Developing Sustainable Food Systems, Policies, and Securities

Abiodun Elijah Obayelu
Federal University of Agriculture, Abeokuta, Nigeria

Oluwakemi Adeola Obayelu
University of Ibadan, Nigeria

IGI Global
PUBLISHER of TIMELY KNOWLEDGE

A volume in the Practice, Progress, and Proficiency in Sustainability (PPPS) Book Series

Published in the United States of America by
IGI Global
Engineering Science Reference (an imprint of IGI Global)
701 E. Chocolate Avenue
Hershey PA, USA 17033
Tel: 717-533-8845
Fax: 717-533-8661
E-mail: cust@igi-global.com
Web site: http://www.igi-global.com

Library of Congress Cataloging-in-Publication Data

Names: Obayelu, Abiodun Elijah, editor. | Obayelu, Oluwakemi Adeola, editor.
Title: Developing sustainable food systems, policies, and securities / Abiodun Elijah Obayelu and Oluwakemi Adeola Obayelu, editors.
Description: Hershey, PA : Engineering Science Reference, 2020. | Includes bibliographical references and index. | Summary: "This book provides multidisciplinary perspectives on how to develop sustainable food systems, policies, and securities"--Provided by publisher".
Identifiers: LCCN 2019042056 (print) | LCCN 2019042057 (ebook) | ISBN 9781799825999 (h/c) | ISBN 9781799826002 (s/c) | ISBN 9781799826019 (eISBN)
Subjects: LCSH: Sustainable agriculture--Nigeria. | Food security--Nigeria. | Women in agriculture--Nigeria. | Land use, Rural--Developing countries.
Classification: LCC S473.N5 D48 2020 (print) | LCC S473.N5 (ebook) | DDC 333.76/1609669--dc23
LC record available at https://lccn.loc.gov/2019042056
LC ebook record available at https://lccn.loc.gov/2019042057

This book is published in the IGI Global book series Practice, Progress, and Proficiency in Sustainability (PPPS) (ISSN: 2330-3271; eISSN: 2330-328X)

British Cataloguing in Publication Data
A Cataloguing in Publication record for this book is available from the British Library.

Practice, Progress, and Proficiency in Sustainability (PPPS) Book Series

Ayman Batisha
International Sustainability Institute, Egypt

ISSN:2330-3271
EISSN:2330-328X

Mission

In a world where traditional business practices are reconsidered and economic activity is performed in a global context, new areas of economic developments are recognized as the key enablers of wealth and income production. This knowledge of information technologies provides infrastructures, systems, and services towards sustainable development.

The **Practices, Progress, and Proficiency in Sustainability (PPPS) Book Series** focuses on the local and global challenges, business opportunities, and societal needs surrounding international collaboration and sustainable development of technology. This series brings together academics, researchers, entrepreneurs, policy makers and government officers aiming to contribute to the progress and proficiency in sustainability.

Coverage

- Global Business
- Technological learning
- Green Technology
- Knowledge clusters
- Innovation Networks
- Global Content and Knowledge Repositories
- ICT and knowledge for development
- Outsourcing
- Strategic Management of IT
- Sustainable Development

IGI Global is currently accepting manuscripts for publication within this series. To submit a proposal for a volume in this series, please contact our Acquisition Editors at Acquisitions@igi-global.com or visit: http://www.igi-global.com/publish/.

The Practice, Progress, and Proficiency in Sustainability (PPPS) Book Series (ISSN 2330-3271) is published by IGI Global, 701 E. Chocolate Avenue, Hershey, PA 17033-1240, USA, www.igi-global.com. This series is composed of titles available for purchase individually; each title is edited to be contextually exclusive from any other title within the series. For pricing and ordering information please visit http://www.igi-global.com/book-series/practice-progress-proficiency-sustainability/73810. Postmaster: Send all address changes to above address.

Titles in this Series

For a list of additional titles in this series, please visit:
https://www.igi-global.com/book-series/practice-progress-proficiency-sustainability/73810

Sustainability in the Entrepreneurial Ecosystem Operating Mechanisms and Enterprise Growth
Yang Gao (Dalian University of Technology, China) Sang-Bing Tsai (Regional Green Economy Development Research Center, Wuyi University, China & Research Center for Environment and Sustainable Development of the China Civil Aviation, Civil Aviation University of China, China) Xiaomin Du (Yingkou Institute of Technology, China) and Chunlin Xin (Beijing University of Chemical Technolog, China)
Business Science Reference • © 2020 • 342pp • H/C (ISBN: 9781799834953) • US $215.00

Digital Solutions and the Case for Africa's Sustainable Development
Albert Ong'uti Maake (UNIVERSITY OF LAY ADVENTIST OF KIGALI, Rwanda) Benard Magara Maake (Kisii University, Kenya) and Fredrick Mzee Awuor (Kisii University, Kenya)
Information Science Reference • © 2020 • 300pp • H/C (ISBN: 9781799829676) • US $195.00

Building an Entrepreneurial and Sustainable Society
Brizeida R. Hernández-Sánchez (University of Salamanca, Spain) José C. Sánchez-García (University of Salamanca, Spain) and Antonio Carrizo Moreira (University of Aveiro, Portugal)
Business Science Reference • © 2020 • 369pp • H/C (ISBN: 9781799827047) • US $235.00

Toward Sustainability Through Digital Technologies and Practices in the Eurasian Region
Gainiya Tazhina (University of International Business, Kazakhstan) and Judith Parker (Teachers College, Columbia University, USA)
Information Science Reference • © 2020 • 294pp • H/C (ISBN: 9781799825517) • US $195.00

Examining the Environmental Impacts of Materials and Buildings
Blaine Erickson Brownell (University of Minnesota, USA)
Engineering Science Reference • © 2020 • 358pp • H/C (ISBN: 9781799824268) • US $245.00

Cross-Border Cooperation (CBC) Strategies for Sustainable Development
Rui Alexandre Castanho (University of Dą browa Górnicza, Portugal)
Engineering Science Reference • © 2020 • 348pp • H/C (ISBN: 9781799825135) • US $225.00

Membrane Technology for Water and Wastewater Treatment in Rural Regions
Rosalam Sarbatly (Universiti Malaysia Sabah, Malaysia)
Engineering Science Reference • © 2020 • 354pp • H/C (ISBN: 9781799826453) • US $205.00

IGI Global
PUBLISHER of TIMELY KNOWLEDGE

701 East Chocolate Avenue, Hershey, PA 17033, USA
Tel: 717-533-8845 x100 • Fax: 717-533-8661
E-Mail: cust@igi-global.com • www.igi-global.com

Samuel Awoniyi, *Department of Agriculture, Economics and Extension, Joseph Ayo Babalola University, Nigeria*
Olusegun. S. Balogun, *Forestry Research Institute of Nigeria (FRIN), Kaduna, Nigeria*
Alaba Dare, *Department of Agricultural Economics and Farm Management, Federal University Agriculture, Abeokuta, Nigeria*
Emmanuel Kanchebe Derbile, *University for Development Studies, Ghana*
Olabisi A. Fawehinmi, *Department of Agricultural Economics and Extension, Bowen University, Iwo, Nigeria*
Rebecca Ibe, *National Horticultural Research Institute (NIHORT), Ibadan, Nigeria*
Mojisola Olanike Kehinde, *Department of Agricultural Economics and Farm Management, Federal University Agriculture, Abeokuta, Nigeria*
Abdul-Azeez Muhammad-Lawal, *Department of Agricultural Economics and Farm Management, University of Ilorin, Ilorin, Nigeria*
Maria Gbemisola Ogunnaike, *Department of Agricultural Economics and Farm Management, Federal University Agriculture, Abeokuta, Nigeria*
Titilope Omolara Olarewaju, *Department of Forest Economics and Extension Services, Forestry Research Institute of Nigeria (FRIN), Ibadan, Nigeria*
Chris Olugbenga Ologbon, *Department of Agricultural Economics, Olabisi Onabanjo University, Nigeria*
Toyin Samuel Olowogbon, *Department of Forest Economics and Extension Services, Forestry Research Institute of Nigeria (FRIN), Ibadan, Nigeria*
Oluseyi Olutoyin Olugbire, *Department of Forest Economics and Extension Services, Forestry Research Institute of Nigeria (FRIN), Ibadan, Nigeria*
Isaac B. Oluwatayo, *Department of Agricultural Economics, Limpopo University, South Africa*
Elizebeth Omolola Oyedepo, *Department of Agricultural Economics and Farm Management, Federal University Agriculture, Nigeria*
Oluwafemi Adebola Popoola, *Nigerian Institute of Social and Economic Research, Ibadan, Nigeria*
Esther Toluwatope Tolorunju, *Department of Agricultural Economics and Farm Management, Federal University Agriculture, Abeokuta, Nigeria*

Table of Contents

Foreword xvi

Preface xviii

Acknowledgment xxv

Section 1
Basic Insights to Sustainable Food Systems, Food Security, Food Policies, and Transformation Options

Chapter 1
Agricultural and Food Policy: Pathways to Sustainable Food Systems and Food Security 1
Abiodun Elijah Obayelu, Federal University of Agriculture, Abeokuta, Nigeria
Simeon Olusola Ayansina, Federal University of Agriculture, Abeokuta, Nigeria

Chapter 2
Seafood Security and Sustainability Through Sustainable Development: A Review of Turkish Seafood Market 16
Sedu Yildirim, Tekirdag Namık Kemal University, Turkey
Merve Kaplan, Kocaeli University, Turkey

Section 2
Food Policy and Agricultural Systems

Chapter 3
Urban Sustainable Growth, Development, and Governance Structures for Revitalization of Open Vacant Spaces in Agriculture and Farming 38
José G. Vargas-Hernández, University Center for Economic and Managerial Sciences, University of Guadalajara, Mexico

Chapter 4
Multi-Stakeholder and Multilevel Food Governance: The Case of the Community of Portuguese-Speaking Countries 60
Sérgio Pedro, Centre for Social Studies, University of Coimbra, Portugal

Chapter 5
The Price Insurance Demand of Rice Producers in the Vietnamese Mekong Delta............................ 73
Huynh Viet Khai, Can Tho University, Vietnam

Section 3
Agricultural Resource Management and Management of Food Resources

Chapter 6
Multiple Scenarios-Based Impact Analysis of Predicted Land-Use Change on Ecosystem Services Value: Evidence From Nigeria.. 83
Aisha Olushola Arowolo, Department of Agricultural Economics and Farm Management, Federal University of Agriculture, Abeokuta, Nigeria
Shakirat Bolatito Ibrahim, Department of Agricultural Economics and Farm Management, Federal University of Agriculture, Abeokuta, Nigeria
Raheem Olatunji Aminu, Department of Agricultural Economics and Farm Management, Federal University of Agriculture, Abeokuta, Nigeria
Abdus Samie, Institute of Agricultural and Resource Economics, University of Agriculture, Faisalabad, Pakistan
Funminiyi Peter Oyawole, Department of Agricultural Economics and Farm Management, Federal University of Agriculture (FUNAAB), Abeokuta, Nigeria

Chapter 7
Implications of Land Grabbing on Sustainable Food Systems.. 109
Agatha Osivweneta Ogbe, Federal University of Agriculture, Abeokuta, Nigeria
Sarah Edore Edewor, Federal University of Agriculture, Abeokuta, Nigeria

Section 4
Management of Agricultural Value Chain Development

Chapter 8
Perspective of Policy Interventions in Rice and Cassava Value Chain Among Women in Nigeria ... 127
Eleojo Grace Adejo, Kogi State University, Anyigba, Nigeria
Olusola Jamiu Saliu, Kogi State University, Anyigba, Nigeria
Patrick Emmanuel Adejo, Kogi State University, Anyigba, Nigeria

Chapter 9
Training Needs Assessment of Palm Oil Processors in Ijebu North Local Government Area, Ogun State... 151
Titilope Omolara Olarewaju, Forestry Research Institute of Nigeria, Nigeria
Felix Oaikhena Idumah, Forestry Research Institute of Nigeria, Nigeria
Agatha Itohan Oseghale, Federal University of Technology, Minna, Nigeria
Lucy Adeteju Orumwense, Forestry Research Institute of Nigeria, Nigeria
Olugbenga Simeon Oke, Forestry Research Institute of Nigeria, Nigeria
Elizabeth Funmilayo Okedeji, National Open University of Nigeria, Nigeria

Section 5
Gender and Its Implications on Sustainable Food Systems and Food Security

Chapter 10
Women Land Rights and Food Security Status of Farming Households in Oyo State, Nigeria......... 163
Abimbola O. Adepoju, Department of Agricultural Economics, University of Ibadan, Nigeria
Rahman A. Adewole, Department of Agricultural Economics, University of Ibadan, Nigeria

Chapter 11
Does Gender Inclusion Really Matter in Sustainable Food Systems?... 181
Sarah Edore Edewor, Federal University of Agriculture, Abeokuta, Nigeria
Agatha Osivweneta Ogbe, Federal University of Agriculture, Abeokuta, Nigeria

Section 6
Migration, Remittances, and Food Security

Chapter 12
Do Foreign Remittances Matter in Attaining Zero Hunger? Evidence From Rural Nigeria.............. 197
Oluwakemi Adeola Obayelu, Department of Agricultural Economics, University of Ibadan, Nigeria
Rebecca Funmi Akinmulewo, Department of Agricultural Economics, Faculty of Agriculture, University of Ibadan, Nigeria

Chapter 13
Impact of Rural-Urban Migration on the Food Consumption Pattern of Farming Households in Ibadan/Ibarapa Agricultural Zone of Oyo State, Nigeria.. 216
Fatai Abiola Sowunmi, Department of Agricultural Economics, University of Ibadan, Ibadan, Nigeria
Funmi Lydia Adeduntan, Department of Agricultural Economics, University of Ibadan, Ibadan, Nigeria

Compilation of References 239

About the Contributors 284

Index 291

Detailed Table of Contents

Foreword xvi

Preface xviii

Acknowledgment xxv

Section 1
Basic Insights to Sustainable Food Systems, Food Security, Food Policies, and Transformation Options

Chapter 1
Agricultural and Food Policy: Pathways to Sustainable Food Systems and Food Security 1
Abiodun Elijah Obayelu, Federal University of Agriculture, Abeokuta, Nigeria
Simeon Olusola Ayansina, Federal University of Agriculture, Abeokuta, Nigeria

Policy plays significant role in defining the food system of any country, and a sustainable food system is necessary for food security. This chapter maps out the causal interactions between food systems, food security and policy, and the challenges in transition to a sustainable food system while respecting the rights of all people to have access to adequate food in Nigeria. Explicit, rigorous, and transparent literature search was undertaken and many articles were assessed and reviewed. Although the results established a mutual relationship between food system and food security, existing literature have widely failed to take interactions between food systems, food security and policy into account. While food production is used as an entry point to improving food system sustainability, the quest for food security are undermining transition towards sustainable food systems. It was found that without right policies in place, it may be difficult to have food systems that are sustainable and ensure food security. This chapter provides a useful contribution to policy, and research on transitions towards sustainable food system. Any policy intervention to address one part of the food systems will impact on other parts and will determine whether a country is food secure or not. Enabling policy environment is therefore essential in ensuring a sustainable food system and for the attainment of food security.

Chapter 2
Seafood Security and Sustainability Through Sustainable Development: A Review of Turkish Seafood Market 16
Seda Yildirim, Tekirdag Namık Kemal University, Turkey
Merve Kaplan, Kocaeli University, Turkey

The world has understood that hunger is one of the most dangerous problems for the future. Accordingly, food security and sustainability are both important issues through sustainable development. This chapter highlights the role of seafood security and sustainability for sustainable development. In this context, seafood security and sustainability for Turkish seafood market was investigated. Turkey is a coastal country, which has accepted 2030 Sustainable Development Goals, with a seafood market and a good sample to investigate seafood sustainability. This study employed secondary data from TURKSTAT and FAO websites to determine Turkish seafood market profile. The study determined seafood security and sustainability based on five dimensions as availability, economic access, physical access, utilization, and stabilization. Seafood sustainability is vital for coastal countries because seafood market brings economic, social, and environmental benefits at the same time.

Section 2
Food Policy and Agricultural Systems

Chapter 3
Urban Sustainable Growth, Development, and Governance Structures for Revitalization of Open Vacant Spaces in Agriculture and Farming 38
José G. Vargas-Hernández, University Center for Economic and Managerial Sciences, University of Guadalajara, Mexico

This chapter analyzes the implications of urban sustainable growth, development, and governance structures for the revitalization of open vacant spaces in agriculture and farming. After reviewing the extensive corpus of literature on the subject, the authors used the critical socio-ecological analysis methodology to determine the main issues, trends, practices, and implications of the urban vacant spaces in relation to the urban sustainable growth and development, the use of urban vacant land in urban agriculture, farming, and gardening, and the collaborative urban governance structures and revitalization of open vacant spaces. It is concluded that transitional use of vacant land and parcels are to be used and utilized for developing a sustainable green city. However, urban vacant land and parcel spaces are required to be utilized for revitalization purposes to be stimulated. Social-ecological analysis focusing on vacant lots in underdeveloped urban spaces hold potential for urban transformation to meet the social needs and improve the ecological services.

Chapter 4
Multi-Stakeholder and Multilevel Food Governance: The Case of the Community of Portuguese-Speaking Countries 60
Sérgio Pedro, Centre for Social Studies, University of Coimbra, Portugal

The contemporary food system, in its global and local dimensions, is a central element of the debate on the sustainability of the planet, a debate that increasingly involves more stakeholders and areas of knowledge in the search for answers to the multiple questions related to the attainment of more sustainable patterns for food and agriculture. The present chapter analyses the participative multi-stakeholder and multilevel model of food governance of the Community of Portuguese Speaking Countries (CPLP), in which stakeholders from different societal and expertise sectors participate in equal manners in the process of co-construction of institutional, technical, and financing measures for the functioning of a given food system. The present chapter has the main goal of sharing and critically analysing the CPLP´s institutional context for the promotion of sustainable food systems as an example of an integrated methodological approach to support the creation of coordinated public policies and institutional conditions to implement a transition to more sustainable food systems and diets.

Chapter 5
The Price Insurance Demand of Rice Producers in the Vietnamese Mekong Delta............................ 73
Huynh Viet Khai, Can Tho University, Vietnam

Using the dichotomous choice contingent valuation method, this chapter helps shed light on the potential for marker-based insurance schemes in Vietnam by empirically exploring the demand for minimum price insurance among rice households. The study showed that the majority of rice farmers accepted the guaranteed price of VND 4,500 per kg, and their accepted insurance fee was about 13% of the guaranteed price and 30% of the break-even price. Farmers growing rice under a monoculture system were less likely to pay for the proposed insurance service, while those with access to any formal credits were more likely to pay for it.

Section 3
Agricultural Resource Management and Management of Food Resources

Chapter 6
Multiple Scenarios-Based Impact Analysis of Predicted Land-Use Change on Ecosystem Services Value: Evidence From Nigeria.. 83
Aisha Olushola Arowolo, Department of Agricultural Economics and Farm Management, Federal University of Agriculture, Abeokuta, Nigeria
Shakirat Bolatito Ibrahim, Department of Agricultural Economics and Farm Management, Federal University of Agriculture, Abeokuta, Nigeria
Raheem Olatunji Aminu, Department of Agricultural Economics and Farm Management, Federal University of Agriculture, Abeokuta, Nigeria
Abdus Samie, Institute of Agricultural and Resource Economics, University of Agriculture, Faisalabad, Pakistan
Funminiyi Peter Oyawole, Department of Agricultural Economics and Farm Management, Federal University of Agriculture (FUNAAB), Abeokuta, Nigeria

This chapter investigates how land-use/land-cover (LULC) changes under different scenarios will affect ecosystem services provisions in Nigeria using multiple data sources. The Markov and dynamics of land system models were integrated to predict future LULC changes while the value transfer methodology was adopted to evaluate the economic value of ecosystem services. The results revealed varying patterns and trends of LULC change under the baseline, forest protection priority, and sustainable economic growth scenarios. Based on the predicted LULC change, the total ecosystem services value in Nigeria will decline under the baseline and forest protection priority scenarios but increase in the sustainable economic growth scenario. The sustainable economic growth scenario showed major positive impacts on the ecosystem service functions of recreation, climate regulation, soil formation, and erosion control. This study concludes that the sustainable economic growth scenario is the best to ensure expected production while safeguarding the environment in Nigeria.

Chapter 7
Implications of Land Grabbing on Sustainable Food Systems.. 109
Agatha Osivweneta Ogbe, Federal University of Agriculture, Abeokuta, Nigeria
Sarah Edore Edewor, Federal University of Agriculture, Abeokuta, Nigeria

Land is one of the most valuable assets required for agricultural production. In Africa, smallholder agricultural producers are faced with a lot of challenges that have highly impacted on productivity and sustainable food systems. The global demand for agricultural land for food and bio-fuel production has increasingly led to the emergence of land grabbing after the 2007-08 food price crisis. The rural poor are the victim of land grabbing as they are faced with declining farmlands, low income generation, and loss of livelihood activities. These have affected the food security status of the rural poor as farmlands are taken from them. The proponents of land grabbing revealed that developing countries are expected to benefit from investments inflow on grabbed land, development of infrastructure, increased income generation, and job creation. They argue that investment in agriculture is necessary to stimulate agricultural production; however, this situation has brought negative effects as most investors failed to keep their end of the transaction on land acquisition deals.

Section 4
Management of Agricultural Value Chain Development

Chapter 8
Perspective of Policy Interventions in Rice and Cassava Value Chain Among Women in Nigeria ... 127
Eleojo Grace Adejo, Kogi State University, Anyigba, Nigeria
Olusola Jamiu Saliu, Kogi State University, Anyigba, Nigeria
Patrick Emmanuel Adejo, Kogi State University, Anyigba, Nigeria

The study investigated trends in rice grain and cassava tuber value addition through processing. Among the staple foods in Nigeria, rice and cassava have gained special prominence and priority attention by the government in terms of their production and value addition. The result indicated that the rice and cassava value chain is affected by different policy regimes. It was also found that women in the north central of Nigeria participated actively in rice and cassava value addition with some challenges. It is recommended that women processors of these commodities should have access to productive resources that can help add value to these commodities, training women on improved value added technologies and innovations by both public and private organizations, and most importantly, making these innovations and technologies affordable, adoptable, and adaptable will go a long way to boost their value added on these commodities through processing.

Chapter 9
Training Needs Assessment of Palm Oil Processors in Ijebu North Local Government Area, Ogun State.. 151
Titilope Omolara Olarewaju, Forestry Research Institute of Nigeria, Nigeria
Felix Oaikhena Idumah, Forestry Research Institute of Nigeria, Nigeria
Agatha Itohan Oseghale, Federal University of Technology, Minna, Nigeria
Lucy Adeteju Orumwense, Forestry Research Institute of Nigeria, Nigeria
Olugbenga Simeon Oke, Forestry Research Institute of Nigeria, Nigeria
Elizabeth Funmilayo Okedeji, National Open University of Nigeria, Nigeria

The study assessed training needs of palm oil processors in Ogun State, Nigeria. A well-structured questionnaire was used to elicit information from 90 palm oil processors. The data was analyzed using both descriptive and inferential statistics. Majority of the respondents were young, married, and experienced in palm oil processing. Women are mostly involved in palm oil processing using manual method of processing with oil palm fruits sourced more from family farms. Respondents require training for manual and mechanized processing methods. Socioeconomic factors have significant influence on different stages of palm oil processing. Poor extension service, high cost of labour, and processing machine were the most perceived constraints to palm oil processing in the study area. The study therefore concludes that there is need for training in oil palm processing. Extension service providers should intensify efforts in this regard so as to boost the palm oil supply both within and outside the country.

Section 5
Gender and Its Implications on Sustainable Food Systems and Food Security

Chapter 10
Women Land Rights and Food Security Status of Farming Households in Oyo State, Nigeria......... 163
Abimbola O. Adepoju, Department of Agricultural Economics, University of Ibadan, Nigeria
Rahman A. Adewole, Department of Agricultural Economics, University of Ibadan, Nigeria

The dominance of men in decision-making processes and leadership positions within the communities has made land allocation, land use, and control skewed in favour of men. This study examined the effects of women's land rights on households' food security status using a sample of 300 representative farmers. Descriptive statistics, household food expenditure, logistic regression, and ordered logit models were the analytical tools used. Results revealed that about 35% of the rural women farmers had land use rights while the remaining 65% had land ownership rights. Women with ownership rights were more food secure, with the majority of the women having residual rights, while only a few had sell rights. Secure women land rights are germane to achieving and sustaining household and national food security. Strategies and instruments for protecting women rights should be developed and implemented, while efforts geared towards designing strategies, assessing multiple dimensions of women empowerment for improved food security status, and welfare of the households should be intensified.

Chapter 11
Does Gender Inclusion Really Matter in Sustainable Food Systems? 181
Sarah Edore Edewor, Federal University of Agriculture, Abeokuta, Nigeria
Agatha Osivweneta Ogbe, Federal University of Agriculture, Abeokuta, Nigeria

Over the past decades, the food systems in developing countries have transformed rapidly. However, the rise in social inequalities has negatively affected, the vulnerable groups as the benefits associated with these transformations are still skewed. This chapter examined the role of gender inclusiveness in promoting sustainable food systems. Employment trends revealed that agricultural employment was higher among males. Five asymmetries (assets, access to agricultural market, access to technology, resilience and risks, and decision making) were identified as limitations to sustainable food systems stemming from the gender differentiated roles. The gender action learning system methodology was adopted using strategies such as empowering men and women through community action learning during catalyst workshops, gender mainstreaming for innovation and institutional change at organizational level, and through advocacy network for policy improvement at the national level. The study concluded that gender inclusion played a crucial role in achieving sustainable food systems.

Section 6
Migration, Remittances, and Food Security

Chapter 12
Do Foreign Remittances Matter in Attaining Zero Hunger? Evidence From Rural Nigeria 197
Oluwakemi Adeola Obayelu, Department of Agricultural Economics, University of Ibadan, Nigeria
Rebecca Funmi Akinmulewo, Department of Agricultural Economics, Faculty of Agriculture, University of Ibadan, Nigeria

Foreign remittance has remained a major source of income and a means to reduce hunger for many poor people in developing countries. The contribution of foreign remittances to food insecurity status of rural households in Nigeria was assessed using data from 2015/2016 Living Standard Measurement Study-Integrated Surveys on Agriculture (LSMS-ISA). Food insecurity status was achieved using the household food insecurity access scale. Data were analysed using descriptive, ordered, and nested logit models. Female-headed households residing in south-east zone with 51 to 70 years old heads and more than six members had greater access to remittances but were severely food insecure. Drivers of food insecurity were age, gender, marital status, education of the household head, membership of cooperatives, access to extension, farm size and per capita income, and living in the north central geo-political zone. Foreign remittances had a positive effect on the food insecurity status of rural households.

Chapter 13
Impact of Rural-Urban Migration on the Food Consumption Pattern of Farming Households in Ibadan/Ibarapa Agricultural Zone of Oyo State, Nigeria.. 216
Fatai Abiola Sowunmi, Department of Agricultural Economics, University of Ibadan, Ibadan, Nigeria
Funmi Lydia Adeduntan, Department of Agricultural Economics, University of Ibadan, Ibadan, Nigeria

The study examined the impact of rural-urban migration on the food consumption pattern of farming households. The study revealed that 73.8% of the households had migrants, while 80.2% of the migrants were male. The highest level of education of most of the migrants was secondary school (71.4%). The study showed that the major reason (63.3%) for migration was for job. The average remittance sent per year was ₦108,119.14. The study revealed that household expenditure on carbohydrate food group accounted for 54.4% of the total households' expenditure on food. The average dietary diversity indices for the migrant (0.345) and non-migrant (0.346) households were low. The study revealed that migration (short and long term) positively influenced per capita food expenditure of respondent. Despite the remittance from some of the migrants, the need to develop the rural areas in terms of provision of basic infrastructures by government is imperative in order to reduce rural-urban migration.

Compilation of References 239

About the Contributors 284

Index 291

Foreword

Food is one of the basic needs of mankind and access to the right quantity and quality of food at the right time on a sustained basis continues to be a major challenge for majority of the world's population. In-spite of rapid advances in technology and technical knowhow in the past several decades, this challenge of hunger and malnutrition has continued to linger. This problem is compounded by the advent of climate change and the growing incidence of unsustainable land use practices, especially across the African continent, which threatens the long-term stability of the environment, with potentially serious long-term food insecurity implications on vulnerable populations.

The question is: How can the nations of the world substantially mitigate this lingering threat of hunger and malnutrition in the coming decade? In addressing this question, we must first understand that food security is a product of sustainable food systems. Thus, appropriate agricultural and food policies must be put into place to drive the emergence of sustainable food systems across the globe, if the global threat of hunger and malnutrition is to be substantially mitigated.

This book titled "Developing Sustainable Food Systems, Policies and Securities" is unique and timely, as it provides a one-stop shop for the most important considerations in understanding the link between food security, sustainable food systems and food policies. The various chapter contributions are original, well thought out, informative, illuminating and incisive.

The book provides a framework for the creation, exploration, or transformation of food systems that will provide solutions to food insecurity challenges through appropriate food policy regimes. It provides an insight into how land use management policies and strategies can help improve agricultural productivity, while minimizing the opening up of new lands and conserving natural ecosystems.

The book extensively explores the difficult question of how to create the enabling policy environment needed to support the development of sustainable food systems, that will in turn guarantee food security. It provides useful insights into how sustainable economic growth policies could help ensure increased agricultural sector output while simultaneously preserving the environment. Furthermore, the book explores the effects of large-scale land acquisition in Sub-Saharan Africa on the local agri-food systems and the implications of emerging land resource utilization patterns for food security and poverty reduction among the vulnerable local populations.

In addition, contributors highlight the importance of gender inclusiveness in achieving sustainable food systems and identifies specific areas of gender asymmetries and key actions required to bridge gender gap so as to achieve sustainable food systems. It advocates the development of gender responsive policies that will factor in gender-specific responsibilities, constraints and resources in agricultural development programmes, recognize and protect women's land ownership rights, and in the long-term ensure that women and men's priorities and needs are reflected at all levels of policy formulation and programme implementation.

In-spite of the numerous challenges faced by farmers in carrying out post-harvest activities, there are opportunities to encourage them in the promotion of value addition to these commodities through equitable and sustainable policies affecting the processing and value addition of these commodities. This book provides some important insights into areas within the agro-processing sector that requires specific policy focus with the aim of achieving food security among rural and urban households.

This book will be an invaluable text for students of agricultural economics and economics in Universities in Nigeria, Africa and other continents of the world. Policy making agencies, planners, consultants, policy researchers, and those interested in issues related to food security and sustainable food systems will also find this book very informative, interesting and useful. I therefore strongly recommend this book for the above-mentioned audience and all other persons that are interested in understanding how appropriate agri-food polices can help to mitigate the threats of food insecurity through the creation of sustainable food systems.

Adebayo B. Aromolaran
Faculty of Agriculture, Adekunle Ajasin University, Nigeria

Adebayo B. Aromolaran *is a Professor of Agricultural & Development Economics and the current Dean of the Faculty of Agriculture, Adekunle Ajasin University, Akungba, Akoko, Ondo State. He is currently, the Country Coordinator for Agricultural Policy Research in Africa (APRA) Workstream 1 Research Project on Agricultural Commercialisation in Nigeria. He obtained his PhD (Agric Economics) from the University of Ibadan, Nigeria (1993) and was appointed Rockefeller Postdoctoral Research Fellow on "Economics of the Family" at Economic Growth Center, Yale University (2001–2003). In 2009, Professor Aromolaran won the UMA LELE award of the International Association of Agricultural Economists (IAAE) for the "Best Contributed Paper on Gender", at the association's 27th Triennial Conference in Bejing, China. The paper titled: "Does Increase in Women Income Relative to Men's Income increase calorie intake among poor households? Evidence from Nigeria" was subsequently Published in Agricultural Economics 41: 239–249, (2010). His primary areas of research are food security, nutrition economics and household economics. He has published in several widely circulating Journals some of which are Food Policy, Oxford Development Studies, Agricultural Economics, International Journal of Development Issues, Food Security Journal, International Economic Journal, African Journal of Agric Resources, African Development Review, Journal of Sustainable Agriculture, Outlook on Agriculture etc. Prof. Aromolaran is a member of the Agricultural and Applied Economic Association (AAEA), Nigerian Association of Agricultural Economists, Farm Management Association of Nigeria, International Association of Agricultural Economics (IAAE), and Nigerian Economic Society (NES). He was Scientific Adviser for the International Foundation of Science (IFS), Stockholm, Sweden (2004– 2010) and the African Institute for Applied Economics (AIAE), Enugu, Nigeria (2010 to 2013). In terms of community engagement and consultancy services, he has worked on several "agricultural development projects" funded by several international organizations as Consultant: UNDP, FAO, World Bank, UNEP, Forum for Agricultural Research in Africa (FARA), USAID, DFID, IFPRI, and UNICEF.*

Preface

INTRODUCTION

The concept of food system is gaining more prominence in recent years amongst both scholars and policymakers because despite the growth in the world food production, a lot of people (more than 800 million) are still chronically malnourished (FAO, 2017). Developing a sustainable food systems can therefore help to find solutions that will provide the world's growing population with a sufficient supply of healthy food within the environmental limits. Sustainable food system is a prism of the present and the future challenges for nutrition, health, sustainable development, community and economic development. It has often been a subject of discussion in nutrition, food, health, community economic development and agriculture. Developing sustainable food system (SFS) is very crucial, though a tough nut to crack, to correcting the world food system's failure. It involves a collaborative network that integrates food production, processing, distribution, consumption and waste management but it remains difficult to engage all these sectors in joint policy actions. The SFS is necessary to achieve Sustainable Development Goal (SDG) 2, which focuses explicitly on food by seeking to "end hunger, achieve food security and improved nutrition and promote sustainable agriculture" by 2030. The SDG 2 is a multiple of other goals that are related to challenges in the food system. SDG 1 focuses on poverty reduction, where agriculture and food has a key role to play. Sustainable agriculture plays a central role in achieving SDG 6 on water, SDG 12 on sustainable consumption and production, SDG 13 on climate change adaptation and mitigation and SDG 15 on land use and ecosystems. Food systems are at the heart of achieving these goals.

Information on policies that would enhance productivity and sustainability of individual agricultural sector is very important but literature is practically devoid of these information and experiences from countries and communities, with respect to comprehensive approach (cross-sectoral policies) to SFS. Hence, this book on "Developing Sustainable Food Systems, Policies and Securities". This book is believed to be the first effort to fill this gap, providing information on proven options for enhancing SFS and how to identify opportunities and actions for exploiting cross-sectoral synergies. The book is a collection of research studies that guide the development of sustainable food systems in practice through an integrated food policy. It explores the nature, extent, and causes of nutrition problems as well as the role that agricultural policy plays on sustainable food systems, in order to tackle food insecurity challenge. This book takes a transdisciplinary approach and considers multi-sectoral actions, integrating health, agriculture, environment, economy, and socio-cultural issues to comprehensively explore develop food systems that is sustainable. Consideration is given to the multi-dimensional nature of food systems, policy, gender and a broader range of scientific topics. Featuring research on topics such as food security, carbon emissions, and nutrition. While the eating patterns are important for building sustainable food and agricultural systems, the development of agricultural and food systems must take into account actions and policies aimed at making food systems more sustainable.

REASONS WHY COUNTRIES IN THE WORLD MUST CHANGE THEIR CURRENT FOOD SYSTEMS

Food systems are experiencing rapid and intense transformations, having to feed a growing global population in a context of persisting economic, environmental and social challenges. Global food is clearly insecure and global food system appears to be approaching its environmental limits. The global food system is escalating environmental challenges through deforestation, pollution, soil degradation, biodiversity erosion, diminishing freshwater resources and greenhouse gas emissions with more than 820 million people facing food insecurity and undernutrition, 672 million people suffering from obesity and 1.3 billion are overweight. Agricultural production and rural livelihoods are being increasingly jeopardized by the impact of climate change and continuing depletion of natural resources (FAO, 2019). This is further exacerbated by challenges such as price volatility, conflicts, crises and migration. Addressing these complex challenges requires integrated and context-specific solutions. The way food is produced, transformed, distributed and consumed are all failing in terms of livelihoods, human health and the environment. A sustainable food system will ensure the production and consumption of sufficiently varied diets containing the right micronutrients (vitamins and minerals) and the fundamental elements of a sustainable food future. Many countries do not produce enough food to feed everyone making such countries to depend on cheap food imports, which works as a quick fix to their food deficit, but does not address the low agricultural productivity that keeps the deficit in place. Studies have established that the challenges countries in the world are facing can be linked to the current food systems. These include excessive tillage of arable land leading to degrading soils, greater release of carbon and locking farmers into unprofitable production systems. People are also eating less diversified diets than before because varieties of food are probably not affordable due to high prices with fewer production by fewer hands. Owing to increasing population and food need, there is the prospect of developing food systems that are sustainable by going on greening of agriculture (increasing use of farming practices and technologies that simultaneously).

Many countries are suffering from pastoral conflict where arable farmers and pastoralists compete between and among themselves for the same natural resources, such as water, fuelwood and fertile land results in a further decline of the ecological resilience of the system and increases tensions and violence which in turn decrease the willingness to invest and food insecurity. Lack of policy formulation or implementation to control the situation serves as a constraint to agricultural productivity, which affects food security and nutrition of the affected countries.

THE CHALLENGES OF FOOD SYSTEMS AND TRANSITIONS TO SUSTAINABILITY WITHOUT CONFRONTATIONS

Today, agriculture stands at a crossroads. There are renewed calls virtually in all the countries for a change in the way food is produced and distributed if the poor and hungry are to be served better and if the world is to cope with a growing population and climate change. Both the conventional and traditional food systems generate substantial pressure on the environment and are currently hard-pressed to meet the food and nutrition requirements of millions of vulnerable people. The cost of externalities caused by current agricultural practices, include those from use of inputs such as pesticides and fertilisers, pollution of waterways and emissions from farm machinery and food-related transport. Expansion of agricultural land at the expense of forests are also enormous. As a result, the challenge for governments is to implement policies that promote not only improved productivity, but also ensure food quality and safety along the value chain.

Food systems that are sustainable are tremendously relevant to growing food security. Large companies are strongly influencing the food system by coming with big supermarkets and food shopping complexes, which harms small and local food systems. Traditional (subsistence) smallholder agriculture is prominent in most developing countries, with limited or no use of off-farm inputs and attendant low productivity, low value added per worker and primarily reliant on extracting soil nutrients with insufficient replenishment by either organic or inorganic fertilisers. The current food system is also susceptible to yield losses due to erratic rainfall, pest and weed infestations and other production related risks, characterized with release of huge amounts of greenhouse gas (GHG) and other toxic pollutants. Degradation of the environment and the natural resources delivers low-cost food at a high cost to the environment is unsustainable owing to the increasing stress on ecosystems. This could jeopardise the achievement of sustainable food systems if a more comprehensive and integrated approach is not factored into the policy actions.

Furthermore, the current traditional food system has limited scope for capital intensive, farm mechanisation and intensive use of agrochemical inputs. Many smallholders' plots, predominantly located in developing countries, are too small to realise the economies of scale required for most of the available commercial farm machinery. In addition, the high cost of purchased inputs, such as chemical fertilisers, pesticides and seeds, generally require that at least some portion of the crops produced must be sold to recover costs. Although, most countries have policies targeted to different components of the food systems to ensure food security, but these policies are typically made in isolation with no specific food plans and strategies that bring them together. Hence, a sustainable food system does not operate in silos but in interconnection of "everything and everybody that influences, and is influenced by all activities involved in bringing food from farm to fork and beyond (Hawkes & Parsons, 2019). In developing sustainable food systems, policymakers and donors must recognize that an important way to support food processing is through increased crop productivity, innovation and technology transfer. Transition to sustainable food systems will require moving from an agriculture-centered to a food systems policy and research framework. In doing that, there is a need to be able to account for and measure externalities produced by these processes. Policies are also needed to increase incentives (and decrease disincentives) for availability, access, and consumption of diverse, nutritious and safe foods through environmentally sustainable production, trade, and distribution. It is therefore not feasible to fix the food system with a few partial measures.

To overcome these challenges, it is important that all actors from input supply to consumers and policy makers see the need for changes and work together to find practical solutions. Development of sustainable food systems should shift from an exclusive focus on boosting production but has its focus on eradicating poverty, increasing resilience, ensuring people's food and nutrition security, promoting good nutrition and health, reducing inequalities, contributing to peace, promoting political stability, regenerating ecosystems, and mitigating climate change.

KNOWLEDGE GAPS IN RESEARCH FROM FOOD SYSTEMS LITERATURE

There is no doubt that there is a growing body of research in food systems literature on the possible ways of transition to sustainable food systems considering the contemporary social, political, health, demographic (growing population) and environmental changes e.g. climate change, resource depletion, biodiversity loss) in recent time. In the course of preparing this book, we identified some knowledge gaps. First, a "whole food systems" approach is still limited in the body of research that makes reference to sustainability transition in food systems in line the findings of Markard Markard, Raven, and Truffer (2012). Second, lack of attitudinal change in embracing gender mainstreaming into the sustainable food systems framework in many countries limits its success. Third, there is a dearth of enough work on how consumer behaviour can be changed towards healthier diets, reduction in food waste and loss. Fourth, there is a paucity of information on what it takes to make agriculture an attractive investment to youths and what the future investment models for agricultural research and development should be, in order to ensure that investments are motivated by facts and priority needs rather than political interests. Fifth, there is still a lack of an operational approach that enables a diagnosis in a given country or region about what are the (a) various dominant food systems (b) inventory of the traditional and modern systems considering the merits and demerits in both cases and (c) the alternatives to dominant ones that may harbour promising innovations for improving or changing unsustainable systems and create new contexts of opportunities for a transition process. Getting all these done will be useful to inform research on transitions to sustainable food system, as well as for policy makers to guide agricultural investments and see how they can orient their innovation policies to support desired transition pathways and respond to any unsustainable dominant food systems.

JUSTIFICATION OF THE BOOK

Food systems are at the heart of many of the major challenges facing the world today because they affect people, and challenges of food systems are the product of many different policies. Real solutions to food systems challenges therefore require an enhanced role of food policy (Hawkes & Parsons, 2019). Sustainable food system is not just about the availability of food, but also about affordability and the preferences (often driven by policy or other incentives) that influence people's access to that food. The focus in recent time has been shifted from activities within the food production system (production, transport, processing) to the outcomes of those activities in the form of the consumption. Because access to affordable, healthy and diverse food depends not only on production but also on factors outside the food production system, a broader approach is required when analysing the impact of policy interventions aimed at enhancing food security. Based on the growing awareness and increase focus on sustainable food systems view as the best to overcome the world's food and nutrition security, this book on 'Developing Sustainable Food Systems, Policies and Securities' presents some major challenges and opportunities in transition from the current traditional food and agricultural systems to a green agriculture; identifies knowledge gaps between food systems, policies and securities; and create a coherent narrative for the necessary sustainability transition.

The book is necessitated to bring together research findings that provide potential solutions to the problems in the development of sustainable food systems and securities as well as agricultural and food policy as the world is bracing up for hard-choice challenges and potentially massive trade-offs around issues related to food and nutrition security in the coming decades. The book is essential for tackling pockets of food and nutrition insecurity; pastoral conflict where arable farmers and pastoralists compete between and among themselves for the same natural resources, such as water, fuelwood and fertile land going on in most countries in the world; reflect on the usefulness of policies in food systems and security;

and draw out some shortcomings, as well as areas of further research and development. The contribution of the book to knowledge is that, it tackles the most contemporary issues in food policy, food security and food systems, while accounting for increased effect of globalisation and growing population. The book is organized into six sections with thirteen chapters disaggregated into different relevant sections after a double blind peer-reviewed and a proper checking that no part of the work was plagiarised or published elsewhere. The book through a systematic review of literature and quantitative data contains basic insights to development of sustainable food systems, food policies and food security with suggestions for transformation options (Chapters 1 and 2), food policy and agricultural systems (Chapters 3, 4 and 5), agricultural resource management, and management of food resources (Chapters 6 and 7), management of agricultural value chain development (Chapters 8 and 9), gender and its implications on sustainable food systems and security (Chapters 10 and 11), as well as migration, remittances and food security (Chapters 12 and 13). The book is expected to deliver new insights for a more accurate process of public policy formulation and private sector investment priorities in food systems and food security. The book is ideally designed for economists, environmentalists, food producers, policymakers, researchers, academicians, and students seeking coverage on agricultural and sustainability issues.

WAY FORWARD

A key message is that food systems are designed by many different decisions and decision-makers beyond what might generally be considered 'agricultural and food policy'. The persistent challenges of food systems and food securities therefore call for better policies. Developing sustainable food systems will require transforming both the content of policies and the processes they are made. It will involve cross-examining the existing policies to see if they are undermining other goals in the system, and whether they align with a vision of what we want the food systems to look like.

ORGANIZATION OF THE BOOK

This book on "Developing sustainable food systems, policies and securities" is made up of 13 chapters distributed across six sections.

Section 1: Basic Insights to Sustainable Food Systems, Food Security, Food Policies, and Transformation Options

This section discusses some contemporary issues and importance of policies on sustainable food system through review of relevant literature and content analysis to examine relationships between sustainable food systems, agricultural and food policy, and food security and draw out some salient lessons.

The first chapter by Obayelu A. E. and Ayansina S. O., sets the scene and identifies critical issues addressed in the rest of the book by bringing out evolving understandings on the links between agricultural and food policies in the development of sustainable food systems, and food security through a systematic literature review. It takes stock of what we know and some of the things we do not know so far about importance of policy in sustainable food systems and food security, outlines the direct or indirect links between agricultural and food policy to food systems and the implications on food and nutrition security. This chapter explores a number of endpoints to create a context for exploration, mitigation, and transformation of food system. The chapter concludes that we need agricultural and food policies that will provide solutions to food systems and food insecurity challenges. Seda Yildirim and Merve Kaplan, in the second chapter using grey literature, explore the importance of seafood security and sustainability in the context of sustainable development using Turkish seafood market as a case study.

Section 2: Food Policy and Agricultural Systems

This second section helps to decipher the importance of agricultural and food policy and how they interact with agricultural systems. José G. Vargas-Hernández, in the third chapter, examines the implications of urban sustainable growth, development and governance structures for revitalization of open vacant spaces in agriculture and farming through a systematic review of the literature. The contribution to the existing literature lies in the fact that the chapter provides, through a systematic review, a comprehensive assessment of the utilization of urban vacant land for sustainable food systems. The fourth chapter by Sérgio Pedro explores the sustainable patterns for food and agricultural controversies in developing and emerging countries. The chapter discusses some contemporary issues through review of relevant literature and content analysis of responses to multiple questions on how to attain sustainable food and agriculture. Huynh Viet Khai in the fifth chapter uses quantitative data and dichotomous choice contingent valuation method to highlight the effect of price of insurance demand on rice producers in the Vietnamese Mekong Delta in Vietnam.

Section 3: Agricultural Resource Management and Management of Food Resources

This section of the book centres on issues relating to resources management and management of food resources. In the sixth chapter, Arowolo, A.O., Ibrahim, S.B., Aminu, R.O. Samie A. and F.P Funminiyi present a multiple scenarios-based impact analysis of predicted land-use change on ecosystem services value on sustainable food system using Nigeria as case study. In Chapter 7, Agatha Osivweneta Ogbe and Sarah Edore Edewor analyse the effects of land grabbing on sustainable food systems.

Section 4: Management of Agricultural Value Chain Development

This section accentuates the issues of agricultural value chain development through exploration of some key innovations in the course of transition to sustainable food systems in developing and emerging countries. Adejo, E.G, Saliu, O.J. and Adejo, P. E., in Chapter 8, use qualitative data to analyse the perspective of policy interventions in rice and cassava value chain among women in Nigeria. In Chapter 9, Idumah, F., Olarewaju, T. O., Oseghale, A.I., Orumwense, L.A., Oke, O. S. and Okedeji, E. use both qualitative and quantitative data to examine training needs assessment in sustainable food system using a case of palm oil processors in Ijebu North Local Government area, Ogun State, Nigeria.

Section 5: Gender and Its Implications on Sustainable Food Systems and Food Security

This section examines the roles of gender in sustainable food systems and food security. The tenth chapter by Adepoju, A.O. and Adewole, R.A. analyse women's land rights and food security status of farming households in Oyo State, Nigeria, while Sarah Edore Edewor and Agatha Osivweneta Ogbe investigate whether gender inclusion really matter in sustainable food systems in Chapter 11.

Section 6: Migration, Remittances, and Food Security

The section examines the roles of remittances in sustainable food systems and food security. In Chapter 12, Obayelu, O.A. and Akinmulewo, R.F. assess the contribution of foreign remittances to attaining zero hunger in rural Nigeria, while Sowunmi, F.A. and Adeduntan, F.L. assess the impact of rural-urban migration on food consumption pattern of farming households in Nigeria using data from Ibadan/Ibarapa agricultural zone of Oyo State in Chapter 13.

Abiodun Elijah Obayelu
Federal University of Agriculture, Abeokuta, Nigeria

Oluwakemi Adeola Obayelu
University of Ibadan, Nigeria

REFERENCES

FAO. (2017). *The State of Food Security and Nutrition in the World 2017*. Rome: FAO.

FAO. (2019). *The State of Food Security and Nutrition in the World 2019*. Retrieved from http://www.fao.org/3/ca5162en/ca5162en.pdf

Hawkes, C., & Parsons, K. (2019). Tackling Food Systems Challenges: The Role of Food Policy. In *Rethinking Food Policy: A Fresh Approach to Policy and Practice*. London: Centre for Food Policy.

Markard, J., Raven, R., & Truffer, B. (2012). Sustainability transitions: An emerging field of research and its prospects. *Research Policy*, *41*(6), 955–967. doi:10.1016/j.respol.2012.02.013

Acknowledgment

The editors appreciate the Almighty God who has made the dream on this book project come true. We wish to acknowledge the concerted efforts of all the contributing authors, the reviewers, and the members of the Editorial Advisory Board (EAB). The idea of the title of the book project stemmed from the fact that the world food production can no longer feed the growing population and production are still taking place in unsustainable manner across many different countries. The questions that therefore came to our mind were: Are there no agricultural and food policies toward sustainable food systems and food security? How do we develop sustainable food systems? Are the policies supportive to sustainable food systems? What are the new solutions to sustainable food systems? We packaged these thoughts and put on a proposal to IGI international. The idea was eventually accepted leading to call for proposals to develop chapters of the book.

We thank the IGI international for the intellectual stimulation and modification of the initial title of the proposal submitted and the various links provided which helped in the widely dissemination of the information. We also thank the African Growth and Development Policy Modelling Consortium (AGRODEP) for putting the proposal on their website.

Abiodun Elijah Obayelu
Federal University of Agriculture, Abeokuta, Nigeria

Oluwakemi Adeola Obayelu
University of Ibadan, Nigeria

Section 1

Basic Insights to Sustainable Food Systems, Food Security, Food Policies, and Transformation Options

Chapter 1
Agricultural and Food Policy:
Pathways to Sustainable Food Systems and Food Security

Abiodun Elijah Obayelu
https://orcid.org/0000-0003-3328-7717
Federal University of Agriculture, Abeokuta, Nigeria

Simeon Olusola Ayansina
Federal University of Agriculture, Abeokuta, Nigeria

ABSTRACT

Policy plays significant role in defining the food system of any country, and a sustainable food system is necessary for food security. This chapter maps out the causal interactions between food systems, food security and policy, and the challenges in transition to a sustainable food system while respecting the rights of all people to have access to adequate food in Nigeria. Explicit, rigorous, and transparent literature search was undertaken and many articles were assessed and reviewed. Although the results established a mutual relationship between food system and food security, existing literature have widely failed to take interactions between food systems, food security and policy into account. While food production is used as an entry point to improving food system sustainability, the quest for food security are undermining transition towards sustainable food systems. It was found that without right policies in place, it may be difficult to have food systems that are sustainable and ensure food security. This chapter provides a useful contribution to policy, and research on transitions towards sustainable food system. Any policy intervention to address one part of the food systems will impact on other parts and will determine whether a country is food secure or not. Enabling policy environment is therefore essential in ensuring a sustainable food system and for the attainment of food security.

DOI: 10.4018/978-1-7998-2599-9.ch001

BACKGROUND TO THE STUDY

Food system is a web of multiple interconnecting elements (Hospes & Brons, 2016; Béné, et al., 2019) and is at the center of global environmental, social, and economic challenges such as resource scarcity, ecosystem degradation, and climate change (Freibauer et al., 2011; Garnett, 2014; Gladek et al., 2016; IPES-Food, 2015; Lang, 2009; Searchinger et al., 2013). It is the interconnected system of everything and everybody that are involved in bringing food from farm to fork and beyond (Parsons et al., 2019). Agriculture is the foundation of all food systems and *food systems* are at the heart of many of the major *challenges* facing the world today. This may not be unconnected to the fact that agricultural produce is the primary source of most nutrients. If agriculture cannot supply all the essential nutrients in the quantity required for good health and productive lives, malnutrition develops (Miller & Welch, 2013). Food systems are linked to the nutritional wellbeing and heath of individuals and populations through the nutrients and other bioactive components contained in the foods they supply. Food system is a term used frequently in discussions about nutrition, food, health, community economic development and agriculture and is influenced by social, political, cultural, technological, economic and natural environments (High-Level Panel of Experts [HLPE], 2014; United Nations Environment Programme [UNEP], 2016; Global Panel, 2016; HLPE, 2017). The concept of food system provides a framework for an integrated description of two-way interaction of food with both natural resources and socio-economic conditions. The food we grow, harvest, process, trade, transport, store, sell and consume is the essential connecting thread between people, prosperity, and planet. A food system involves all processes and infrastructure involved in satisfying a population's food security (Porter et al., 2014). Food systems include food security outcomes (availability, accessibility, utilization and stability) of food as well as other socioeconomic and environmental factors (Ericksen, 2008; Ericksen et al., 2010; Ingram, 2011; Kearney, 2010).

Current food systems in Nigeria is observed not to be delivering food security and healthy food for everyone, nor are they sustainable using the limited natural resource inputs. While food production has more than doubled and diets have become more varied (and often more energy-intense); a lot of people are still hungry, suffering from micronutrient deficiencies (in particular of vitamin A, iodine, iron and zinc) and overweight or obese. The complexity of the food system creates challenges for identifying and incorporating food security supporting policy opportunities. Due to the breadth of policies, there are possibilities of some policies negating another policy's effectiveness. For example, the benefits of local community incentives for more fruits and vegetables in convenience stores may be overwhelmed by federal or state policies that create a favourable business environment for the production of highly processed foods. Food related studies seem not to have paid due attention to the linkages between sustainable food systems, food security and food policy.

This chapter therefore synthesizes the interrelationships between food systems, food security and food policy goals, the challenges, and how to overcome them in the case of Nigeria. The chapter addresses the following questions:

1. What are the challenges facing food systems?
2. How do food systems link to food security and food policy?
3. What are the missing links in connecting food systems, food security and food policy in Nigeria?
4. How can food systems and policy help to address the problem of food insecurity in Nigeria?

The main objective of the chapter is to explain the term 'food systems' in greater detail, and show linkages between food systems, food security and policy, analyze the major issues associated with food systems, food security and policy in Nigeria, and draw out the kinds of support that can be given for sustainable food systems that can leads to a food secure nation.

Justification of the Study

The 2030 Agenda for Sustainable Development clearly shows that move towards sustainable food systems is crucial to achieving sustainable development (Table 1). Such a transition is vital to achieving sustainable food and nutrition security for present and future generations. Only a few studies however address the relation between food systems sustainability and food security. A better understanding of the linkages between sustainable food systems and food security is necessary to achieve the second Sustainable Development Goal (SDG 2), 'Zero Hunger' (End hunger, achieve food security and improved nutrition and promote sustainable agriculture) in the context of the 2030 Agenda. The food systems are changing rapidly around the world especially in terms of production, dietary patterns and nutrition habits. The role of policy in tackling food systems challenges in a connected way therefore requires more attention now ever than before (Parsons & Hawkes, 2018; Global Panel on Agriculture and Food Systems for Nutrition, 2014). The sustainability of the food system is critical not just to those working in the food sector, but to everybody. Findings from this chapter will help to understand the food system complexities, connections between food security policy and food systems, and the challenges of sustainable food systems. The findings will help on how to explore, design innovation and policy options for overcoming food system challenges, and identify the most important issues regarding natural resources, as well as the opportunities for effective policy, social and/or technical interventions.

MATERIAL AND METHODS

The analysis builds on an in-depth review of the recent literature. More than sixty documents were identified on 'food system(s)', using two research engines: Science Direct and Google Scholar. The research scanning included academic research documents, journal articles, books and book chapters, government and international institution studies, reports, working papers, and other gray literature sources, mostly published between 2000 and 2019.

RESULTS AND DISCUSSION

An Overview of the Interrelationships Between Food Systems and Food Security

Food system is "the full set of actors, resources, processes and activities that encompass the domains of food production, processing, distribution, consumption and the disposal of waste, and the outcomes of these activities, including nutrition and health, socioeconomic wellbeing and environmental quality, as well as the tradeoffs and synergies between the various outcomes (Melesse et al., 2017). By implication, food system is made up of a number of activities (see Figure 1) whose primary aim is to increase food

Table 1. The place of food systems in the 2030 Sustainable Development Goal Agenda

	Goal	Relevance to Food Systems
1	No poverty	Almost 80% of poor people live in rural areas
2	Zero hunger	We produce enough food for everyone, yet about 800 million go hungry
3	Good health and well-being	Good health starts with nutrition
4	Quality education	Nutritious food is critical to learning
5	Gender equality	Women produce half the world's food, but have much less access to land
6	Clean water and sanitation	Sustainable agriculture holds potential to address water scarcity
7	Affordable and clean energy	Modern food systems are heavily dependent on fossil fuels
8	Decent Work and Economic Growth	Agricultural growth in low-income economies can reduce poverty by half
9	Industry, innovation and infrastructure	Agriculture accounts for a quarter of gross domestic product (GDP) in developing countries
10	Reduced inequalities	Land reforms can give fairer access to rural land
11	Sustainable cities and communities	Rural investment can deter unmanageable urbanization
12	Responsible consumption and production	One third of the food we produce is lost or wasted
13	Climate action	Agriculture is key in responding to climate change
14	Life below water	Fish gives 3 billion people 20% of their daily animal protein
15	Life on land	Forests contain over 80% of the world's terrestrial biodiversity
16	Peace, justice and strong institutions	Ending hunger can contribute greatly to peace and stability
17	Partnerships for the goals	Partnerships help raise the voice of the hungry

Source: Adapted from FAO (2017)

security (van Berkum et al., 2018). Food system is the network of activities connecting people to their food. It covers all stages of the value chain from growing and harvesting agricultural products through processing, packaging, transporting, selling, cooking, consuming, and the disposal of waste food and packaging (UNEP, 2016). The food system is the biggest user of key natural resources, such as terrestrial and marine biodiversity, soils, freshwater, minerals and fossil fuels (UNEP, 2016). Issues concerning the food system include governance and economics of food production, its sustainability, the degree to which food is wasted, how food production affects the natural environment and the impact of food on individual and population health. A key characteristic of the food system is the extensive linkages, interdependencies and feedback loops between value chain stages and the wider environment, society and economy.

The term food security on the other hands has an important history and represents a key concept for policymakers (Bureau & Swinnen, 2018; Candel & Biesbroek, 2018). It means many things to many people but despite this variability, the concept continues to features in local, national and international food policy discourse. Food security has been used in nearly 200 different ways across the world (Smith et al., 1993). As early as 1992, Maxwell and Smith (1992) reviewed more than 180 items discussing concepts and definitions of food security. Many of these earlier definitions however centered on food production. More definitions have in recent times emerged of food security highlighting access to food (Department for Environment, Food and Rural Affairs [DEFRA], 2006) as one of the key components in keeping with the 1996 World Food Summit definition (FAO, 1996) that defined food security as a situation when "all people, at all times, have physical and economic access to sufficient, safe, and nutritious food to meet their dietary needs and food preferences for an active and healthy life." The worldwide attention on food access was given impetus by the food "price spike" in 2007-2008, triggered by a complex set of long and short-term factors (FAO, 2009; von Braun & Torero, 2009). Today, food security is built on four pillars (Committee on World Food Security, 2012; Ericksen, 2008; FAO, WFP, & IFAD, 2013; United Nations System High Level Task Force on Global Food Security, 2011); food availability (that is, sufficient quantities of food available on a consistent basis); food access (that is, having sufficient resources to obtain appropriate and nutritious foods); food use/utilization (that is, appropriate use, based on knowledge of basic nutrition and care); and stability in food availability, access and utilization.

Increase in the efficiency and productivity of food systems have resulted in successes around the world in reducing the prevalence of hunger and improving nutrition. To ensure food security, all components of food systems need to be sustainable, resilient, and efficient. Where food systems do not perform adequately, food security and nutrition are threatened. Linkages between food security and food system are therefore both functional and operational.

The food security status of any group can be considered as the principal outcome of food systems, if these systems are defined broadly and generically. Attaining food security will require ensuring the sustainable use of natural resources and lowering the environmental impacts of food system activities. Changes towards sustainable food systems affect food availability (e.g. Ely et al., 2016; Jurgilevich et al., 2016; Kuokkanen et al., 2017; Levidow, 2015; Pant, 2016), food access (e.g. Audet et al.,2017; Kuokkanen et al., 2017), food utilization (e.g. Davies, 2014; Ely et al., 2016; Jurgilevich et al., 2016) and stability (e.g. Marsden, 2013) either positively or negatively. Thus the recent literature about the relationship between food security and food systems emphasizes the need for the food system to be not only sustainable but also resilient enough to cope with threats and uncertainties (often in relation to climate change) (Global Panel, 2016; World Wide Fund for Nature, 2016; UN, 2015; HLPE, 2017): concepts like 'adaptability' and 'transformability' play a key role in enhancing the resilience of the food system. Resilience thinking is a form of systems thinking that embraces the need to change in order to survive (Ge et al., 2017; Tendall et al., 2015). Sustainable solutions for a sufficient supply of healthy food requires changes in the food system, such as more efficient use of natural resources and less waste.

Relationship Between Food Systems and Food Policy

Food policy implies the setting of goals for the food system, such as production, environmental impact and nutrition, and "determining the processes of achieving these goals" (Pinstrup-Andersen & Watson, 2011). Food policies are designed to influence the operation of the food and agriculture system. Policies are made around production and processing techniques, marketing, availability, utilization and consumption of food. Policy shapes who eats what, when, where and at what cost (Lang et al., 2009). Food policymakers engage in activities such as regulation of food-related industries, establishing eligibility standards for food assistance programs for the poor, ensuring safety of food supply, food labeling, and even the qualifications of a product to be considered organic. Policies can be used to address problems of food system or used as a threat to food system. There can be policy interventions that aim at promoting foods which are generally good for people's health and also protect nature.

For instance, policies on agricultural land affect who farms. Animal welfare policy influences how animals reared for food are treated. Food trade policies have bearing on what is imported and exported, as well as on transport costs, tariffs and the global competitiveness of national food businesses. Food aid policies affect the type and effectiveness of assistance to countries experiencing conflict and instability, and food safety policies establish the mechanisms for reducing food contamination. Food waste policies have implications for environmental resources. Policies on food education impact on people's cooking skills. Food labelling guidance affects consumers' knowledge about what they eat. Nutrition policies can influence the standards of food on sale and food advertising. Social policy can limit how much money people have for food. Policies on labour influence how much time and energy people have to shop for and prepare meals. Rules on occupational health direct how farms and food businesses support the health of their staff. Energy policy affects land use and the cost of fuel for food production. Policies on migration determine who works where in the food system. Exchange rate policy affects the profitability of food imports and exports. Planning and tax rules can encourage or discourage investments by farmers. Government investment in research influences food industry innovation.

Changing Food Systems in Nigeria

Nigeria is one of the countries where transition nutrition is existing leading to compounding problems of either under-nutrition or over-nutrition. Urban food demand is different from that in rural areas (Brauw & Herskowitz, 2019). It is now well established that rising household incomes lead to unhealthier diets by favouring an increased consumption of fats, oils, sugar, animal-based products and processed foods unlike most industrial countries (for example, the United States or the United Kingdom), the effects of increased income have generally been considered as beneficial, resulting in better quality diets, better healthcare, lower morbidity and mortality from infectious diseases and lower risk of obesity. Consumers are seeking for more convenient foods (processed food) resulting in a quickly evolving food system. Food consumers are more sedentary in their lifestyles making them susceptible to malnutrition in the form of overweight and obesity. At the same time, undernutrition also remains a substantial concern, particularly in rural areas, while micronutrient deficiencies are a persistent problem throughout the country. The poorest families have the worst diets from a nutritional standpoint because lower-income people have less access in their immediate neighborhood to affordable healthy foods like fruit and vegetables that are relatively more expensive than high-calorie foods.

Production methods used by smallholder Nigerian farmers who are mainly involved in production of food have not changed significantly. Some of the farmers now sell through an increasingly complex array of food chains such as food shopping complex and supermarkets. Food shopping centres and supermarket chains are gaining power and influence in the food system with heavy use of advertising. Adegboye et al., (2016) in their study identified evidence of the dietary changes in Nigeria from 'traditional' to more processed foods, due to the difference in the nutrition transition across states and differences in the food systems, including production, processing, distribution, trade, food environments and consumer behaviour (HLPE, 2017). This is because the food systems determine availability, affordability, convenience, and desirability of various foods.

Causes of Food Systems Failure and the Effect on Food Security

The current food systems in Nigeria are not delivering a huge range of foods to the citizens because they are associated with multiple challenges. A review of these challenges shows that the Nigeria food systems are shaped by a multitude of factors, such as geography, demography, urbanization, globalization; socioeconomic status, income, marketing, consumer attitude; religion and culture. These factors impact also on food security at national, local, and household level. Food system failure is a significant disruption in the provision of food such that the food security of the community is compromised. Intermediate level events that result in food system failure can be broadly classified into failures of accessibility, availability, and acceptability. Food systems fundamentally depend on natural resources, such as land, soil, water, biodiversity, minerals, biomass and fossil fuels (Figure 1). Food system failures could occur through resource depletion; contamination or other damage; disease; and shortages or high prices. Production failure may also occur due to events that result in a season with production below expectations due to weather, plant or animal disease infestation, business failure, or ineffective business management practices. Key statistics show that in many cases these resources are not currently managed sustainably or efficiently: an estimated 33% of soils are moderately to highly degraded due to erosion, at least 20% of the world's aquifers are overexploited, over 80% of the input of minerals (e.g. phosphate) do not reach consumers' plates and 29% of 'commercial' fish populations are overfished. Due to population growth, changes in dietary patterns driven by growing wealth (more meat, dairy and fish consumption) and climate change, the pressures on natural resources are expected to increase over the coming decades. External food systems challenges are related to rapid migration towards (peri-)urban settlements that may increase the likelihood of food losses and waste in longer-distance food chains.

Since food systems involve the ways in which food is farm, process, sold and consume, the environment affects food systems and food systems also affect the environment including the depletion of soil quality and biodiversity (loss of plant, bird and insect life), as well as increasing greenhouse gas emissions (e.g. from transportation). As a challenge, some of the Nigerian agricultural lands are gradually being used for industrial and residential purposes leading to limited amount of agricultural land offered for sale or rent and the poor access to capital. This is making it difficult for the younger generation to enter farming, while the high proportion of older farm operators moderates long-term return investments.

All aspects of food systems potentially resulted to climate change and the security of food is affected by climate change. This is severe in Nigeria where more than 70% of agriculture is rain fed. The effect of drought is becoming worse in some parts of the country because more greenhouse gases such as carbon dioxide, methane and nitrous oxide are released into the atmosphere thereby causing air temperatures to rise and more rapid evaporation of moisture from water bodies. Investment in irrigation system of farming will allow farmers to augment the amount of water supplied to plants at regular intervals to aid crop growth and re-vegetation of dry areas during periods of inadequate rainfall.

Food Unavailability

Food availability is a situation where food exist with no barrier of getting such. It depends on the level of local production, imports, stock levels and net trade in food items. It becomes unavailable when there are barriers to prevent its acquisition by the community. Food supply requires both purchased food and donated food to reach sufficiency and failure of either system will lead to hunger. Food is unavailable in Nigeria due to supply chain failure and failure of the food donation system (emergency government assistance programs). Food supply chain failures result at points of production, processing, wholesale, distribution, retail, or food donation source points.

Food Inaccessibility

This is a situation where people have no physical and economic ability to acquire food. Food inaccessibility in Nigeria is observed to be as a results of two major factors. These are economic and physical barriers.

Economic barriers make food inaccessible or unaffordable to the community due to very high food prices or decreases in net income. High food prices have multiple effects, including decreased supply (caused by production disruption); and increases in production, processing, distribution or retail cost that are passed on to the consumer, such as increases due to higher fuel costs (Chodur et al., 2018). The amount of income available to purchase food are influenced through changes in the amount of the population making a living wage, unemployment, or the failure of safety nets that supplement earned income. Physical barriers also affect ability to travel to the provisioning point. Provisioning points are inaccessible due to events including transportation barriers, lack of proximity to a provisioning point, or interruptions to normal means of transit.

Food Unacceptability

Food acceptability encompasses food security component of utilization (or biological use of the nutrients contained in food) as well as the cultural acceptability of the food, which is available and accessible (Hammelman & Hayes-Conroy, 2015). Some foods are accessible to the community and available at a vending point but the acceptability of such food is still a challenge. Most foods are not acceptable if the food supply is not nutritionally adequate, not medically good, not religiously/culturally appropriate, or food is distasteful to consumers for reasons including flavor, appearance, or actual or perceived quality. A lot of people are beginning to know that one of the primary effects of unhealthy food system is the high rates of diet related ill-health such as obesity and overweight. Cultural beliefs and taboos as well as religious beliefs are also found to influence the food choices of consumers (Ijewere & Odia, 2012).

Challenges of Food Systems and Food Security in Nigeria

Although Nigeria has the largest and most elaborate National Agricultural Research Systems (NARS) in sub-Sahara Africa (Table 2) operating under the umbrella body known as Agricultural Research Council of Nigeria (ARCN) to address the challenges facing agriculture, the country food systems is still being confronted with some challenges. Nigeria is still a food-deficit country. The high urban demand is met through cheap food imports, which further lowers the incentives for investments in Nigerian agriculture. As such, the mismatches between supply and demand at many levels and in many dimensions seem to be mutually reinforcing each other. Environmental trends, such as soil degradation, climate change, water scarcity, deforestation and decreasing biodiversity pose further threats to the food system. Over 40% of people are reported to be food insecure, 50% malnourished with rising overweight and disease burdens. The small-scale farmers use low inputs, fertilizers and the pastoralists compete for access to land with high level of erosion, flood and water pollution, high food import bill, food waste and loss.

Achieving food security had failed in Nigeria due to the food systems commonly in used. The use of crude technology rather than investment in improved technologies is clearly a dominant factor affecting food security in Nigeria. Other major pitfall are poor storage systems and lack of processing capacity. For instance, in Nigeria, a tomato farmer who finds it difficult to sell his entire baskets of vegetables after harvesting may end up losing a huge sum of money because he has no way of storing or processing them to avoid spoilage. This loss of food accounts for some of the increase in price of food commodities, as farmers look to cover their losses one way or another. Staple food such as beans, which are common sources of plant protein are lost to insects. Hence, farmers and traders use pesticides to try to prevent weevil infestation in crops post-harvest and before taking them to the market with residual effects on the health of the consumers.

Figure 1. Links between food system, food security and policy goals
Source: Adapted from Melesse et al., (2017).

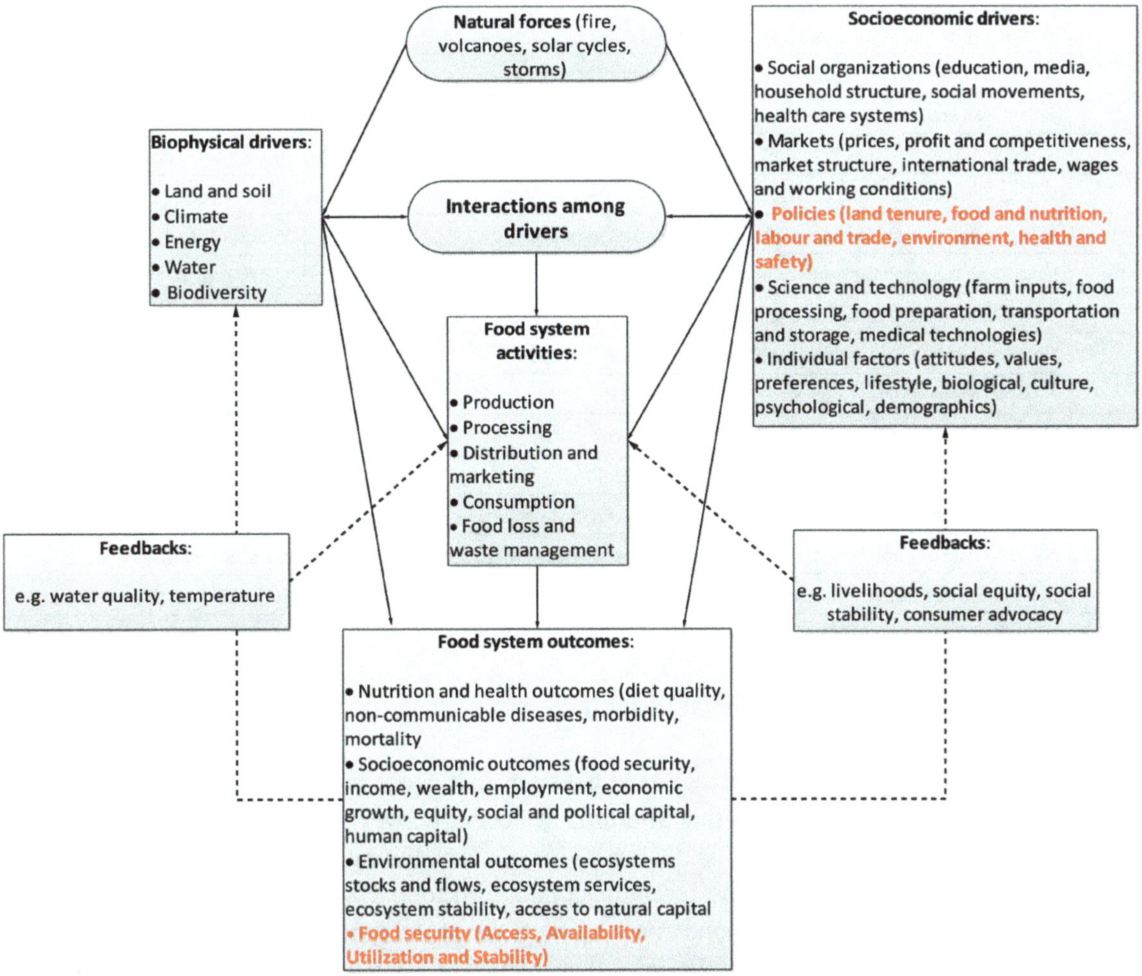

Communities depending on livestock, those who practice mixed farming on a small scale, and consumers in cities all have specific demands regarding livestock production systems and food and nutrition security implying that pastoralism (and livestock keeping) is an integral part of the food system and constitutes an important component of the agricultural economy in Nigeria. It is also closely linked to the social and cultural lives of pastoralists and of resource-poor farmers for whom animal ownership ensures varying degrees of sustainable farming, economic stability and food and nutrition security. The scarcity of resources has also been identified as one of the major drivers of the pastoral-farmer conflict in the Northern of Nigeria affecting sustainable food systems and food security.

Table 2. List of national agricultural research systems in Nigeria

S/N	Name of the Institutes	Location
1	National Root Crops Research Institute (NRCRI)	Umudike, Abia State
2	National Horticultural Research Institute (NIHORT)	Ibadan, Oyo State
3	Cocoa Research Institute of Nigeria (CRIN) Ibadan	Ibadan, Oyo State
4	Nigerian Institute for Oil Palm Research (NIFOR)	Benin-City, Edo State
5	Rubber Research Institute of Nigeria (RRIN)	Benini-City, Edo State
6	Nigerian Institute for Oceanography and Marine Research (NIOMR)	Victoria Island, Lagos State
7	Lake Chad Research Institute (LCRI)	Maiduguri, Borno State
8	National Veterinary Research Institute (NVRI)	Vom, Plateau State
9	National Institute for Fresh-Water Fisheries Research (NIFFER)	New Bussa, Niger State
10	Nigerian Stored Products Research Institute (NSPRI)	Ilorin, Kwara State
11	National Cereal Research Institute (NCRI)	Badeggi, Niger State
12	Institute for Agricultural Research & training (IAR&T)	Ibadan, Oyo State
13	National Animal Production Research Institute (NAPRI) Shika Zaria	Zaria, Kaduna State
14	National Agricultural Extension & Research Liason Services (NAERLS)	Zaria, Kaduna State
15	Institute for Agricultural Research (IAR)	Zaria, Kaduna State

CONCLUSION AND RECOMMENDATIONS

This chapter illustrates the interactions that exist between food systems, food security and agricultural and food policy. The chapter concludes that interlinkages occur between the multiple components of the food system and policy which must not be dealt with in isolation, but rather be addressed as an interconnected system. Food-related policies are rarely designed with food systems as their primary objective or their primary concern. Policy decisions affect many parts of the food system, including agriculture, consumer knowledge and food choices. Policies that influence food systems' performance also affect food security. Agricultural and food policy is therefore essential for a sustainable food system. Policy that affect a components of food system affects the others components because of the interdependencies of the components which directly or indirectly influence food security. Food systems are directly linked to issues on food security with interactions between policies and processes at local, national, regional and global levels. Challenges in food system affect food security outcomes. This systematic review represents a useful contribution to research on transitions towards sustainability in agriculture and food sectors, and provides insights into how such research can contribute to addressing the challenges of food insecurity and malnutrition. Findings from the chapter suggests the need to move beyond silos by fostering cross-sectoral collaboration and the integration of food and agricultural policy makers, those in agro-food sustainability transitions and food security research fields. Thus, more attention needs to be paid to understanding and promoting policy actions that support nutrition-enhancing food systems as a whole. Policy solutions that will protect food system and ensure food security will also require greater political and social will from multiple sectors of our society.

Pragmatic way forward to overcome the challenges of agricultural and food policy- sustainable food systems-food security nexus in Nigeria include:

- Mass education of people on the understanding of food systems, the failures, system boundaries and spatial dimensions.
- Fixed investments for adaptation of production systems (provided by international banks) and/or climate-smart finance for climate mitigation purposes (provided by NGOs) are required for food system responses to climate change.
- Elimination of subsidies that encourage unsustainable food production or practices but stimulate the demand for healthy food and encouraging producers to invest in more sustainable production methods.
- Creation of adequate legal frameworks to secure property rights and land tenure and regulate access to, and use of water, biodiversity and ecosystems services.
- Creation of adequate legal frameworks to regulate environmental impacts on food systems, for example to prevent nutrient losses at all stages, especially in the livestock sector.
- Investment in technology and research development for locally suitable seeds and breeds (with proper infrastructure, distribution system, quality assurance and certification schemes) need to be strengthened.
- Adoption of consumption-oriented policies to stimulate healthy and sustainable eating patterns (for example the creation of a healthy food environment). Governments, in collaboration of course with the private sector and civil society (including NGOs), can directly and indirectly encourage the consumption of healthy food through subsidies, taxes, dietary guidelines, labelling, information, research and other measures.

REFERENCES

Adegboye, O. R., Smith, C., Anang, D., & Musa, H. (2016). Comparing and Contrasting Three Cultural Food Customs from Nigeria and Analyzing the Nutrient Content of Diets from These Cultures with the Aim of Proffering Nutritional Intervention. *Critical Reviews in Food Science and Nutrition*, *56*(15), 2483–2494. doi:10.1080/10408398.2013.862201 PMID:25830896

Audet, R., Lefèvre, S., Brisebois, É., & El-Jed, M. (2017). Structuring tensions and key relations of Montreal seasonal food Markets in the Sustainability Transition of the Agri-food sector. *Sustainability*, *9*(3), 320. doi:10.3390u9030320

Bene, C., Prager, S. D., Achicanoy, H. A., Toro, P. A., Lamotte, L., Cedrez, C. B., & Mapes, B. R. (2019). Understanding food systems drivers: A critical review of the literature. *Global Food Security*, *23*(1), 149–159. doi:10.1016/j.gfs.2019.04.009

Brauw, A., & Herskowitz, S. (2019). *Consumer demand and processed foods in Nigeria: a complex, evolving picture*. Retrieved from https://a4nh.cgiar.org/2019/02/28/consumer-demand-and-processed-foods-in-nigeria-a-complex-evolving-picture/

Bureau, J. C., & Swinnen, J. (2018). EU policies and global food security. *Global Food Security*, *16*, 106–115. doi:10.1016/j.gfs.2017.12.001

Candel, J. J. L., & Biesbroek, R. (2018). Policy integration in the EU governance of global food security. *Food Security*, *10*(1), 195–209. doi:10.100712571-017-0752-5

Chodur, G. M., Zhao, X., Biehl, E., Mitrani-Reiser, J., & Neff, R. (2018, July 3). Assessing food system vulnerabilities: A fault tree modeling approach. *BMC Public Health*, *18*(1), 817. doi:10.118612889-018-5563-x PMID:29970077

Committee on World Food Security. (2012). *Coming to terms with terminology: Food security, nutrition security, food security and nutrition, food and nutrition security*. Retrieved online on 10 Nov 2019 at http://www.fao.org/fsnforum/sites/default/files/file/Terminology/MD776(CFS___Coming_to_terms_with_Terminology).pdf

Davies, A. R. (2014). Co-creating sustainable eating futures: Technology, ICT and citizen-consumer ambivalence. *Futures*, *62*, 181–193. doi:10.1016/j.futures.2014.04.006

Department for Environment, Food and Rural Affairs (DEFRA). (2006). *Food security and the UK: An evidence and analysis paper.* Retrieved from http://archive.defra.gov.uk/evidence/economics/foodfarm/reports/documents/foodsecurity.pdf

Ely, A., Geall, S., & Song, Y. (2016). Sustainable maize production and consumption in China: Practices and politics in transition. *Journal of Cleaner Production, 134*(Part A), 259–268. doi:10.1016/j.jclepro.2015.12.001

Ericksen, P., Stewart, B., Dixon, J., Barling, D., Loring, P., Anderson, M., & Ingram, J. (2010). The value of a food system approach. In *Food Security and Global Environmental Change* (pp. 25–45). London, UK: Earthscan.

Ericksen, P. J. (2008). Conceptualizing food systems for global environmental change research. *Global Environmental Change*, *18*(1), 234–245. doi:10.1016/j.gloenvcha.2007.09.002

Fanzo, J., Cogill, B., & Mattei, F. (2012). *Metrics of sustainable diets and food systems*. Technical Brief - Biodiversity International. Retrieved from https://www.bioversityinternational.org/uploads/tx_news/Metrics_of_sustainable_diets_and_food_systems_1572.pdf

FAO, WFP, & IFAD. (2013). The State of Food Insecurity in the World 2013: The Multiple Dimensions of Food Security. Rome: Author.

FAO. (2017). *Food and Agriculture: Driving action across the 2030 Agenda for Sustainable Development.* Rome: Food and Agriculture Organization of the United Nations (FAO). Retrieved from http://www.fao.org/3/ai7454e.pdf

Freibauer, A., Mathijs, E., Brunori, G., Damianova, Z., Faroult, E., Gomis, J. G., & Treyer, S. (2011). *Sustainable food consumption and production in a resource-constrained world. European Commission - Standing Committee on Agricultural Research (SCAR).* The Third SCAR Foresight Exercise. doi:10.2777/49719

Garnett, T. (2014). Three perspectives on sustainable food security: Efficiency, demand restraint, food system transformation. What role for life cycle assessment? *Journal of Cleaner Production*, *73*, 10–18. doi:10.1016/j.jclepro.2013.07.045

Ge, L., Anten, N. P. R., van Dixhoorn, I. D. E., Feindt, P. H., Kramer, K., Leemans, R., ... Sukkel, W. (2016). Why we need resilience thinking to meet societal challenges in biobased production systems. *Current Opinion in Environmental Sustainability*, *23*, 17–27. doi:10.1016/j.cosust.2016.11.009

Gladek, E., Fraser, M., Roemers, G., Sabag, M. O., Hirsch, P., & Kennedy, E. (2016). *The global food system: An analysis*. Retrieved from https://www.metabolic.nl/publications/global-food-system-analysis

Global Panel on Agriculture and Food Systems for Nutrition. (2014). *How can Agriculture and food system policies improve nutrition?* Technical brief, London, UK: Global Panel on Agriculture and Food Systems for Nutrition. Retrieved from http://www.glopan.org/sites/default/files/Global%20Panel%20 Technical%20Brief%20Final.pdf

Global Panel on Agriculture and Food Systems for Nutrition. (2016). Food systems and diets: Facing the challenges of the 21st century. London, UK: Author.

Hammelman, C., & Hayes-Conroy, A. (2015). Understanding cultural acceptability in urban food policy. *Journal of Planning Literature*, *30*(1), 37–48. doi:10.1177/0885412214555433

HLPE. (2014). Food losses and waste in the context of sustainable food systems. A report by the High Level of Experts on Food Security and Nutrition of the Committee on World Food Security.

HLPE. (2017). Nutrition and food systems. A report by the High Level Panel of Experts on Food Security and Nutrition of the Committee on World Food Security, Rome.

Hospes, O., & Brons, A. (2016). Food system governance: A systematic literature review. In Food Systems Governance: Challenges for Justice, Equality and Human Rights (pp. 13-42). Abingdon-on-Thames: Routledge.

Ijewere, A. A., & Odia, E. O. (2012). Cultural Influences on Product Choice of the Nigerian Consumer. *Indian J. Econ. Bus.*, *11*, 139–155.

Ingram, S. J. I. (2011). *From Food Production to Food Security: Developing interdisciplinary, regional-level research*. Wageningen University. Retrieved from edepot.wur.nl/176450

IPES-Food. (2015). *The new science of sustainable food systems: Overcoming barriers to food systems reform*. Retrieved from http://www.ipes-food.org/images/Reports/IPES_report01_1505_web_br_pages.pdf

Jurgilevich, A., Birge, T., Kentala-Lehtonen, J., Korhonen-Kurki, K., Pietikainen, J., Saikku, L., & Schösler, H. (2016). Transition towards circular economy in the food system. *Sustainability*, *8*(1), 1–14. doi:10.3390u8010069

Kearney, J. (2010). Food consumption trends and drivers. *Philosophical Transactions of the Royal Society of London. Series B, Biological Sciences*, *365*(1554), 2793–2807. doi:10.1098/rstb.2010.0149 PMID:20713385

Kuokkanen, A., Mikkilä, M., Kuisma, M., Kahiluoto, H., & Linnanen, L. (2017). The need for policy to address the food system lock-in: A case study of the Finnish context. *Journal of Cleaner Production*, *140*, 933–944. doi:10.1016/j.jclepro.2016.06.171

Lang, T. (2009). *Food security and sustainability: The perfect fit.* Retrieved from http://www.sd-commission.org.uk/data/files/publications/SDCFoodSecurityPositionPaper.pdf

Lang, T., Barling, D., & Caraher, M. (2009). *Food policy: integrating health, environment and society.* Oxford: Oxford University Press. doi:10.1093/acprof:oso/9780198567882.001.0001

Levidow, L. (2015). European transitions towards a corporate-environmental food regime: Agroecological incorporation or contestation? *Journal of Rural Studies*, *40*, 76–89. doi:10.1016/j.jrurstud.2015.06.001

Marsden, T. (2013). From post-productionism to reflexive governance: Contested transitions in securing more sustainable food futures. *Journal of Rural Studies*, *29*, 123–134. doi:10.1016/j.jrurstud.2011.10.001

Maxwell, S., & Frankenberger, T. (1992). *Household Food Security: Concepts, Indicators and Measurements: A Technical Review*. New York: UNICEF and IFAD.

Melesse, M. B., Berg, M., Béné, C., Brauw, A., Brouwer, I. D., & Remans, R. (2017). *Improving Diets through Food Systems: Metrics and Methods for Analysis*. Unpublished Article.

Miller, D. D., & Welch, R. M. (2013). Food system strategies for preventing micronutrient malnutrition. *Food and Agriculture Organization of the United Nations*. Available online at http://www.fao.org/fileadmin/templates/esa/Papers_and_documents/WP_13_06_Miller_Welch.pdf

Pant, L. P. (2016). Paradox of mainstreaming agroecology for regional and rural food security in developing countries. *Technological Forecasting and Social Change*, *111*, 305–316. doi:10.1016/j.techfore.2016.03.001

Parsons, K., & Hawkes, C. (2018). *Connecting food systems for co-benefits: How can food systems combine diet-related health with environmental and economic policy goals? Policy Brief 31*. Copenhagen: World Health Organisation. Retrieved from http://www.euro.who.int/__data/assets/pdf_file/0007/387070/policy-brief-31-austria-eng.pdf?ua=1

Parsons, K., Hawkes, C., & Wells, R. (2019). What is the food system? A Food policy perspective. In *Rethinking Food Policy: A Fresh Approach to Policy and Practice*. London: Centre for Food Policy. Available online at https://www.city.ac.uk/__data/assets/pdf_file/0008/471599/7643_Brief-2_What-is-the-food-system-A-food-policy-perspective_WEB_SP.pdf

Pinstrup-Andersen, P., & Watson, D. D. (2011). *Food policy for developing countries: the role of government in global, national, and local food systems*. Ithaca: Cornell University Press.

Porter, J. R., Xie, L., Challinor, A. J., Cochrane, K., Howden, S. M., Iqbal, M. M., ... Travasso, M. I. (2014). Food security and food production systems. In *Climate Change 2014: Impacts, Adaptation, and Vulnerability. Part A: Global and Sectoral Aspects. Contribution of Working Group II to the Fifth Assessment Report of the Intergovernmental Panel on Climate Change* (pp. 485–533). Cambridge, UK: Cambridge University Press.

Searchinger, T., Hanson, C., Ranganathan, J., Lipinski, B., Waite, R., Winterbottom, R., & Heimlich, R. (2013). *Creating a sustainable food future: Interim findings. A menu of solutions to sustainably feed more than 9 billion people by 2050*. Retrieved from https://www.wri.org/sites/default/files/wri13_report_4c_wrr_online.pdf

Smith, M., Pointing, J., & Maxwell, S. (1993). *Household Food Security, Concepts and Definitions: An Annotated Bibliography*. Brighton: Institute of Development Studies, University of Sussex.

Tendall, D. M., Joerin, J., Kopainsky, B., Edwards, P., Shreck, A., Le, Q. B., ... Six, J. (2015). Food system resilience: Defining the concept. *Global Food Security*, *6*, 17–23. doi:10.1016/j.gfs.2015.08.001

UNEP. (2016). *Food Systems and Natural Resources*. A Report of the Working Group on Food Systems of the International Resource Panel.

United Nations. (2015). *Zero Hunger Challenge. Pathways to zero hunger: Transforming our Food Systems to Transform our World*. Rome: Committee on World Food Security.

United Nations System High Level Task Force on Global Food Security. (2011). *Food and nutrition security: Comprehensive framework for action. Summary of the Updated Comprehensive Framework for Action*. Rome: UCFA.

van Berkum, S., Dengerink, J., & Ruben, R. (2018). *The food systems approach: sustainable solutions for a sufficient supply of healthy food*. Wageningen: Wageningen Economic Research, Memorandum 2018-064, 32.

World Wide Fund for Nature (WWF). (2016). *The Global Food System: An Analysis*. WWF Netherlands & Metabolic Lab.

Chapter 2
Seafood Security and Sustainability Through Sustainable Development:
A Review of Turkish Seafood Market

Seda Yildirim
Tekirdag Namık Kemal University, Turkey

Merve Kaplan
https://orcid.org/0000-0001-5096-5546
Kocaeli University, Turkey

ABSTRACT

The world has understood that hunger is one of the most dangerous problems for the future. Accordingly, food security and sustainability are both important issues through sustainable development. This chapter highlights the role of seafood security and sustainability for sustainable development. In this context, seafood security and sustainability for Turkish seafood market was investigated. Turkey is a coastal country, which has accepted 2030 Sustainable Development Goals, with a seafood market and a good sample to investigate seafood sustainability. This study employed secondary data from TURKSTAT and FAO websites to determine Turkish seafood market profile. The study determined seafood security and sustainability based on five dimensions as availability, economic access, physical access, utilization, and stabilization. Seafood sustainability is vital for coastal countries because seafood market brings economic, social, and environmental benefits at the same time.

DOI: 10.4018/978-1-7998-2599-9.ch002

INTRODUCTION

The rising world population (7.7 billion currently – UN Department of Economic and Social Affairs, 2019) needs more food, fiber and fuel. These unlimited needs and wants of people mostly cause extinction of biodiversity, destruction of forest lands, farming areas, oceans and seas (Aladjadjiyan, 2012). The continuation of humanity is subject to environmental consciousness and awareness of settled economic and social life as fully as possible. over the last decade, environmental, social and economic policies made sustainable integration and sustainable development possible. Although sustainable development history is not so old, both developed and developing countries have transformed their economic policies and development strategies based on sustainability principle recently (Yıldırım et.al., 2016). The idea of sustainability began in 1960s (International Institute for Sustainable Development, 2012) but the formal adventure of sustainable world began with World Commission on Environment and Development (WCED) in 1987. Sustainable development term was introduced to the world by the report of WCED (1987) called as "Our Common Future". This report identified population and human resources, food security and sustainability, biodiversity, energy (sustainable energy resources), industry (sustainable production) and the urban challenge (sustainable city planning) as key action plans towards achieving a sustainable world (United Nations, 1987). Then, the United Nations (UN) (1972) determined the relationship between economy, society and environment through United Nations Conference on the Human Environment (UNCED). The Rio Declaration and Agenda 21 at the The Earth Summit (1992) and UN members emphasized the need for a sustainable development and action plan (Klarin, 2018). It can be said that real action plan was set up by Rio+20. The first sustainable development goals, Millennium Development Goals, were launched in 2012 and main themes of sustainable development were presented to countries. After Rio+20, the United Nations (UN) accelerated the process and Sustainable Development Goals in 2015 (IISD, 2012).

Food shortage will cause a big problem in the future, if there aren't any effective policies for food security and food sustainability (FAO, 2017). According to sustainable development goals, both developed and developing countries try to make their agricultural methods and technologies sustainable. For example, raising organic livestock and organic agriculture have become important movements for food security recently. However, these food security movements or action plans generally don't include seafood. Although international seafood trade is so high, seafood security has been ignored for a long time (Smith et.al., 2010). Seafood is an important and almost fundamental protein source for developing countries but an optional food for developed countries (Tlusty, 2012). It can be said that coastal developing or coastal emerging countries are more interested in seafood security and sustainability. Coastal countries generally prefer seafood as a fundamental food, just like crop or animal source foods. Seafood export and domestic sales also contribute substantially to coastal countries' economy (FAO, 1996). Turkey is a developing country surrounded on three sides by sea and can also open to ocean by Turkish Bosporus and Gibraitar Strait. Turkey a part of the coastal areas bordering the Mediterranean Sea and the Black Sea (TÜDAV, 2017). This study assessed the seafood security and seafood sustainability in Turkey.

FOOD SECURITY AND SUSTAINABLE DEVELOPMENT GOALS

There are different definitions of food security in the literature (FAO, 2002b). Food security was first formally defined by World Food Summit in 1974. FAO (2002b) summarized the historical development of food security concept as focusing on volume and supply of food, food security was accepted as "*the availability of adequate global food supplies to sustain food consumption and production with proper price*" in 1970s (ODI, 1997). A view of food security changed in 1980s and it was defined as "*access of all people at all times to adequate food for healthy life*" (World Bank, 1986). By 1990s, food security became a more important issue for the world leaders and World Food Summit (1996) defined it as "*when all people, at all times, have physical and economic access to sufficient, safe and nutritious food to meet their dietary needs and food preferences for an active and healthy life*" (FAO, 2002b).

In 2000s, food security was accepted as a multi-dimensional term with many indicators to determine global food security. The most common definition for food security came from World Food Summit (1996) as when people can acquire sufficient quantities of food anytime and anywhere and maintain their health by food, then food security has been achieved. (FAO, 2006). Food and Agriculture Organization of the United Nations (FAO) (2008a) defined food security through four dimensions as "*physical availability, economic and physical access, food utilization and stability of other dimensions*" (Table 1).

Table 1. Four dimensions of food security

Dimensions	Definitions
Physical availability:	This dimension presents quantity of food production and stock levels and other numerical amounts of food as food availability.
Economic and physical access:	The supply of food at the national or international level should be globally sufficient. In addition, household can afford food easily.
Food utilization:	The physical and economic access of food is not enough for preventing hunger. Also healthy and rich variety food should be provided.
Stability:	This dimension determines that people should not have any risk of being hunger based on climate change, economic factors or political instability.

Source: (adapted from FAO, 2008a)

Food security is mostly related with issues of sustainability and availability as much as agricultural trends to meet the world's food demand. Focusing on just food production or agricultural technology will not be enough for food security and sustainability in the long term (Sunderland et.al., 2013). In this point, food security is thought with sustainability to prevent hunger in the long term. The world has been agreed that food security and hunger are important issues for the humanity since 1990s (Duncan, 2017). Fighting food insecurity sustainably with the sustainable development goals prove that food security and ending hunger should be main goals for sustainable world (FAO, 2018a). In September 2000, world leaders accepted the United Nations Millennium Declaration and set out some goals which were aimed till 2015. These first sustainable goals were called the Millennium Development Goals (MDGs). The eight MDGs includes were: "reducing poverty and hunger; achieving global primary education; achieving global gender equity; reducing child mortality; improving health; fighting with HIV/AIDS and other diseases; providing environmental sustainability; and developing global partnership" (Asian development Bank, 2009). Halving the proportion of people suffering from hunger and poverty was the first goal of the MDG. Although the proportion of poverty decreased in low-income countries, the proportion of hunger did not decrease according to MDG 2011 report. This proved that fighting hunger and food insecurity need a holistic approach.

Food security became a formal goal for the world by 2012. However, fighting hunger has been argued and it is a main problem like poverty. It was found that the MDGs were not enough to improve poverty, hunger, environmental pollution or other global problems, therefore the United Nations prepared a new action plan as 2030 Agenda for Sustainable Development in September 2015. This Agenda includes new sustainable development goals that are expected to be achieved by 2030 (UN General Assembly, 2015).

The 2030 Sustainable Development Goals with 17 main goals, extended new themes like energy, clean water, lands, sustainable cities, oceans and seas (Table 2). The SDGs have fundamental changes and it has more detailed sub-goals (Sustainable Development Goals Knowledge Platform, 2015). For example, poverty and hunger are determined in separate titles and both of them have different sub-goals (Yıldırım & Bostancı, 2018).

Being a vital need for humanity, food security will provide human existence that food security and sustainability will remain a focus for sustainable world forever; because poverty causes hunger, environmental pollution causes hunger, global warming causes hunger and food insecurity. Goal 2 directly aims at providing sufficient food for every people in the world (Hepburn & Bellmann, 2018). When countries maintain food security, hunger will disappear as soon as possible. Thus, ending hunger is dependent on food security but food security is also depended on other SDGs in practice (CFS, 2015a). Table 2 guides us to understanding the relationship between food security and 2030 SDGs.

Table 2. The relationship between food security and 2030 SDGs

Goals	Aims[a]	The Relationship Between Food Security and SDGs
SDG 1: No Poverty	Its aims is to reduce poverty for all people over the worldwide. Most of people still live below the international poverty line. By 2030, countries aims to reduce poverty for every men, women, children and the proportion of poor citizens should fall by half.	Poverty and hunger is related together. Poverty always brings hunger or inadequate access to food. Low economic conditions cause inadequate access to food or inadequate food utilization (CFS, 2017)
SDG 2: Zero Hunger	It aims at providing quality, healthy and sufficient nutrition and food for people in everywhere. There are so many people in the world that still can't get healthy food or sufficient food in a day. By 2030, countries should provide food security and sustainability for humanity.	Food security means that people can have an economic and physical access to food. Rate of people or citizens who have inadequate food in a day, should be decreased by 2030. When countries provide economic and physical access to food for everyone, there will be food security. Goal 2 is directly related with food security (Perez-Escamilla, 2017).
SDG 3: Good Health and Well-Being	It aims at improving global health for people. Especially, under-5 mortality rate should be decrease in the least developed countries. Fatal diseases and contagious diseases should be reduced by effective treatments.	Food security includes access to healthy foods and various nutrition as much as physical or economic availability. With this goal, food utilization can determine health and well-being of people.
SDG 4: Quality Education	Equitable and quality education is a fundamental right for every girls and boys in the world. Providing free and quality education is important by 2030.	Poverty and lack of education increase hunger and food insecurity. Urban food insecurity mostly occurs because of urban poverty, unemployment, higher food prices and lower education (Mutisya et.al., 2016). In addition, rural countries which have mostly food insecurity and higher hunger rates, need higher education. Education can be vital factor for fighting with hunger in rural countries (De Muro & Burci, 2007)

continues on following page

Table 2. Continued

Goals	Aims[a]	The Relationship Between Food Security and SDGs
SDG 5: Gender Equality	By 2030, it is targeted at reducing every kinds of discrimination against women and girls and it is important to provide equal work, equal education and equal life-standards for women and girls in the world.	According to The Committee on World Food Security (2011), achieving food security and sufficient nutrition for women, men and children are related with sustainable development efforts. It is important to ensure that women have equal rights in health, education and food as men. (CFS, 2011).
SDG 6: Clean Water and Sanitation	Providing clean water for all people is an important aim. Rising population is a threat for clean drinking water in the future that protection of water and related eco-systems is important.	Water is key for life. Every kind of food and nutrition is based on water availability. For example, crops and livestock need water and water products all come from clean water. Keeping clean water sources and drinking water are important goals to achieving food security as much as sustainable development. (CFS, 2015b)
SDG 7: Affordable and Clean Energy	Energy demand should be met by sustainable energy resources. Energy consumption and energy need are increasing day-to-day. Most of energy resources are harmful for environment. So, sustainable energy resources can meet energy demand.	The agrifood chain needs energy to complete the process. Production of crops, fish, livestock and etc. need energy. Accordingly, energy type will influence food security. In addition, energy consumption causes environmental pollution, which is so important to prevent climate change in the long term (Sims et.al., 2015)
SDG 8: Decent Work and Economic Growth	Improving economy and providing job and income for every people in the world is an important goal.	Decent employment is important to providing food security and reducing poverty. Because buying food or reaching nutrition is related to income or purchasing power. People will buy food up to their income (FAO, 2012).
SDG 9: Industry Innovation and Infrastructure	Infrastructure system should be sustainable be of good quality and support economic and social development.	Industry innovation must include food industry with new and sustainable technologies in food production to achieve food security globally. For example, food production systems should be sustainable and healthy and rural and urban areas should be re-designed to protect forests, seas and agricultural areas (UNCTAD, 2017).
SDG 10: Reduce Inequalities	Every kind of discrimination should be prevented by 2030.	Unfortunately, most of the hungry people are women and girls. The index of global hunger shows that poverty and hunger are higher in countries with severe inequalities. In this context, reducing inequalities will support food security.(Asian Development Bank, 2013)
SDG 11: Sustainable Cities and Communities	By 2030, sustainable and green communities should be increased and city life of people should be re-designed based on principles of environmental protection and social well-being.	For sustainable world, urban life of people is a big challenge. Most urban population lives in polluted environment and consume basic resources such as energy, drinking water or forest much more than rural areas. Promoting sustainable cities will affect sustainable agriculture or fishing positively in the long run (Bostancı & Yıldırım, 2018).
SDG 12: Responsible Consumption and Production	Consumers should take responsibility and ethical or green consumption behaviour should be supported.	Sustainable consumption trends and sustainable production models enhances a sustainable world. Less and green consumption are less harmful to the environment Consumers can help improve the environment and also sustainable production technologies can stop or reduce the effects of climate change and environmental pollution.

continues on following page

Table 2. Continued

Goals	Aims[a]	The Relationship Between Food Security and SDGs
SDG 13: Climate Action	To prevent damages of climate changes, countries should get action plan for natural disasters.	Climate changes mostly influence food availability and food stability. Especially, global warming and weathers, food sources such as agricultural foods, animal origin foods or seafood are affected negatively. (HLPE, 2012).
SDG 14: Life Below Water	Marine and coastal ecosystems should be protected for the future. To maintain fishing and other profits of the oceans and seas, protection of oceans, seas and their ecosystem are important aims.	Oceans and seas are important resources for humanity. So many lives are depended on oceans and seas. In general, developed countries without coastlines and developing countries with coastlines are dependent on oceans and seas for food security. Because major proportion of their food supply comes from the oceans or seas in coastal countries.
SDG 15: Life on Land	Forest areas decreased last decades. For sustainable world and natural ecosystem, protecting forest areas is an important goal.	Protection of forest areas is important for food security in the long term. Forest areas provide food sources for economic, social and environmental sustainability. Hunting or gathering forest foods contribute to food security. Forests also provide food for animals too. (HLPE, 2017)
SDG 16: Peace, Justice and Strong Institutions	Ending violence in the society is an important issue. Well-being and health of people are based on peaceful society.	Peace and justice are important issues for well-being of life in a country. Highest proportion of hunger people live in war areas. In addition, strong institutions to achieve or support food security is necessary.
SDG 17: Partnerships for the Goals	Global goals bring countries together to achieve a sustainable world in the long term. Specifically, less developed countries need developed and developing countries' support for economic and social development.	The world has understood that there must be an international coordination for water and food security. For example, the MDB Working Group on Food and Water Security co-chaired by the European Bank for Reconstruction and Development (EBRD) and the Asian Development Bank (ADB) has cooperated together since 2010. (EBRD, 2011). Global coordination is necessary to achieve global food security.

[a]The information is based onSustainable Development Goals Knowledge Platform, (2015)
Source: Compiled by authors

Both academic studies and public reports have proved that food security is a prior issue for the sustainable world. Food security and fighting with hunger is an old goal since 1990s. The relationship between food security and sustainable development goals has been realized over time (Fan et. al., 2018). The last Sustainable Development Goals includes many goals and sub-goals that are connected to agriculture and food. The SDG 2 directly focuses on food security and ending hunger but other SDGs are related to food security significantly. For example, SDG 1 aims to reduce poverty in which food security plays a key role. Sustainable agriculture also plays an important role in achieving SDG 6, SDG 12, SDG 13 and SDG 15 (Brooks, 2016). Perez-Escamilla (2017) posited that food insecurity was mostly associated with poverty. It also caused poor physical and mental health as well as children's learning problems. However, unemployment, socioeconomic inequalities, dirty drinking water, lack of housing security increased food insecurity (Perez-Escamilla, 2017). Mendes (2015) analysed how 2030 SDGs affected food security by using the toolkit part of OECD's Framework for Policy Coherence for Sustainable Development. Mendes (2015) categorized potential impacts of 2030 SDGs on food security using the toolkit part of OECD's Framework for Policy Coherence for Sustainable Development (Table 3).

Table 3. Potential impacts of 2030 SDGs on food security

<table>
<tr><th colspan="4">Food Security Dimensions</th></tr>
<tr><th>Accessibility</th><th>Utilization</th><th>Sustainability</th><th>Availability</th></tr>
<tr><td>SDG 1 – positive impact</td><td>SDG 1 – positive impact</td><td>SDG 4– positive impact</td><td>SDG 6 negative and positive</td></tr>
<tr><td>SDG 3 – positive impact</td><td>SDG 3 – positive impact</td><td>SDG 5– positive impact</td><td>SDG 7 negative and positive</td></tr>
<tr><td>SDG 4 – positive impact</td><td>SDG 4
– positive impact</td><td>SDG 8– positive impact</td><td>SDG 9 negative</td></tr>
<tr><td>SDG 5 – positive impact</td><td>SDG 5– positive impact</td><td>SDG 12– positive impact</td><td>SDG 10– positive impact</td></tr>
<tr><td>SDG 6– positive impact</td><td>SDG 12– positive impact</td><td>SDG 13– positive impact</td><td>SDG 11– positive impact</td></tr>
<tr><td rowspan="2">SDG 8– positive impact</td><td rowspan="2">SDG 16– positive impact</td><td>SDG 14– positive impact</td><td>SDG 12– positive impact</td></tr>
<tr><td>SDG 15– positive impact</td><td></td></tr>
<tr><td>SDG 9– positive impact</td><td rowspan="7"></td><td>SDG 16– positive impact</td><td>SDG 13 negative and positive</td></tr>
<tr><td rowspan="2">SDG 10– positive impact</td><td rowspan="2">SDG 17– positive impact</td><td>SDG 14 negative and positive</td></tr>
<tr><td>SDG 15 negative and positive</td></tr>
<tr><td>SDG 11– positive impact</td><td rowspan="4"></td><td>SDG 16– positive impact</td></tr>
<tr><td>SDG 12– positive impact</td><td>SDG 17– positive impact</td></tr>
<tr><td>SDG 16– positive impact</td><td rowspan="2"></td></tr>
<tr><td>SDG 17– positive impact</td></tr>
</table>

SDG 2 is directly related to food security and it is excluded from the table
Source: (adapted from Mendes, 2015)

When economic and social conditions get better, people can access more food. In addition, environmental pollution and climate change are big challenges for food security and sustainability. Sustainable city planning, sustainable industry, sustainable production and sustainable economic policies and implementations can improve natural environment and reduce pollution or impacts of climate change (Bostancı & Yıldırım, 2018). For food security, natural resources and natural environment should be protected (Mendes, 2015). Consequently, 2030 Sustainable Development Goals aim at improving economic, social and environmental life for the future. Achieving food security depends on 2030 SDGs, such that food security policies should be multi-faceted in addressing poverty, hunger, energy usage, inequalities and pollution (Mendes, 2015; Diaz-Bonilla & Hepburn, 2016; Ismail, 2018).

SEAFOOD SECURITY AND SUSTAINABILITY

The fisheries and aquaculture sector are pivotal to achieving food security, nutrition and health and they also play a major role in preventing hunger for many countries in the world. Fish meet fundamental protein need and provides useful nutrition sources as long-chain omega-3 fatty acids, iodine, vitamin D and calcium, thus making seafood security and sustainability an important issue for actualizing a sustainable world (FAO, 2018b). Seafood products can be provided from either capture fisheries or aquaculture (Asche et al., 2015). Food and Agriculture Organization of the United Nations (FAO) refers to about 2500 species for captured fisheries resources that people can provide various seafood products from oceans and seas to meet hunger need. However, marine capture fisheries resources are so close to full exploitation globally ("Capture fisheries resources," 2019). Aquaculture includes farming plants and animals in three basic areas as inland, coastal and marine. With aquaculture, countries can provide various animal and plant species and this industry is one of the fastest growing agricultural industries in the world ("Aquaculture resources," 2019). Marine capture fisheries' landings include small pelagic fish and aquaculture producers generally prefer to use small pelagic fish to reduce feed cost. There is always a conflict between human consumption and animal consumption (including aquaculture) of small pelagic fish such as herring, sardine, anchovy et cetera. (Allison, 2011). Accordingly, sustainable aquaculture, reducing overfishing and protecting rare species of fish are crucial to achieving seafood security and sustainability.

Fisheries provide jobs, incomes and nutrition to millions of people. In poor coastal areas, many people's life is dependent on fisheries (Ye, 2015). Thus, seafood sustainability is an important concern for coastal countries in maintaining economic growth and human well-being simultaneously (Brooks, 2016). The developing countries derive more economic benefits through fisheries than developed countries (Martini & Lindberg, 2013). Some studies show that fisheries trade contributes to food security and poverty reduction in developing countries (Gillett & Lightfoot, 2002; Horemans & Kébé, 2006; Chuenpagdee et.al., 2016; Ababouch, 2015). For instance, Kawarazuka and Bene (2011) studied 30-year time series of exports and per capita domestic availability of fish in 14 countries (Maldives, Sierra Leone, Solomon Islands, Comoros, Kiribati, Bangladesh, Cambodia, Indonesia, Gambia, Senegal, Seychelles, Sa Tome & Principe, Sri Lanka, Lao PDR, Japan, Togo, Philippines, Congo DR, Vanuatu, Guinea, South Korea, Thailand, Malaysia, Myamar, Cameroon, Malawi, Cote d'lvoire, Nigeria, Uganda, Viet Nam) and found that availability of fish increased when fish export increased in 50% of the countries. This result can be related to growth of aquaculture and captured fisheries. On the other side, countries with high fish consumption (domestic), high population growth and poverty, can have a trouble with fish supply. Economic and physical access to seafood therefore determine food security and poverty. In aquaculture and fisheries industries, food insecurity generally appears because of inequalities in purchasing power between buyers and sellers or small fisheries and factory owners and between women and men (Allison, 2011).

The relationship between food security and the SDGs can be linked into seafood security (Table 4). Bennett et.al. (2018) summarized the contribution of fisheries to food and nutrition security to seafood security in 2030 Sustainable Development Goals (Table 4).

Beyond reducing poverty and food insecurity, aquaculture provides food (fish) for low-income consumers, income and employment for small producers, as well as support domestic development in developing countries. The cases of Malawi, Bangladesh and the Philippines showed that small-pond or cage aquaculture systems contributed income and employment to poor and smallholder households. In this context, the improved policies for aquaculture industry are pivotal to achieving food security and economic support for the middle and lower class households (Allison, 2011).

By 2050, world population will be 9.2 billion with most of the population growth expected in South Asia, the Middle East and Africa (Un, 2012; Grabar, 2013). According to United Nations (2012). Rising population will bring big poverty and hunger together like today. Almost 2 billion people will suffer from hunger in the future (United Nations, 2012). Although the future seems gloomy, there are some optimistic scenarios to prevent hunger, poverty. The rising seafood industry has the potential to solve the future's hunger problem and contribute to global nutrition need (FAO, 2002a; Naylor et.al., 2009; Tlusty, 2012; Tacon & Metian, 2013; Kobayashi et.al., 2015; El Sheikha & Xu, 2017).

Table 4. The Potential Impacts of Seafood for the Sustainable World

Contribution of Seafood and Fish Consumption	
Playing a major role in food production system	Since 1945, FAO has recognized the importance of capture fisheries to end hunger but recent times show the biggest interest for seafood in achieving food security
By 2014, 167,2 million metric tons of fish was produced	This means that the global population consumed 17% of animal protein.
The global supply of fish (capture fisheries and aquaculture) provides a great animal protein for more than 3.1 billion people	In some countries such as Island and coastal countries (Maldives, Cambodia, Sierra Leone, Kiribati, Solomon Islands, Sri Lanka, Bangladesh, Indonesia, and Ghana), protein need depends on seafood
Seafood provides the most healthy food for people	Fish comprise vitamins A, D, B, calcium, phosphorus, zinc, iron and iodine. These micro-nutrients improve and support human health, especially in children, whose need for protein (which can be obtained from fish) is so critical for their growth.
With the global fish supply, omega-3 polyunsaturated fatty acids is provided for cardiovascular and brain health	To protect heart health, fish is a fateful food.
Fish consumption is necessary for women, infants and children	In low-income countries and less developed countries, there is a big need for adequate protein and nutrition to maintain health, especially to reduce child mortality among under-five children owing to inadequate nutrition. At this point, seafood and fish are both great rescuers.
Risk of Seafood and Fish Consumption	
Consumption of fish can cause some healthy problems	Fish consumption may cause mortal problems because fish food can include toxic substances such as polychlorinated biphenyls, dioxins, methylmercury, microplastics etc. This problem mostly depended on environmental pollution. For seafood security and human health, oceans and seas should be protected and keep clean as much as possible.

Source: (Adapted from Bennett et.al., 2018)

SEAFOOD SECURITY AND SUSTAINABILITY IN TURKISH SEAFOOD MARKET

Turkey is a fast growing OECD country (OECD, 2019) and the economic and social development performance of Turkey has been impressive since 2000s (The World Bank, 2019). Turkey has known sustainable development concept for a long time. The tenth National Development Programme (NDP) of Turkey was based on Rio summit in 1992. Turkish NDP encouraged sustainable growth and development. Between 2010 and 2015, Turkey accelerated its efforts for sustainable development by working on Millennium Development Goals and also accepted the vision of 2030 Sustainable Development Agenda. Since 2016, Turkey has been working towards the realization of the 17 sustainable development goals by 2030 (Republic of Turkey, Ministry of Development, 2016).

The aquaculture sector is one of the four sub-sectors of the agricultural sector, along with crop production, animal production and forestry in Turkey. Turkish seas, which have a coastline of 8333 km, show different characteristics in terms of average temperature and salinity. In the North is the low temperature and saline (0.17-0.18%) Black Sea; in the West and South are the Aegean and Mediterranean with high temperature and salinity (0.33-0.39%) with a mixture of straits; and the third coast is the Sea of Marmara. There is a decrease in the number of seafood species in the transition from the Mediterranean to the Black Sea, but an increase in the population size. The different characteristics of these seas affect not only Turkish hunting but also the aquacultural activities in the country (ZMO, 2017).

Turkey and FAO (Food and Agriculture Organization of the United Nations) have cooperation together. The FAO helps Turkey to address food security and food safety, sustainable use of natural resources and institutional capacity enhancement (FAO, 2018b). Being a coastal country, Turkey also recognizes the importance of keeping oceans and seas clean for the future. As determined in 2030 SDGs, marine life underwater is necessary issues in public legal arrangements in Turkey. For instance, The Ministry of Environment and Urbanization of the Republic of Turkey has provided up-to-date data on how measures are taken in the Marmara, Aegean, Mediterranean and Black Sea Regions to combat marine pollution and what is done to combat pollution (Table 5).

Table 5. Protection of Turkish Seas

Seas	Current Profile
Sea of Marmara	According to the recent data from Turkish Ministry of Environment and Urban Planning (2018), pollution monitoring is carried out at 47 points in the Marmara Sea. In the works carried out in accordance with the swimming water quality regulation, 168 out of 174 swimming areas were found to be in compliance with the regulations and inspections in the inappropriate areas.
The Black Sea	According to a recent data from Turkish Ministry of Environment and Urban Planning (2018), pollution for the Black Sea was monitored in 69 places. Necessary monitoring was made at 258 swimming water points and 251 points were approved in the swimming water areas. With respect to ship wastes, there are waste procurement services at 32 coastal facilities in the Black Sea.
The Mediterranean Sea and the Aegean Sea	Bathing water analysis was performed at 342 points in the Aegean Region and 323 groundwater analyzes were performed in the Mediterranean Region (Turkish Ministry of Environment and Urban Planning, 2018). With respect to ship waste, a total of 87 coastal facilities in the Mediterranean and Aegean Sea regions provided waste reception services. For ship and coastal security, three of them are in Aegean and Mediterranean Regions and six Regional Emergency Response Plan offices were prepared.

Source: (Adapted from Turkish Ministry of Environment and Urban Planning, 2018; Yıldırım & Kaplan, 2019)

The protection of the Seas and the prevention of pollution in the seas necessitate international cooperation. Turkey also participated in international agreements such as Protection of the Black Sea Against Pollution (Bucharest) Convention and Mediterranean Conservation of the Marine Environment and Coastal Area (Barcelona) Convention. According to the relevant agreements, participating countries prepare a report every year on their work on the prevention or reduction of pollution in the seas (Turkish Ministry of Environment and Urban Planning, 2018). With respect to legal arrangements and law, Turkey protects its seafood and fish. In terms of protection of the sea creatures and underwater life in Turkey, with the "Commercial Fisheries 4/1 Communiqué Regarding Regulation of Hunting (Communiqué No: 2016/35)" information relating to the ban on hunting the species like sea bream, red star, great coyote (sand) shark, red coral, black coral, sun shark, sea horse, shark, minaret, dwarf shark, devil minaret, ears, spurred camgöz, pina, sea turtles, seal, mayan, seal, dolphin, whales, sea meadows, sturgeon, grouper, oily fish, lagos, and mushrooms (Legislation Information System, Official Gazette, 2016). In the same communiqué, the size and weight of the fish and other sea creatures to be hunted, as well as the rules and prohibition related to hunting were specified. Hunted Fish includeanchovy, sardine and horse mackerel, bonito-toric, shield, tongue and flounder, blue fin tuna, written tuna, tapestry (tombic) and long wing tuna (tulina), leer and lambuka, grouper and lagos, shrimp, bivalve mollusk, sea snail and sea eggplant, insect, lobster and blue crab, octopus hunting, seaweed and sponge production (Legislation Information System, Official Gazette, 2016).

Having four Seas, Turkey is also rich in wetlands and ranks first among the Middle East, Eastern and European countries (Frenken, 2009). Turkey is therefore endowed with various kinds of aquaculture resources. Table 6 shows Turkish aquaculture and seafood production. It can cconcluded that seafood products are an integral part of Turkish food security.

Table 6. Turkish seafood production

Year	Sea Products (Captured)	Aquaculture Production	Freshwater Products (Captured)
	(Tonnes)	(Tonnes)	(Tonnes)
2000	460,521	79,031	42,824
2001	484,410	67,244	43,323
2002	522,744	61,165	43,938
2003	463,074	79,943	44,698
2004	504,897	94,010	45,585
2005	380,381	118,277	46,115
2006	488,966	128,943	44,082
2007	589,129	139,873	43,321
2008	453,113	152,186	41,011
2009	425,046	158,729	39,187
2010	445,680	167,141	40,259
2011	477,658	188,790	37,097
2012	396,322	212,410	36,120
2013	339,047	233,394	35,074
2014	266,078	235,133	36,134
2015	397,731	240,334	34,176
2016	301,464	253,395	33,856
2017	322,173	276,502	32,145
2018	283,955	314,537	30,139

Source: (Turkish Ministry of Agriculture and Forestry, 2019)

The availability dimension shows the current amount of food in a country. To achieve food availability, the food must be found in abundance through domestic production, food aid or importing activities (FAO, 2008b; Napoli, 2011; Simon, 2018; Haug, 2018). In this context, we can conclude higher quantity of seafood and aquaculture food will provide higher economic and physical access to seafood.

It can concluded that over time, Turkey has a potential to keep seafood security with various types of fishes, Turkey can provide different types of fish-food and nutrition for people (Table 7). In addition, Turkey is rich in other seafood products such as octopus, spiny lobster, sea snail et cetera. (see Table 8). Turkey therefore provides adequate seafood products both for domestic and foreign markets.

Captured fisheries and aquaculture provide so many benefits for Turkish economic and social development. Other kinds of seafood (except fish) also provides food and nutrition for Turkey. Turkey has various kind of seafood in the market that people have many alternative fish food in Turkish seafood market.

Table 7. Captured Fisheries Types in Turkey

Type of Fish	2016	2017	2018
Total (tonnes)	263,724,5	269,676,4	222,023,6
Leer fish	186,6	211,9	181,7
Greater amberjack	7,0	8,5	8,4
Albacore	25,2	44,0	37,8
Hake-European hake	783,8	1,011,3	1,019,3
Red mullet	1,453,6	1,406,4	1,399,3
Goldon banded	78,7	69,4	49,9
Sprat	50,224,9	33,949,5	20,056,6
Seabream	495,1	590,0	544,1
Common sole	352,2	486,4	432,4
John dory	47,1	48,3	52,1
Common seabream	25,3	28,8	44,7
Angler fish	176,0	185,2	219,9
Shore rockling	10,4	11,7	15,8
Frigate mackerel	406,8	474,1	367,0
Meagre	23,6	10,1	55,9
Silverside	516,5	489,3	591,5
Anchovy	102,595,2	158,093,8	96,451,7
Painted comber	17,9	12,1	11,0
Eurepean barracude	115,7	96,2	75,3
Black skorpion fish	138,6	306,0	208,2
Annular bream	84,2	86,6	45,9
Horse mackerel	8,859,8	8,065,6	14,221,8
Scad	2,288,6	4,919,3	6,456,1
Brown mearge	4,5	3,0	4,3
Picarel	328,9	285,9	255,3
Turbot	221,1	167,4	139,2
Two banded bream	125,0	210,9	128,4
Gobies	50,5	2,8	12,8
Grey mullet	1,825,7	2,313,6	1,592,4
Angelshark	2,8	0,9	0,3
Sword fish	76,5	441,0	427,0
Red gurnard	54,3	56,6	43,8
Trigla lineata	4,2	7,5	6,8
Chup mackerel	1,602,0	2,043,0	1,503,5
Topeshark	22,3	23,2	21,0
Bogue	2,795,1	3,175,0	3,559,3
Waker	230,6	32,6	111,1

continues on following page

Table 7. Continued

Type of Fish	2016	2017	2018
Seabas	131,7	135,1	151,4
Small-Scalled	28,2	20,4	40,6
Blue fish	9,573,6	1,935,7	5 767,4
Saddled seabream	90,2	92,3	63,7
Striped bream	980,0	1,171,7	1,062,7
Whiting	11,540,8	8,248,0	6,813,9
European coger	2,6	0,3	-
Conger eel	123,7	152,3	181,7
Croaker	30,9	26,7	25,3
Dusky grouper	11,0	3,1	2,6
Bluefin tuna	1,324,0	1,514,7	1,283,7
Little tunny	184,1	479,8	616,6
Piper	3,4	2,5	2,2
Atlantic bonito	39,459,6	7,577,6	30,920,4
Large-eye dentex	32,9	9,4	8,2
Flounder	8,8	7,1	6,3
Pilchard	18,162,1	23,425,7	18,854,0
Black sea bream	51,4	19,8	52,0
Saupe	127,6	144,6	120,0
Dentex	53,9	47,2	69,4
Sharpsnout seabream	2,1	2,1	2,0
Striped red	3,047,0	2,074,4	2,914,9
Twaite shad	1,642,0	1,576,2	1,605,3
Blue spatled bream	13,5	17,2	26,2
Mackerel	61,9	728,2	368,8
Thornback ray	116,1	183,0	82,6
Gar fish	267,8	252,8	263,6
Saury	131,3	152,9	139,3
Other	266,0	307,7	227,2

Source: (TURKSTAT, 2019a)

Table 8. Other Seafood Production

Types of Seafood Products	2016	2017	2018
Total (tonnes)	263,724,5	269,676,4	222,023,6
Octopus	245,9	162,7	223,7
Spiny lobster	1,1	5,0	1,9
Norway lobster	0,1	1,4	2,3
Sea snail	10,353,7	9,194,1	9,672,3
Common lobster	1,5	1,8	4,7
Oyster	-	-	-
Long finned squid	389,0	421,9	523,6
Speckled shrimp	50,1	54,1	46,0
Green tiger prawn	719,8	728,6	758,8
Caramote prawn	252,4	208,0	219,4
Giant gamba prawn	1,669,1	1,382,8	299,0
Carpet Shell	4,8	-	0,8
Striped venus	20,931,7	34,941,1	44,532,8
Mediterranean mussel	77,5	535,6	603,8
Bearded horse mussel	-	-	-
Warty venus	-	-	-
Cuttle fish	925,1	986,0	1,041,9
Common shore crab	6,0	1,3	14,9
Great Scallop	-	-	-
Blue crab	2,0	8,8	10,5
Other	299,8	1 506,1	761,9

Source: (TUIK, 2019a)

According to 2018 Turkish Fisheries data, the last profile of Turkish seafood market included these statistics (TURKSTAT, 2019b):

- Aquaculture production decreased in 2018 (- 0,3%) compared to the previous year and reported as 628,631 tons. Sea fish accounted for 35.3% of the total production, 9.9% for other seafood, 4.8% for domestic aquaculture and 50% for aquaculture.
- Captured fisheries decreased by 11.4% in 2018, while aquaculture increased by 13.8%
- 314,094 tons of production made by hunting, while the production of aquaculture was realized as 314,537 tons. Marine fisheries decreased by 11.9% and domestic fisheries decreased by 6.2% compared to the previous year.
- 33.4% of aquaculture production took place in Inland waters and 66.6% in the Seas
- Eastern Black Sea Region took the first place with 31.5% in sea fish production. This region was followed by the Western Black Sea with 30.6%, Marmara with 18.4%, Aegean with 15% and Mediterranean with 4.5%.

- the average fish consumption per capita was 5.49 kg in 2017, the average fish consumption per capita was reported as 6,14kg (increased by 11.8%) in 2018.

Further, Turkey has a major role in the global seafood market. GLOBEFISH Highlights (2019) reported that Turkey exported seabass and seabream to the European Countries (EU28) greatly that Greek exporters have to reduce their prices. In 2018, Turkish seafood market had its record production level and also turbulent financial conditions. Unfortunately, the weakening Turkish Lira and other financial conditions affected domestic market demand negatively. Having lower prices, Turkey was the biggest exporter of seafood in the EU28, the Middle East market as Lebanon and also in the Russian Federation in 2018 (FAO, 2019). Thus, the most important competitive advantage is low priced seafood and geographic position of Turkey.

CONCLUSION

The outcome of this chapter provides useful information on seafood security and sustainability in the context of sustainable development goals for developing coastal countries. Sustainable development goals aim to provide sustainable solution for main problem areas such as poverty, hunger, inequality and environmental pollution. Fast consumption and environmental pollution decrease agricultural food products, seafood products and food of animal origin. Seafood also plays a major role in food security. Low-income and less developed coastal countries, largely dependent on seafood sustainability to achieve seafood security. Seafood sustainability enhances seafood security by keeping oceans and seas clean, preventing pollution, as well as ending global warming. If there are cleaner oceans and seas, they can provide higher quantities of seafood for a long time. The availability and utilization of seafood will promote access of seafood in the market. Seafood security and sustainability profile of Turkey are summarized as:

- **Seafood Availability:** Turkey can produce seafood sufficiently for both domestic and foreign markets. With keeping cleanness of seas and protecting biodiversity by law, Turkey can maintain its availability of seafood products for a long time.
- **Economic Access:** The availability of seafood will bring lower prices in the domestic markets. Turkey can produce its own seafood and citizens prefer to buy cheap and available proteinous seafood. Fish products are consumed seasonally and people generally purchase fresh fish from the market. Although the number of fish consumers is so low, the high price of other animal-origin food encourages Turkish households to consume fish products (FAO, 2008b)
- **Utilization:** Higher species of seafood products will bring different types of nutrition for people. Turkey has lots of diverse species of captured seafood and aquaculture that consumers can choose from .
- **Stabilization and Sustainability:** Turkey is a developing country and follows 2030 Sustainable Development Goals to achieve its development. In this context, Turkey keeps environmental sensibility in its economic and social policies. Turkey also has its own legal system for fisheries and has collaboration with international organization. Seafood security and sustainability can be achieved if Turkey keeps sustainable strategy and policy in fisheries and aquaculture..

Consequently, Turkish seafood security seems almost sustainable in the long term. The availability of seafood products is sufficient both for domestic and foreign markets. Developing countries generally need more resources to complete their economic development. The increasing population also requires more food, energy, goods and services,. and this makes achieving sustainable development goals so hard in developing countries. Relating food security the 17 SDGs, will be a critical issue for being a sustainable country. Seafood security is therefore crucial for Turkey's long run economic and social development.

REFERENCES

Ababouch, L. (2015). *Fisheries and Aquaculture in the Context of Blue Economy*. UN Economic Commission for Africa. Retrieved from https://www.afdb.org/fileadmin/uploads/afdb/Documents/Events/DakAgri2015/Fisheries_and_Aquaculture_in_the_Context_of_Blue_Economy.pdf

Aladjadjiyan, A. (2012). *Food Production: Approaches, Challenges and Tasks*. InTech. Retrieved from https://www.intechopen.com/books/food-production-approaches-challenges-and-tasks/physical-factors-for-plant-growth-stimulation-improve-food-quality

Allison, E. H. (2011, Dec.). *Aquaculture, Fisheries, Poverty and Food Security*. Working Paper 2011-65. The WorldFish Center.

Aquaculture Resources. (2019). *Aquaculture resources*. Retrieved from http://www.fao.org/fishery/resources/aquaculture/en

Asche, F., Bellemare, M. F., Roheim, C., Smith, M. D., & Tveteras, S. (2015). Fair Enough? Food Security and the International Trade of Seafood. *World Development*, *67*, 151–160. doi:10.1016/j.worlddev.2014.10.013

Asian Development Bank. (2013). *Gender Equality and Food Security. Women's Empowerment as a Tool against Hunger*. Retrieved from https://www.adb.org/sites/default/files/publication/30315/gender-equality-and-food-security.pdf

Bennett, A., Patil, P., Kleisner, K., Rader, D., Virdin, J., & Basurto, X. (2018). *Contribution of Fisheries to Food and Nutrition Security: Current Knowledge, Policy, and Research*. NI Report 18-02. Durham, NC: Duke University.

Bostancı, S. H., & Yıldırım, S. (2018). *From Sustainable Communities to Ecologıcal Citizenship: Searching For Ecological Environment-Sensitive Life*. ICOMEP'18-Spring, International Congress of Management, Economy and Policy, Turkey.

Brooks, J. (2016, Apr 6). *Insights, New Approaches to Economic Challenges Food Security and the Sustainable Development Goals*. Academic Press.

Capture Fisheries Resources. (2019). *Capture fisheries resources*. Retrieved from http://www.fao.org/fishery/resources/capture/en

CFS. (2015b). *Water for Food Security and Nutrition. Policy recommendations*. CFS 42 Final Report. Retrieved from http://www.fao.org/3/a-av046e.pdf

CFS. (2017). *Global Strategic Framework for Food Security and Nutrition (GSF)*. Retrieved from http://www.fao.org/3/MR173EN/mr173en.pdf

CFS (Committee on World Food Security). (2015a). *Framework for Action for Food Security and Nutrition In Protracted Crises*. Retrieved from http://www.fao.org/3/a-bc852e.pdf

Chuenpagdee, R., McConney, P., Munro, G., Ferreira, B., Marschoff, E., Rice, J., & Rosenberg, A. (2016). *Social and Economic Aspects of Sea-Based Food and Fisheries*. United Nations.

De Muro, P., & Burci, F. (2007). *Education for Rural People and Food Security A Cross Country Analysis*. Rome: Food and Agriculture Organization of the United Nations.

Diaz-Bonilla, E., & Hepburn, J. (2016). Trade, Food Security, and the 2030 Agenda. Geneva, Switzerland: International Centre for Trade and Sustainable Development (ICTSD), International Environment House 2.

Duncan, J. (2017). Global Food Security Governance: Civil society engagement in the reformed Committee on World Food Security. London: Routledge.

EBRD (European Bank for Reconstruction and Development). (2011). *Private Sector for Food Security Initiative*. Retrieved from https://www.ebrd.com/downloads/research/factsheets/foodsecurity.pdf

El Sheikha, A. F., & Xu, J. (2017). Traceability as a Key of Seafood Safety: Reassessment and Possible Applications. *Reviews in Fisheries Science & Aquaculture*, *25*(2), 158–170. doi:10.1080/23308249.2016.1254158

FAO. (1996). Fisheries and Aquaculture In Sub-Saharan Africa: Situation And Outlook In 1996. FAO Fisheries Circular No. 922 FIPP/C922. Rome: FAO.

FAO. (2002a). Reducing Poverty and Hunger: The Critical Role of Financing for Food. *Agriculture and Rural Development. International Conference on Financing for Development*, Monterrey, Mexico.

FAO. (2002b). *Trade and Food Security: Conceptualizing the Linkages Expert Consultation*. Rome: FAO.

FAO. (2006). *Food Security, Policy Brief, June 2006, Issue 2*. FAO.

FAO. (2008a). *An Introduction to the Basic Concepts of Food Security*. Retrieved from http://www.fao.org/3/al936e/al936e00.pdf

FAO. (2008b). Food Security Information for Action, Availability Assessment and Analysis, Lesson, Assessing Availability, Learners'. *Notes*.

FAO. (2012). *Decent rural employment for food security: A case for action. Knowledge Platform. Rural Employment. Gender, Equity and Rural Employment Division*. Rome, Italy: Economic and Social Development Department.

FAO. (2016). FAO/INFOODS Global Food Composition Database for Fish and Shellfish Version 1.0-uFiSh1.0. Rome, Italy: FAO.

FAO. (2017). *Strengthening Sector Policies for Better Food Security and Nutrition Results*. Policy Guidance Series.

FAO. (2018a). Transforming Food and Agriculture to Achieve the SDGs. 20 interconnected actions to guide decision-makers. Rome: FAO.

FAO. (2018b). *Achieving Blue Growth Building Vibrant Fisheries and Aquaculture Communities.* Retrieved from http://www.fao.org/3/CA0268EN/ca0268en.pdf

Frenken, K. (2009). *Irrigation in The Middle East Region in Figures AQUASTAT Survey – 2008.* FAO Water Reports 34. Rome: FAO.

Gillett, R., & Lightfoot, C. (2002). *The Contribution of Fisheries to the Economies of Pacific Island Countries.* A report prepared for the Asian Development Bank, the Forum Fisheries Agency, and the World Bank, Asian Development Bank.

Grabar, H. (2013). *Could Earth's Population Peak in 2050?* Retrieved from https://www.citylab.com/life/2013/04/could-earths-population-peak-2050/5208/

Haug, R. (2018). *Food Security Indicators: How To Measure And Communicate Results.* Noragric Report No. 83 January 2018. Department of International Environment and Development Studies, Noragric Faculty of Landscape and Society, Norwegian University of Life Sciences.

HLPE. (2017). *Sustainable Forestry for Food Security and Nutrition.* A report by The High Level Panel of Experts. HLPE Report 11.

HLPE (High Level Panel of Experts). (2012). Food security and climate change. A report by the High Level Panel of Experts on Food Security and Nutrition of the Committee on World Food Security.

Horemans, B. & Kébé, M. (2006). Enhancing The Economic Contribution Of Fisheries To West And Central African Nations. *IIFET 2006 Portsmouth Proceedings.*

IISD (International Institute for Sustainable Development). (2012). *Sustainable Development Timeline.* Retrieved from https://www.iisd.org/sites/default/files/publications/sd_timeline_2012.pdf

Ismail, M. (2018). Food security is key to SDGs success. *The ASEAN Post.* Retrieved from https://theaseanpost.com/article/food-security-key-sdgs-success

Kawarazuka, N., & Béné, C. (2011). The potential role of small fish species in improving micronutrient deficiencies in developing countries: Building evidence. *Public Health Nutrition, 14*(11), 1927–1938. doi:10.1017/S1368980011000814 PMID:21729489

Klarin, T. (2018). The Concept of Sustainable Development: From its Beginning to the Contemporary Issues. *Zagreb International Review of Economics & Business, 21*(1), 67–94. doi:10.2478/zireb-2018-0005

Kobayashi, M., Msangi, S., Batka, M., Vannuccini, S., Dey, M. M., & Anderson, J. L. (2015). Fish to 2030: The Role and Opportunity for Aquaculture. *Aquaculture Economics & Management, 19*(3), 282–300. doi:10.1080/13657305.2015.994240

Leggett, J. A., & Carter, N. T. (2012). *Rio+20: The United Nations Conference on Sustainable Development.* Congressional Research Service.

Legislation Information System, Official Gazette Date: 13.08.2016 Number of Official Gazette: 29800, No. 4/1 On Communication for Regulation of Fishing Fishing (Communication No: 2016/35). (n.d.). Retrieved from https://www.mevzuat.gov.tr/Metin.Aspx?MevzuatKod=9.5.22750&MevzuatIliski=0&sourceXmlSearch=ticari%20ama%C3%A7l%C4%B1

Martini, R., & Lindberg, C. (2013). *Fishing for Tomorrow: Managing fisheries for sustainable development. Coherence for Development, Better Policies for Better Lives, (2)*. OECD.

Mendes, C. (2015). *Policy Coherence for Sustainable Development Toolkit Module 2: Food security*. Retrieved from https://www.oecd.org/gov/pcsd/Thematic%20module%20on%20Food%20Security_DRAFT.pdf

Mutisya, M., Ngware, M. W., Kabiru, C. W., & Kandala, N. (2016). The Effect of Education on Household Food Security in Two Informal Urban Settlements In Kenya: A Longitudinal Analysis. *Food Security, 8*(4), 743–756. doi:10.100712571-016-0589-3

Napoli, M. (2011). Towards a Food Insecurity Multidimensional Index (FIMI). *Master*. In *Human Development And Food Securıty, (2010/2011)*. Roma: Roma Tre University.

Naylor, R. L., Hardy, R. W., Bureau, D. P., Chie, A., Elliott, M., Farrell, A. P., ... Nichols, P. D. (2009, Sept. 8). Feeding Aquaculture in An Era of Finite Resources. *Proceedings of the National Academy of Sciences, 106*(36), 15103-15110. 10.1073/pnas.0905235106

ODI (Overseas Development Institute), (1997, Feb.). *Global hunger and food security after the World Food Summit.* Briefing Paper 1997, (1). London: ODI.

OECD. (2019). OECD Environmental Performance Reviews: Turkey 2019. OECD Environmental Performance Reviews. Paris: OECD Publishing.

Perez-Escamilla, R. (2017). Food Security and the 2015–2030 Sustainable Development Goals: From Human to Planetary Health. *Current Developments in Nutrition, 1*(7), e000513. doi:10.3945/cdn.117.000513 PMID:29955711

Republic of Turkey, Ministry of Development. (2016). Report on Turkey's Initial Steps towards the Implementation of the 2030 Agenda for Sustainable Development. Ankara: Author.

Ryder, J., Iddya, K., & Ababouch, L. (2014). Assessment and Management of Seafood Safety and Quality Current Practices and Emerging Issues. *FAO Fisheries and Aquaculture Technical Paper*, 574.

Simon, G. A. (2018). Definition of Food Security, The Four Dimensions of Food Security, Food Insecurity, Food Sovereignty, Food Systems (Master in human Development and Food Security). Roma Tre University.

Sims, R., Flammini, A., Puri, M., & Bracco, S. (2015, Nov.). *Opportunities for Agri-Food Chains To Become Energy-Smart*. FAO and USAID. Retrieved from http://www.fao.org/3/a-i5125e.pdf

Smith, M. D., Roheim, C. A., Crowder, L. B., Halpern, B. S., Turnipseed, M., Anderson, J. L., ... Selkoe, K. A. (2010). Sustainability and Global Seafood. *Science, 327*(5967), 784–786. doi:10.1126cience.1185345 PMID:20150469

Sunderland, T., Powell, B., Ickowitz, A., Foli, S., Pinedo-Vasquez, M., Nasi, R., & Padoch, C. (2013). *Food Security and Nutrition: The Role Of Forests*. Discussion Paper. CIFOR. Retrieved from https://www.cifor.org/publications/pdf_files/WPapers/DPSunderland1301.pdf

Sustainable Development Goals Knowledge Platform. (2015). *Transforming our world: the 2030 Agenda for Sustainable Development*. Retrieved from https://sustainabledevelopment.un.org/post2015/transformingourworld

Tacon, A. G. J., & Metian, M. (2013). Fish Matters: Importance of Aquatic Foods in Human Nutrition and Global Food Supply. *Journal Reviews in Fisheries Science*, *21*(1), 22–38. doi:10.1080/10641262.2012.753405

The World Bank. (2019). Retrieved from https://www.worldbank.org/en/country/turkey/overview

Tlusty, M. F. (2012). Environmental improvement of seafood through certification and ecolabelling: Theory and analysis. *Fish and Fisheries*, *13*(1), 1–13. doi:10.1111/j.1467-2979.2011.00404.x

TÜDAV (Turkish Marine Research Foundation). (2017). *2017 Turkish Seas Report*. Retrieved from Retrieved from http://tudav.org/wp-content/uploads/2018/04/TUDAV_2017_Denizler_Raporu_s.pdf

Turkish Ministry of Agriculture and Forestry. (2019). *Fishery Statistics*. Retrieved from https://www.tarimorman.gov.tr/sgb/Belgeler/SagMenuVeriler/BSGM.pdf

Turkish Ministry of Environment and Urban Planning. (2018). *Sık Sorulan Sorular, Deniz Ve Kiyi Yönetimi* [Frequently Asked Questions, Sea and Marine Management]. Retrieved from https://csb.gov.tr/sss/deniz-ve-kiyi-yonetimi

TURKSTAT. (2019b, June). *Aquaculture (Amsterdam, Netherlands)*, *2018*, 12. Retrieved from http://tuik.gov.tr/PreHaberBultenleri.do?id=30697

TURKSTAT (Turkish Statistical Institute). (2019a). *Fishery Statistics*. Retrieved from http://www.tuik.gov.tr/PreTablo.do?alt_id=1005

UN Department of Economic and Social Affairs. (2019). News, Growing at a slower pace, world population is expected to reach 9.7 billion in 2050 and could peak at nearly 11 billion around 2100. Author.

UN General Assembly. (2015). *Resolution adopted by the General Assembly on 25 September 2015*. A/RES/70/1. Retrieved from https://www.un.org/en/development/desa/population/migration/generalassembly/docs/globalcompact/A_RES_70_1_E.pdf

UNCTAD. (2017). The Role Of Scıence, Technology And Innovatıon In Ensurıng Food Securıty By 2030. United Nations.

United Nations. (1987). *Report of the World Commission on Environment and Development: Our Common Future*. Retrieved from https://sustainabledevelopment.un.org/content/documents/5987our-common-future.pdf

United Nations. (2012). *Back to Our Common Future Sustainable Development in the 21st century (SD21) project Summary for policymakers*. United Nations. Retrieved from https://sustainabledevelopment.un.org/content/documents/UN-DESA_Back_Common_Future_En.pdf

United Nations (UN). (1972). *Report of the United Nations Conference on the Human Environment.* Stockholm, 5-16 June 1972, A/CONF.48/14/Rev.1. UN.

World Bank. (1986). *Poverty and Hunger: Issues and Options for Food Security in Developing Countries.* Retrieved from http://documents.worldbank.org/curated/en/166331467990005748/pdf/multi-page.pdf

World Food Istanbul. (2018). *Infografik: Türkiye'de Balık ve Deniz Mahsülleri* [Infographics: Fish and Seafood in Turkey]. Retrieved from https://www.worldfood-istanbul.com/Fuar-Hakk%C4%B1nda/Sektorel-Haberler/Infografik-Turkiye-de-Bal%C4%B1k-ve-Deniz-Mahsulleri

Ye, Y. (2015). Global Fisheries, Current Situation and Challenges. In H. D. Smith, J. L. Suárez de Vivero, & T. S. Agardy (Eds.), *Routledge Handbook of Ocean Resources and Management, 1th. Ediiton.* Routledge. doi:10.4324/9780203115398-15

Yıldırım, S., Yıldırım, D. Ç., & Gedikli, A. (2016). Sustainable Consumption Trends in the World in the Context of Green Economy and Sustainability. In M. M. Erdoğdu, T. Arun, & I. H. Ahmad (Eds.), *Handbook of Research on Green Economic Development Initiatives and Strategies* (pp. 66–84). IGI Global. doi:10.4018/978-1-5225-0440-5.ch004

ZMO (The Chambers of Agricultural Engineers). (2017). *Genel Merkez, e-kütüphane, Su Ürünleri Raporu* [Headquarters, e-library, Fisheries Report]. Retrieved from http://www.zmo.org.tr/genel/bizden_detay.php?kod=27302&tipi=17&sube=010.02.2017

KEY TERMS AND DEFINITIONS

2030 Sustainable Development Goals: The last accepted global sustainable development goals that including 17 main goals to achieve until 2030.

Committee on World Food Security (CFS): It is an organization that including international members and governments to ensure food security globally. CFS runs a report regularly and determines the relationship between food security and sustainable development goals.

FAO-The Food and Agriculture Organization: It is an organization of the United Nations that leading international efforts to fight with hunger and achieve food security.

Food Security: It can be defined as sustainable economic and physical access of various food ingredients globally.

Millennium Development Goals: First accepted globally sustainable development goals Including 8 basic goals to achieve until 2015.

Seafood Security: Like as food security, availability, economic and physical access to adequate and quality seafood by people.

Seafood Sustainability: To keep food security, protecting and maintaining availability of seafood and biodiversity of oceans and seas for the future.

Sustainable Development: A development approach that aims to balance between economy, environment, and society.

Section 2

Food Policy and Agricultural Systems

Chapter 3
Urban Sustainable Growth, Development, and Governance Structures for Revitalization of Open Vacant Spaces in Agriculture and Farming

José G. Vargas-Hernández
https://orcid.org/0000-0003-0938-4197
University Center for Economic and Managerial Sciences, University of Guadalajara, Mexico

ABSTRACT

This chapter analyzes the implications of urban sustainable growth, development, and governance structures for the revitalization of open vacant spaces in agriculture and farming. After reviewing the extensive corpus of literature on the subject, the authors used the critical socio-ecological analysis methodology to determine the main issues, trends, practices, and implications of the urban vacant spaces in relation to the urban sustainable growth and development, the use of urban vacant land in urban agriculture, farming, and gardening, and the collaborative urban governance structures and revitalization of open vacant spaces. It is concluded that transitional use of vacant land and parcels are to be used and utilized for developing a sustainable green city. However, urban vacant land and parcel spaces are required to be utilized for revitalization purposes to be stimulated. Social-ecological analysis focusing on vacant lots in underdeveloped urban spaces hold potential for urban transformation to meet the social needs and improve the ecological services.

DOI: 10.4018/978-1-7998-2599-9.ch003

INTRODUCTION

Cities losing population have to deal with high levels of land and structures vacancies. Population changes have an impact on the elasticity of towns that may have high levels of urban vacant land, and in turn, on the housing market. Population trends and urban land use and vacancy patterns have to consider their impacts on vacant land and structures. Deflating cities losing population and the territory is not expanding, have less vacant land, and more vacant structures.

Urban vacant land may stand out in some exceptional cities among the post-industrial cities in the developed world from where residents migrated following the new patterns of industrial jobs. City policies, growing local economy, and immigration tend to reduce the amount of urban vacant land. Urban farming and agriculture in empty and vacant lots and other urban spaces of post-industrial landscapes are reminiscent of the urban streetscape with vegetable gardens and livestock enclosures.

Vacant spaces are part of the holistic study on the relationships between urban ecology and urban patterns related to temporal and spatial dynamics. Urban vacant land, buildings, and structures are considered temporary and transitory, albeit long-term, in some cases, conditions. Urban ecology has emerged from vacant urban land, where abandoned urban landscapes have the potential to serve as ecological laboratories. Vacant landscapes are laboratories providing critical resources for data collection and observations in urban ecology subject to the city regulations of land and natural resources. High vacancy areas may begin to represent the city as an ecology scale development of new emergent ecosystems. Urban ecology laboratories well managed to provide vacancy solutions related to biodiversity, land management, ecosystem services, education, employment, etc.

Transitional use of vacant land and parcels are to be used and utilized for developing a sustainable green city. Social-ecological analysis focusing on vacant lots in underdeveloped urban spaces hold potential for urban transformation to meet the social needs and improve the ecological services.

This paper opens with a socio-ecological analysis of urban vacant spaces and their interrelationships with urban sustainable growth and development. After determining the relevant implications, it is analyzed the current trends of using urban vacant land in urban agriculture, farming, and gardening. Finally, the article analyses the collaborative urban governance structures and revitalization of open vacant spaces. The goal is to present the main conclusions on the proposal that urban spaces hold potential for urban green transformations to meet social needs and improve ecological services of communities.These transformations from urban vacant land and parcel spaces to be utilized for revitalization purposes require a collaborative governance structure to provide support to be stimulated.

URBAN VACANT SPACES

Urban vacant land, as a broad concept, include diverse forms and sizes from greenfields, greenbelts, brown-fields, wastelands, abandoned land, derelict and uncultivated land, etc., all of them that have different types of land in urban areas which have subtle differences (Bowman & Pagano, 2000; Bowman & Pagano, 2004; Kremer et al., 2013). Vacant land has been described as wasteland (Mathey & Rink, 2010) dead space (Coleman, 1982), derelict landscape (Jakle & Wilson, 1992). Dead spaces are derelict land, vegetated wastelands, and abandoned buildings, construction sites, materials dumps, etc. Another typology of urban vacant land was identified as post-industrial, derelict, unattended with vegetation, natural, and transportation-related (Kim, Miller, & Nowak, 2018).

Urban vacant space is defined as unused and unexploited spaces in the city, as well as the parking lots (Communauté Urbaine de Montréal, 1996). Open land private or public have been intrinsically associated with the underdeveloped nature of the land, vegetated or not. A building or lot is vacant if it has been vacant for two years or more (Accordino & Johnson 2000). Small public and private-owned vacant and unused land plots and open green spaces located in inner and dense urban areas are usually easy to be utilized. Housing and Urban Development (HUD) (2015) defines vacant address the one that mail delivery staff on urban routes have identified as not collecting for 90 days or longer, which increases the accuracy of data.

A strategic typology of urban vacant land and structures is based on changes in population and land areas. Changes in urban vacant land have different causes such as types, physical characteristics, urban boundaries, disinvestments, growth patterns, deindustrialization, suburbanization, annexation, pollution, contamination, etc. Cities develop an interactive effect between changes in population and urban land area. Inflating cities are the ones that experience at the same time growth in population, characterized by economic growth (Pettit, 2014), and land area and the opposite; cities dilute when they lose.

Inflating and diluting cities have different built environments, although the inflating cities report a below-average urban vacancy rate land while diluting cities have a higher urban vacancy land. For example an average of 16.7% of large US cities' land area is considered vacant (Newman, Bowman, Jung Lee, & Kim, 2016). The differences between the inflating and diluting cities respond to causes of changes, conditions, and urban land vacancy determination. The most relevant causes of vacancy are disinvestment, suburbanization, annexation, contamination, and deindustrialization.

Bowman and Pagano (2004) found that expanding city boundaries develop hag levels of vacant land. A compressing city is when the population increase, but the land area remains the same, and the town diluting is when the land area increases, but the population diminish. The interactions between population and urban boundaries change the urban conditions in terms of vacant land and structures (Cohen, 2006; Wiechman, 2007).

Urban vacant premises can be classified by type, size, and vacancy property expressed in terms of area. The vacancy by property type can be warehouses, factories, offices, and shops. The vacancy rates expressed in terms of area are for urban planning policy-making and statistical monitoring. Usually, retail has the lowest vacancy rate while warehousing, factories, and offices have the highest.

Designations, conditions, amounts, causes of increases, and decreases as characteristics of urban vacant land are relevant to define and classify the various types in a typology that enable to facilitate the analysis and the implications. Determining the types of urban land is the starting point to evaluate the potential of urban areas for the redevelopment of vacant spaces for residential, commercial, and industrial development.

Underused and unpaired developed land have received different terms such as brownfield sites, vacant land and buildings, derelict and previously developed land and buildings, etc. (Alker et al., 2000; Syms, 1994, 1998, 1999, 2001). Brownfield is extended to a dynamic view of vacant and partially vacant property and buildings in a constant state of flux. Vacant premises and brownfield sites and buildings have some attributes associated with vacant land available for redevelopment as resources for regeneration challenging to identify.

Actors with a role in the use of vacant land property can identify and utilize it to scale-up urban agriculture, farming, and gardening. Vacant land can be used for urban agriculture (Draus, Roddy, & McDuffie, 2014), green infrastructure (Schilling & Logan, 2008), and implemented with public space (Gough & Accordino, 2013). Urban land and buildings become vacant over time, and they may be reused.

Urban vacant land that has reached their limits of employment and population in some cities, the amount of buildable is small proportion, and it has a reverse relationship with the population (Northam, 1971). Urban land has large percentages of vacant land spaces that have large interests to many stakeholders. Urban sustainable development and regeneration policies are prompting the study and analysis of urban vacant land and property and their efficient refurbishment and reuse.

The wide range of urban vacant land distribution patterns is dependent on population density variations and urban structural configurations. Urban vacant land is a relevant issue almost in any city and any country showing different differences and trends in the types and amounts and requiring different prescriptions and actions to fit specific situations and conditions. The urban studies prescribe abandoned structures as declining neglected urban vacant land areas (Accordino & Johnson, 2000). The land is vacant if no structure exists on it, and humans did not utilize the property excluding underutilized parcels (Németh & Langhorst, 2014).

Data from National Commission on Urban Problems (1968) and also, according to Bowman and Pagano (2000) cities have between 12.5-15% vacant land and lots tending to be concentrated in low-income poor neighborhoods (Brulle & Pellow, 2005; Kremer et al., 2013). Urban space growth is highly related to meet the needs of housing and infrastructure development for population and corresponding to the population growth. A large amount of urban vacant land is in cities with population growth and boundary expansion. Conditions of vacant land restrict land redevelopment, and the amounts vary with regions (Bowman & Pagano, 2004). Some urban areas are less likely to develop vacant lands.

Otherwise, if there is no correspondence between population and city growth, the results can lead to vacant urban land and buildings (Bowman & Pagano, 2004; Bontje, 2004). Vacant urban land in cities is increasing in city centers where the land use densities are declining due to manufacturing activities declines after having been reached their limits on employment creation and population density. Cities that have declined in the manufacturing sector have large quantities of vacant land.

Vacant landscapes improve urban biodiversity and may provide environmental education, although when these spaces are unmanaged contributes to the perception of urban failures. Vacant landscapes afford a range of possibilities, and they have impacts upon one another. Multiple urban land-use scenarios are a tool that helps to allocate vacant urban spaces to vegetable production, including residential green areas, gardens, rooftops, etc.

Vacant landscapes with highly-vegetated scenarios may bring a more clean air effect and cooling effect (Smith, Li, & Turner, 2017). Multiple vegetable production scenarios in portions of urban vacant land spaces such as rooftops can be promoted to be used as greenhouse hydroponics to produce enough vegetables to meet the demand of inhabitants. Vacant residential yard spaces can be used to produce hydroponics both low- and high-intensity yields in order to meet the vegetable demand. Vacant residential spaces use of hydroponics are minimized because their practices have high financial and labor costs. Vacant landscapes are also designed as parks requiring the inputs of communities for a positive function or utility.

Data sets often created for different purposes might provide a better platform for information on vacant buildings. Accurate and updated information on building vacancy should be available and comprehensive and reliable. For instance, in Saint Paul Minnesota - Vacant Buildings Database contains all of the registered vacant buildings in the City, as well as their type. Ecosystem services are delivered through the management of vacant land in the urban ecosystem composed of several spheres and Vacant Building Category. Vacant buildings should be considered in terms of value and units as a proportion of total building stock of the area. Fostering value for places in vacant land requires improvements in physical design based on legibility and order of landscape functions and utility, including the inputs of the stakeholders. Social-ecological conditions of urban landscapes and their spatial patterns in vacant land requires new tools for assessment and monitoring. The spatial patterns of multiple ecosystem services of urban vacant land should be mapped with urban social needs.

The stock of vacant industrial premises in the context of the local economy involves the characteristics of the industrial development model as a conditioned effect leading to vacancy. The condition of premises and building vacancies are an outcome of the collective decisions of participants, occupiers, and owners. At the level of vacant industrial floor space, there are some variations in redevelopments and reoccupations across the urban spaces. Some cities located in the American Rust Belt with high vacancy rates look like woodlands but containing a significant number of urban residents (Burkholder, 2012).

Later were recorded newly change components of emergent vacancies, out of which some were contentious of the local industrial property market. Ball (2002) focuses on vacant industrial premises or brown buildings in the local urban industrial property market emphasizing a framework for the analysis of the re-use and refurbishment evaluated in terms of reoccupation.

Acquisition, subdivision, and re-use of vacant industrial buildings and premises via letting is an essential mechanism through which they are returned to use. Development companies and agencies are directly involved in self-managed letting with the acquisition, subdivision, and reopening of vacant industrial premises and buildings. Costs, benefits, and constraints analysis of re-use vacant industrial buildings versus new build are lower. Vacancy in industrial premises and buildings lacking strategic direction may take an incrementalist approach of re-using.

Re-using and reoccupation of vacant industrial premises and buildings in the city avoided spaces that can be transformed into more useful spaces that may contribute to reclaiming the redundant city. Revitalization of vacant urban lots to be transformed into green spaces can become a valuable resource for local economies, social communities, environments, and ecosystem services (Bowman & Pagano, 2004; Burkholder, 2012; Kremer et al., 2013; Rupprecht & Byrne, 2014).

The vacant and derelict built environment should be recycled rather than use undeveloped greenfield sites. Data related to empty and vacant buildings on the type, location, value, size, length, and age should be recorded, monitored, and aggregated to local, regional, and national levels. The implementation of the strategies is relevant to create a typology of community land-use types based on vacancy characteristics, incomes of residents in the area and vacancy market rates to convert these vacant spaces into urban green spaces with their various inherent functions (Akers, 2013; Detroit Works Project, 2014, p. 51). A significant amount of vacant urban land is not devoted to any functional use, out of which some are considered buildable. Still, some other vacant land is deemed to be unbuildable due to physical constraints. Vacant land increases as the population decreases.

URBAN SUSTAINABLE GROWTH AND DEVELOPMENT

Urban vacant land is an asset for cities growing in population and expanding boundaries and urban areas depopulated, creating opportunities for encouraging urban sustainable growth and development, such as the smart growth approach. Many of the city planning departments cannot provide records on vacant or underutilized land, premises and buildings, so much remain to be achieved to have a record on the available city vacant spaces. Large scale urban sustainable planning agenda should address and consider all the possible vacancies.

Buildings remaining total or partially vacant are an inactive resource to be reused and redevelopment. The use and reuse of vacant spaces for housing and industrial buildings, including factory complexes, are linked to the notions of the urban environment and sustainable city. A current debate on the reuse of vacant industrial buildings and premises area is concerned with the related costs, benefits, and constraints against the new buildings (Ball, 1998; Hall, 1998).

Not always the costs and sustainability of reusing vacant dwellings or industrial buildings and premises and lower than the costs and sustainability of new buildings. However, the economic consideration of the appropriate refurbishment and reusing seems to be more attractive beyond the sustainability considerations. Industrial building policies tend towards an opportunistic approach of financial resources, although buildings at risk analysis and other local initiatives are required to deal with vacant buildings.

Vacant industrial parcels located along watercourses have the opportunity to manage large quantities of water. Unoccupied buildings and vacant property or partly used in heterogeneous both function and form, represent a wasted capacity and unemployed resource. The high number of vacant parcels and the increasing unemployment characterizes any city in decay, as it is the case of Detroit.

Property data from particularly floor space, combined with particularly occupation and rates merge to provide consistent data on vacant buildings. Mature vacant building and premises stock and floor space of vacant factories, shops, and offices aged more than one Century or older lead to the conclusion than the age is related to the vacancy to a certain extent.

Small parcels, under-used plots and vacant land areas, back yards, container gardening, etc., are equally significant for urban planning and development to undertake green activities and clean-up operations of land contamination and air and water pollution, or provide land areas as community gardens for recreation and education of housing and community developments.

Urban communities implementing urban farming and agriculture in its vacant and residential spaces, hydroponic vegetables, can be considered marginal in terms of the relative population density. Urban vacant land can be available by tax incentives and provision of user permits to private landowners and the inclusion of urban farming and agriculture land in sustainable urban planning. Vacant lands are reservoirs of biomass and plant materials which supports to achieve higher urban sustainability standards. Urban vacant parcels can be turning into urban farming to secure food and increase sustainability.

Local urban planning authorities must have an urban planning framework to analyze site-specific data on developed vacant and derelict land and buildings that might be available for redevelopment. Mapping of urban vacant land that can be used for urban agriculture and farming, gardening and aquaculture should be defined and classified as urban and peri-urban vacant land to establish a land bank. This policy should be clearly stated in the urban development plan and be enhanced with tax incentives and users' permits to use vacant urban loans available on a temporary basis to the urban disadvantaged and poor.

Developmental city densities are related to the elasticity of urban vacancy issues where more elastic cities have large portions of vacant land, most of which are left as non-productive.

Vacant lots agglomerated in areas with a low concentration of green urban green spaces on most densely populated areas overlapping high social needs of ecological spaces resulting in low ecological value.

The urban territory considered vacant land, and lots vary in each of the cities, which is determined by their conditions and causes, increasing and decreasing their elasticity. Elastic cities have more vacant land than inelastic cities, despite the opposite abandonment relationship where low vacant land proportion tend to have high rates of abandonment (Bowman & Pagano, 2000). The vacant land areas are very comparable for both types of inelastic cities, compressing and deflating cities. Inelastic urban areas report more than twice as many vacant lands as elastic cities.

A high population density community can dedicate all the vacant and lawn spaces to grow vegetables instead of using hydroponics. The temporal occupation of used vacant lots develops the ability to transform the conditions of the socio-ecological systems. Management and design of vacant land use contribute ecologically to sustainable cities by creating webs of sustainable urban spaces to provide ecosystem services. One of the potentials of vacant land is its ability to provide ecosystem services to the urban population. Vacant land should be addressed to be classified considering the scenarios with ecosystems and cultural services included.

The unit vacancy is not a ubiquitous condition as opposed to land vacancy almost in abandonment in city-specific aggregations of surface area in urban environments containing a large amount of biomass and biodiversity and infrastructural integration to harness ecosystem services.

Urban vacant, residential yard, and industrial rooftop spaces are considered different land uses. Some land-use maps do not distinguish between parking lots and other vacant spaces.

Vacant land delivering ecosystem services has some characteristics to be managed as collective assets requiring an intimate understanding of conditions, uses, and colorization forces. Management and design of the provision of ecosystem services take into consideration the categorization of vacant lands. Vacant land spaces in urban environments support the coexistence and evolution of species in associative ecosystems with ecosystems service implications for the urban settlements.

Ecosystem services are delivered through the management of vacant land in the urban ecosystem composed of several spheres. The provision of ecosystem services to urban vacant land may potentially become a natural resource for shrinking cities for sustainable development under a framework of urban agroecology (for example, the urban-to-rural gradient in the city of Leipzig, Germany). The characteristics of vacant land define the type and quality of ecosystem services such as food, freshwater, urban air, etc., also tied to the conditions of resource management.

Urban vacant areas leverage ecosystem services over the long term assuring management and planning driven by cost-effectiveness, sanitation, and public safety to the city and its population. Urban gardening as a low-cost vacant land management optimization approach is used through greening and production, reducing the risk of failure. Vacant landscapes provide not only ecosystem services but also socio-cultural services such as aesthetics, recreation, education, etc. and should be considered in planning.

The growing green areas of urban vacant land are related to the benefits of the increasing quality of life for residents. Vacant and other derelict lands can produce food, and at the same time, the greening effects are positive, sustainable outcomes in terms of human health and well-being for the entire population and opportunities to cooperate and environmental awareness benefiting the communities from ecosystems.

Usually, urban planning and sustainable development decisions on vacant buildings are made in an information vacuum. Obsolete is a negative connotation of older vacant industrial spaces implicit in the word refurbishment instead of re-use. Land vacancy planning should consider and provide advantages for future development. The spatial analysis is relevant to determine the geographical composition of the vacant premises to the built environment, which may be used in sustainable urban planning, policy-making, monitoring, and control.

Analysis of the vacant industrial floor space of old industrial urban areas with the expectation to be re-used in the local industrial property market is crucial to plan better cities. Some cities have a land market of high value, growing housing demand, and are almost entirely built out with few vacant or abandoned properties. The research of Ball (1994), based on direct survey work, focuses on the property market on the condition and character of vacant floor space on refurbishment and re-use processes. The market for refurbished reopened, and reoccupied with good quality industrial premises and buildings involves local authorities, developers, real estate agents, and others that have the opportunity to profit in the property market.

A survey on vacant buildings in industrial premises conducted by Ball and Bord (1994) in sustainability analysis identified their development status as still vacant, reoccupied, and redeveloped. The reoccupancy rate is measured as the percentage of reoccupied industrial floor space of total vacant. Some vacant buildings are retained in good infrastructure conditions, while some reoccupations occur in bad conditions. Vacant buildings remaining for a long period of time can be redeveloped for more suitable uses or can be reoccupied, and still, others stay as persistent vacancies.

The flat earth and floor space of derelict sites include vacant buildings and property below street level and above the shop and high-rise complex, within non-domestic buildings. Vacant industrial buildings and premises bring the attention of urban sustainable development and policy communities for actions in terms of reoccupation, refurbishment, and reusing as activities of good practice, policy, and strategy options with sustainability implications.

Vacant land and empty, no occupied buildings are usually perceived negatively and correlated with deteriorated low property values and a place for violence and crime (Hoffman et al., 2012). Greening urban vacant lots can reduce stress and violence (Branas, Cheney, MacDonald, Tam, Jackson, & Ten Have, 2011). A difference-in-differences analysis of health, safety, and greening vacant urban space. Am. J. Epidemiol. 2011, 174, 1296–1306. Urban vacant land, buildings, and structures usually are described from a negative perspective.

There has been scarce research on the refurbishment and reusing of vacant urban dwellings and urban industrial premises property concerning urban sustainable development. Persistent vacancies can have engaged in major refurbishment activities. Long-term changes in building oblige to shift from new building construction to refurbishment and maintenance of existing vacant spaces (Kohler & Hassler, 2002).

The urban local authorities should identify the location, value, size, and age of vacant premises and how long they have been vacant. Urban local planning authorities should specify data on previously developed, lying, occupied, derelict, currently in use, and allocated vacant land and buildings for redevelopment. Local planning sustainable development should focus on restoring biodiversity in vacant urban lots assessed as economic, social, and ecological resources with an emphasis on the intersection of human development and wildlife needs.

Urban vacant land restoration has challenges and benefits in economic, socio-ecological systems and the ecosystem services supplied to the urban population living in cities. Vacant land restoration should be an interdisciplinary approached, combining economic, socio ecosystem concerns from a holistic urban land use perspective (Anderson & Minor, 2017). Restoration techniques implemented in urban vacant land must be suited to increase the urban green spaces aimed to improve poor and low-income urban areas.

Evaluating to restore vacant lots and their biodiversity must be considered using methods to assess the economic, social, and environmental sustainability outcomes (United Nations, 2002; Andersson, 2006). As usual, some of the vacant industrial premises are not being actively marketed, and reoccupations may occur where assumptions are being actively marketed. The transition from vacant premises to re-use and reoccupation may occur with different conditions of high or marginally refurbishment, or not, subdivided or reconfigured.

Urban local planning authorities might be underestimating the non-domestic premises that make up vacant buildings and premises due to the identification of reporting on building footprint rather than vacant floor space within the building (Myers & Wyatt, 2003). Some of these vacant buildings are reoccupied over some time, depending on the local market and involving small and uncertain businesses operating in non-expensive working spaces and areas.

In some urban areas, there is a volume of disused industrial floor space and vacant premises, as it has been reported by Ball (1996) involving persistently vacant buildings of persistent disuse in local property in Stoke-on-Trent in the British West Midlands, addressing an issue of reuse to urban developers. Some vacant premises and properties have deteriorated to a degree in structural terms, which only required a limited refurbishment to return to some level of sustainable use.

Vacant urban spaces can provide residential accommodation over other commercial urban premises on designated urban areas supported by some fiscal incentives. Tax incentives may encourage the use of vacant urban spaces in designated areas with specialized treatment to house building, as it was successfully introduced, for example, in Ireland (Norris & Winston, 2009). Vacant urban spaces can be utilized for urban agriculture, horticulture, and vegetable cultivation using new technologies and methods to guarantee food security and sustainability. Vacant dwelling can be temporarily and permanently, being this one the unoccupied inhabitable dwellings. In some urban settlements, there are high level of permanently and also temporally vacant dwellings. Persistently and permanently vacant premises and property tend to be in poorer condition than both newly, temporally vacant and reoccupied premises.

URBAN VACANT LAND IN URBAN AGRICULTURE, FARMING, AND GARDENING

Urban vacant land represents an opportunity for declining cities to become more sustainable, offering local food sources and food systems to the inhabitants of urban communities. Empty urban vacant lots project negative image unless they are used for gardening, farming, and agriculture to obtain healthy food and improve retail food landscape. Using urban vacant land for growing food in the high quality of produce, more natural, healthy despite concerns for soil toxicity and economic savings. Local economic and population growth contributes to decreasing land vacancy. Vacant lots and parcels are difficult to regenerate due to the small size, odd shape, and location disconnection. Urban land-use policies and regulations are deemed as methods for increasing and decreasing urban land vacancies.

Home and community farming, gardening, agriculture, and aquaculture on vacant urban public and private land can be promoted for domestic food production. Urban agriculture is related to the one that takes place within the built-up city and peri-urban agriculture in the areas surrounding the city (Nugent, 2000). Reclaim vacant land to scale food production and 'de-alienating' urban dwellers helps to overcome the forms of metabolic rift. Economic crises and their consequent foreclosures of homes may result in the rise of vacant lots and land which can be used for creating urban farming, agriculture and gardens through the empowerment of more self-reliant local communities, increasing food security, better nutrition and reducing the crime rates (Metcalf & Widener, 2011).

The structural urban vacant land crisis based on urban industrial and brownfield land can be gradually replaced by housing and the real estate market. Urban land management practices support the use of vacant land for urban gardening, farming, and agriculture, enhancing the attractiveness for alternative urban development and increasing the value in the real estate market. Urban gardens and farms arise on vacant urban lots in the margins of the food system, and the built environment may provide some food when the market fails. Urban vacant spaces should be considered as possible locations to produce food, which may be decompensating the loss of rural land dedicated to agricultural activities.

Urban agriculture is practiced on public and private vacant land, gardens, rooftops, etc. There is revamping food production in urban settings transitioning unutilized vacant spaces and rooftops into productive areas. Urban vacant spaces can be used for urban horticulture in rooftops, private balconies, fallow land, roadsides, etc. Non frequently used roads in the cities can be converted into green space for a pedestrian and bicycle path (Bohn & Viljoen, 2011). Urban vacant spaces and resources for urban farming and agricultural production are costlier than the one in rural areas, one of the reasons why urban farming is put on the roofs of buildings and other infrastructures. Available rooftops and vacant spaces can be used for hydroponic greenhouses.

The vacant roof spaces of large public, community, commercial, and industrial buildings are sites that can be used potentially for vegetable and ornament plants that can be exposed to sunlight. A good example has been the use of vacant abandon land used to plant flowers, ornamental plants, herbs, and vegetables, next to a government building, as reported by Hui (2011). Growing of vegetables and ornament plants around housing compounds and other vacant lands not used for public green spaces and neglected by the owners and local authorities improving the microclimate.

Non-profit urban farming and agriculture activities can be located in the poorest urban area, where vacant lots are numerous, socioeconomic conditions, food access issues, and the structural inequities are more severe. Declined cities with decreasing socioeconomic conditions and vacant land have been able to utilize it to cultivate food and reinvest in neighborhoods (McGuire, 2007; McMillan, 2008; Sterpka, 2009; Herzog, 2011).

Urban farming and agriculture projects are often rooted in the vacant "lumpengeography" (Walker, 1978) of the city in capitalist redevelopment. Urban gardens have reclaimed the commons by cultivating vacant lots based on new normative conceptions of urban space (Crane et al., 2012). Commons can take over a vacant lot moving to subvert its exchange value with other use values to support the community provisioning of food. The use-value of the urban vacant land is exceeded by the exchange value when there are redevelopment initiatives, a housing bubble, and the arrival of new incomers.

Urban agriculture under the neoliberal urbanism framework (Peck, Theodore, & Brenner, 2009) urban is a common strategy of community groups to encourage neighborhood social interaction, reclaim vacant urban space and urban green areas aimed to foster livelihoods for the unemployed. Green urban gardens emerged from undervalued vacant land may contribute to the rising adjacent land property values (Voicu & Been, 2008), and ultimately threaten their tenure. Vacant lots, open land spaces, and underutilized green spaces with agricultural potential can be identified using aerial technology.

Transformation of vacant lots into gardens may be motivated by poor low-income neighborhoods for food security contributing to the gentrification process (Crouch, 2012; Tortorello, 2012). The gentrification process of New York in the 1980s led to rising land values, including leasing vacant land for squatter gardens, but often racialized tensions arose whether to use vacant land as space for gardens for low-income housing (Schmelzkopf, 1995).

Opportunities for urban agriculture on vacant land supported by local capital impose some obstacles to its expansion such as funding, changes in land values, externalities from environmental pollution, moving gentrifiers, etc.

Urban vacant spaces can be of low and high farming intensity in terms of monitored parcels and plots. The care received, the yields produced of traditional and organic agriculture. High- and low-intensity urban vacant space practiced on a large parcels yields derive from yield ratios of community and collective gardens of publicly owned land. Urban vacant and residential yard spaces in a community can be used for urban farming and agriculture with low-intensity fresh vegetable yield estimates, to meet the demand of the entire urban population. Urban residential and business vacancies as a type of vacant addresses are usually more abundant than other types of vacancies.

Vacant spaces city-owned can be transformed into a privately managed urban farming. Urban farming and agriculture may be creatively achieved for food growing considering the compact city (Barr, 1997) making use of the derelict, under-utilized and interstitial spaces by overlapping uses in rooftops, vacant plots and roadside verges (1997, p. 16). Risky areas of urban agriculture can be prevented making some urban vacant spaces and land available for urban farming by local authorities and other involved stakeholders of the food system (Kremer et al., 2013; Taylor & Lovell, 201; Hamilton et al., 2013; Lovell, 2010; Mok et al., 2014; Bohn & Viljoen, 2011).

The use of high and low-intensity urban farming and agriculture in vacant and residential space have an impact on the population to feed. An urban community with a high population density may be supported by farming and agriculture in vacant and residential areas complemented by hydroponics. Urban communities with large amounts of vacant space can produce for the more consumptive population relying upon the demand of more populated areas that do not have vacant and roof spaces to feed their own residents. Despite that urban communities may have a surplus of vacant land available, they can supplement hydroponics to meet the population demand on fresh vegetables. Urban vacant and residential spaces may produce vegetables using low-intensity yields in combination with industrial rooftops implementing hydroponics to meet the demand.

Low-intensity yields are applied to vacant spaces in residential yards, and sometimes high-intensity yields can be applied to these areas but more clearly privately-owned vacant spaces dedicated to urban farming and agriculture. Front yards of public and private buildings might be as not registered vacant land, which can be dedicated to urban gardening, farming, and agriculture following strategies aligned with urban densification. Local governments may aim to dedicate urban vacant spaces, to community and collective gardens available to agriculture and vegetable production allocated are considered a low-risk plan requiring small financial investments.

Most of the time, urban vacant spaces devoted to producing food also aim to foster socio ecosystems services, social and environmental benefits. Urban agriculture using vacant spaces for community and neighborhood government plays a positive role in improving the local environment by serving poor low-income people (Hough & Barrett, 1987).

City initiatives may enable residents to grow food on vacant lots that aids subsistence and are donated to local community organizations (Bearre, 1976). Urban agriculture works on public land and in private plots, and the city encourages property tax reassessment to use vacant land (Schutzbank, 2012; Williams, 2010). The urban food production system in vacant public spaces involving local residents to grow personal gardens on lots or in private lots and industrial rooftops. Urban agriculture contributes to building community, providing fresh and healthy food, recreation, making more beautiful neighborhoods (Stewart et al., 2019), and productive use of vacant land (Hodgson et al., 2011).

Urban farming and agriculture to make use of idle vacant spaces that project a negative image possibly will be the best decision undertaking minimal resources to promote city development and growth. ZFarming (Zero-acreage Farming that means agriculture in or on urban structures) is a subtype of urban farming that uses some of the principles of ground-based urban farming. At the same time it differs from it by using the building stock instead of vacant land and parcels, brownfields and farmland. ZFarmingpresents diverse opportunities for resource-efficiency and challenges, regulatory and technical frameworks and restraints. ZFarming is a grassroots initiative of a bottom-up engagement or volunteers and residents organized on loosely groups that have access to buildings and rooftop spaces, occupant and transform, which differs from ground-based urban farming in vacant urban spaces.

ZFarms use technologies of industrial farming and green roof in vacant buildings and rooftops. ZFarming integrates food production security into the urban land scarcity exploiting vacant buildings, rooftops, and unused urban spaces. Urban farming and agriculture are spreading across urban vacant land as a source of ecosystem services and food, supported by local civil society and government. Urban farming may be better suited to be implemented in vacant land, although others consider vacant land as an opportunity to have a natural source to provide themselves better quality vegetables and fruits.

The city government may have to make decisions to have some options to buy vacant land and parcels and also may work with non-profit organizations to develop urban green areas, community and neighborhood gardens, etc. Urban green space projects may vary across neighborhoods but reduce vacant land and parcel properties and revitalize neighborhoods and communities by increasing property values and community's socio-economic status, creating green jobs, improving public health, etc. (Douglas, 1975; Vaughan, Kaczynski, Wilhelm Stanis, Besenyi, Bergstrom, & Heinrich, 2013; Heynen, Perkins, Roy, Brown, & Jameton, 2000; Schilling & Logan, 2008).

Urban vacant spaces have the potential to produce vegetables to meet the urban vegetable demand at lower operating costs. The fresh vegetable demand can be achieved if all urban vacant space, industrial rooftops, and residential gardens space are utilized. Urban farming and agriculture initiatives take place on large urban vacant parcels and in city green areas and parks, from small private and community gardens to large urban farms run by non-profit organizations and business market gardeners.

A good example is the city of Detroit has more than 60 square miles of vacant urban lots. The Potato Patch plan adopted across the USA was an initiative launched in Detroit during the Depression of 1893 when more than 1500 families farmed on vacant lots used for gardening to provide food. During the recession of the 1970s, an urban gardening program known as 'inflation gardens' took over vacant lots in American cities (Brown & Jameton, 2000; Lawson, 2005; Schmelzkopf, 1995). Urban gardening initiatives installed in adapted urban space of temporary nature "on vacant lots and formerly or future built-up sites" (Fuhrich & Goderbauer, 2011, p. 53), can be developed to serve for educative and recreational purposes.

Many residents envision urban gardening with eco-roofs, converted parking and vacant lots, and any other underutilized spaces that may provide food (City of Portland, 2007). In 1934 in New York, more than 5000 vacant lots were transformed into an urban gardening program (Brown & Jameton, 2000; Lawson, 2005).

COLLABORATIVE URBAN GOVERNANCE STRUCTURES AND REVITALIZATION OF OPEN VACANT SPACES

New forms of urban gardening, farming, and agriculture can re-use and re-appropriate urban vacant land and adaptive open green space to shift functions under new forms of governance framework offering new possibilities of sustainable urban development. Vacant urban spaces may be facilitated the micro-scale urban agriculture integrated into social housing developments such as neighborhoods and communities greening, farming and gardening, as well as small-scale animal raising (Dubbeling et al., 2009). Vacant land has always been at the center stage of local governments, non-profit organizations, collectives and community members concerned about urban agriculture and farming initiatives for food production in the city.

Local government can hold the governance and management of public urban vacant space controlled by a wide range of public services supported by citizen and grass-root involvement and participation non-bureaucratic (Kul-ke et al., 2011, p. 222) building consensus for all stakeholders on regulating and controlling the use of vacant spaces. Collaborative governance structures can give support to a framework for more collaborative and sustainable urban planning practices. It canoffer synergistic possibilities of new socio-economic transactions and functions, and serve as a platform for exchanges and meeting spaces to engage residents in the participative process for revitalizing vacant open spaces.

Urban planning local authorities identify non-domestic vacant buildings that make up most of the amount of vacant building stock. Vacant buildings are those unoccupied for more than one year, structurally sound, and in a reasonable state of repair (DETR, 2000, p. 34, 1998). Partially occupied vacant buildings are more difficult to redevelop because it may include vacant floor space to be reused. Vacant floor space and premises can be classified by type, size, postcode, length of vacancy, age, and any other attribute. However, vacant commercial buildings have been considered more problematic for urban planning.

Long-term vacancy requires more oriented regeneration policies (National Land Use Database, 2000). Vacant buildings require recording reliable and comprehensive information. Consistent information of long-term vacant floor space in buildings within non-domestic premises is relevant for urban planning purposes. Vacant property availability for urban developers require to have reliable and comprehensive data and information available on the previous uses of empty buildings, accessibility of service provision, etc., difficult to extrapolate (Kohler & Hassler 2002, p. 229).

Local authorities are usually the larger urban landowners, much of which this land is vacant and derelict in making a good use as suitable land for urban farming and food production in urban green spaces. An urban land inventory is the first step, determining how much vacant land can be used for urban agriculture is depending on the site assessments and negotiation with the involved stakeholders. No accurate inventory and policy on unused urban plots, vacant land, and existing open green spaces on public or private land can set out the potential for urban farming, agriculture and gardening, and the impact on economic growth, social equality, and environmentally sustainable city. Urban vacant land inventories monitored on databases create more holistic management and regulatory system.

Local authorities usually own suitable vacant land for suitable development such as the allotment gardens (Crouch, 1998) to enable and actively be promoted as areas of redevelopment for community gardens of urban food-production projects which may be likely to be sold to developers, or left the vacant land when maintenance costs are high. Good use of vacant and derelict lands increases urban local food production providing environmental benefits and opportunities to recycle organic waste. To mitigate the effects of the high rise cityscape and urban spaces created by increasing building densities can change of UHI and microclimate\. To improve the urban landscape, local authorities are promoting the use of any available vacant plot of land, rooftops, window boxes (Vandermeulen, et al., 2009).

This urban governance structure and sustainable urban planning framework provide a multifunctional space for participative consensus-oriented management and they support re-using and revitalizing urban vacant public and private open spaces. Urban environmental management supported collective practices may enhance the sustainable use of vacant land and lots to stimulate the use of public and private green spaces. A high vacancy of lots and buildings have negative connotations that can overshadow other positive community assets (Garvin et al., 2013). Vacant land is an asset for urban green areas and agriculturally based city.

Urban communities and neighborhoods focus on most problematic types, conditions, and causes, such as deindustrialization, disinvestment, sectoral, technological, and demographic changes of vacant land, in order to develop urban and locational policies and regulations (Greenstein & Sungu-Eryilmaz, 2004). Most parcels are difficult to revitalize and regenerate because they are small and disconnected which is mostely caused by suburbanization and disinvestment.

There is no clear consensus to determine the condition of land and structure vacancies, which vary by time and city government criteria. City officials are responsible for determining, designing, and deeming vacancy of spaces, lots and structures based on other methods such as informal feedback of residents, duration of vacancies, building and health inspections, delinquency, and other vacancy conditions.

Vacant municipal and semi-public urban land might be accessed through policies and programs promoting the social inclusion of disadvantaged urban households as a strategy to enhance nutrition and food security. Urban disadvantaged and poor households can have access to urban vacant land supported by policies and programs implemented to secure their rights to produce food aimed to improve their livelihoods. Vacant land offers opportunities for vulnerable groups to reclaim the source and ownership of a better food system. Vacant urban land can be converted to agricultural food-based production and creating employment for the most vulnerable and poor people.

The green and organic food and verdure emerging from vacant urban lots and marginal spaces are related to a movement in favor of the creation of new commons from redeveloping industrial brownfields into scale-up some new urban green areas (DeSousa, 2004; Rosol, 2005). Most of the commons are the vacant spaces and wastelands of urban spaces, including the natural and agricultural resources which have been commodified, such as biodiversity, seeds, water, etc. Urban local authority data must identify the location, type, and size of vacant premises. The vacancy identification is the notification of premises becoming unoccupied and recorded by the urban local authorities.

Management of vacant land and green areas co-populated with other vegetation on once-occupied vacant lots in residential spaces is important. Plants in green vacancy areas require resources for some kind of maintenance required to survive. Some of the benefits of urban agroecology in vacant lots as a natural resource are the species growing, the decrease in crime, reduction of poverty, etc. These benefits are also related to urban watersheds, green networks, and provide a framework for city integration and incorporation. Uses and conditions of vacant lots develop a network in the urban environmental and socio-ecological system that continues evolving due to disturbances and disasters of vacant areas.

Management scenarios may involve increasing the soil permeability and nutrient on vacant urban land to increase surface biomass for carbon sequestration. Vacant land located in urban watersheds has nature well-distributed in residential parcels. Land abandonment and vacancy give opportunities to new compositions of nature with areas of opportunistic species and adaptive vegetation leading to a certain type of ecosystem services and referring as indicators of the surviving site uses. Food produced on local vacant land is an important survival strategy for migrants, refugees, and displaced people.

Cities at the forefront of vacant land management are focusing more on the tax revenue lost than on the ecological benefits. Myers & Wyatt (2004) have developed a methodology drawing on property taxation relating to vacant non-domestic buildings to identify the type, size, and location of a long-term vacant property. The market for vacant and partially vacant buildings is imperfect and based on incomplete information with the not efficient allocation of resources. A vacant parcel has multiple use values exceeding the exchange value on the market. Urban agriculturalists cultivate community and collective farms or gardens on vacant land.

As the owner's strategy to selling or to find tenants, some vacant buildings can persistently be retained in good conditions for reoccupations, even though some have occurred in poor conditions to sell or rent at a lower price and release some capital. Vacant buildings are a fundamental resource (Kohler, 1999, p. 318) of social and cultural worth that encourages efficient use involving some environmental impacts. The composition of building stock requires knowledge improvement (Kohler & Hassler, 2002)

From investments, some dwellings may be left vacant to await capital appreciation to take also the advantage of incentives (Fitz Gerald & Winston, 2005). Property investments and specialized construction organizations can be involved in the portfolio acquisition and reuse of vacant industrial premises and develop working industrial buildings or designing and building residential areas for housing. Occupancy status of premises, occupied or vacant, has no impact on their valuation with the assignment of a rate. Vacant and empty premises could be affected by the normal level of rates of listings classified as industrial or housing prevalent with these types of property.

CONCLUSION

The conclusion of this chapter is that urban local spaces hold potential for urban green transformations to meet the social needs and improve the ecological services of communities. These transformations from urban vacant land and parcel spaces to utilized for revitalization purposes require a collaborative governance structure to provide support. Transitions from vacant land and parcel spaces to be utilized need to be stimulated.

Vacant land becomes a valuable asset in supplying the ecosystem services as essential resources such as biodiversity, food, and clean, fresh water, Etc. at sustainable rates. Vacant land plays an essential role in the provision of ecosystem services interconnected and independent as well with the ecological spheres.

REFERENCES

Accordino, J., & Johnson, G. T. (2000). Addressing the Vacant and Abandoned Property Problem. *Journal of Urban Affairs*, *22*(3), 301–315. doi:10.1111/0735-2166.00058

Akers, J. (2013). Making markets: Think tank legislation and private property in Detroit. *Urban Geography*, *34*(8), 1070–1095. doi:10.1080/02723638.2013.814272

Alker, S., Joy, V., Roberts, P., & Smith, N. (2000). The definition of brownfield. *Journal of Environmental Planning and Management*, *43*(1), 49–69. doi:10.1080/09640560010766

Anderson, E. C., & Minor, E. S. (2017, January 1). Vacant lots: An underexplored resource for ecological and social benefits in cities. In *Urban Forestry and Urban Greening*. Elsevier GmbH. Doi:10.1016/j.ufug.2016.11.015

Anderson, E. C., & Minor, E. S. (2017, January 1). Vacant lots: An underexplored resource for ecological and social benefits in cities. In *Urban Forestry and Urban Greening*. Elsevier GmbH. Doi:10.1016/j.ufug.2016.11.015

Anderson, E. C., & Minor, E. S. (2017, January 1). Vacant lots: An underexplored resource for ecological and social benefits in cities. In *Urban Forestry and Urban Greening*. Elsevier GmbH. Doi:10.1016/j.ufug.2016.11.015

Andersson, E. (2006). Urban landscapes and sustainable cities. *Ecology and Society*, *11*(1), 34. doi:10.5751/ES-01639-110134

Ball, R. M. (1996). *The use and re-use of vacant industrial buildings, Proposal for funding under the EPSRC Sustainable Cities Managed Programme*. Division of Geography, Staffordshire University.

Ball, R. M. (1998). *The use and re-use of vacant industrial buildings*. EPSRC Research Grant GR=K93464, Final Project Report, EPSRC.

Ball, R. M. (2002). Re use potential and vacant industrial premises: Revisiting the regeneration issue in Stoke-on-Trent. *Journal of Property Research*, *19*(2), 93–110. doi:10.1080/09599910210125223

Ball, R.M. & Bord, D. (1994, Sept.). *Vacant industrial premises in Stoke-on-Trent: a survey and analysis of the situation in August 1994*. Consultancy Report submitted to Stoke-on-Trent City Council and the Staffordshire Training and Enterprise Council.

Bearre, C. D. (1976, April 8). Farm-a-Lot: Detroiters turn vacant lots into gardens. *Ann Arbor Sun*. Retrieved from http://oldnews.aadl.org/node/201103

Bohn, K., & Viljoen, A. (2011). The edible city: Envisioning the continuous productive urban landscape. *Field J*, *4*(1), 149–161.

Bontje, M. (2004). Facing the challenge of shrinking cities in East Germany: The case of Leipzig. *GeoJournal*, *61*(1), 13–21. doi:10.100710708-005-0843-2

Bowman, A. O. M., & Pagano, M. A. (2000). *Vacant Land in Cities: And Urban Resource*. Brookings Institution, Center on Urban and Metropolitan Policy.

Branas, C. C., Cheney, R. A., MacDonald, J. M., Tam, V. W., Jackson, T. D., & Ten Have, T. R. (2011). A difference-in-differences analysis of health, safety, and greening vacant urban space. *American Journal of Epidemiology*, *174*(11), 1296–1306. doi:10.1093/aje/kwr273 PMID:22079788

Brown, K. H., & Jameton, A. L. (2000). Public health implications of urban agriculture. *Journal of Public Health Policy*, *21*(1), 20–39. doi:10.2307/3343472 PMID:10754796

Brulle, R. J., & Pellow, D. N. (2005). Environmental justice: Human health and environmental inequalities. *Annual Review of Public Health*, *27*(1), 103–124. doi:10.1146/annurev.publhealth.27.021405.102124 PMID:16533111

Burkholder, S. (2012). The new ecology of vacancy: Rethinking land use in shrinking cities. *Sustainability*, *4*(6), 1154–1172. doi:10.3390u4061154

Burkholder, S. (2012). *The New Ecology of Vacancy: Rethinking Land Use in Shrinking Cities*. Academic Press.

City of Portland. (2007). *Voices from the community: The vision PDX community input summary* (Abridged Version). Retrieved from https://www.visionpdx.com/downloads/visionPDX%20CIS%20 Abrdg%20Versi.pdf

Cohen, B. (2006). Urbanization in Developing Countries: Current Trends, Future Projections, and Key Challenges for Sustainability. *Technology in Society*, *28*(1), 63–80. doi:10.1016/j.techsoc.2005.10.005

Colasanti, K., Hamm, M., & Litjens, C. (2012). The city as an "agricultural powerhouse?" Perspectives on expanding urban agriculture from Detroit, Michigan. *Urban Geography*, *33*(3), 348–369. doi:10.2747/0272-3638.33.3.348

Coleman, A. (1982). Dead Space in the Dying Inner City. *The International Journal of Environmental Studies*, *19*(2), 103–107. doi:10.1080/00207238208709976

Communauté Urbaine de Montréal. (1996). [Montreal Urban Community]. *Carte d'Occupation du Sol* [Land Occupancy Card]. Retrieved from http://cmm.qc.ca/biotopes/docs/volets_1_et_2.pdf

Crane, A., Viswanathan, L., & Whitelaw, G. (2012). Sustainability through intervention: A case study of guerrilla gardening in Kingston, Ontario. *Local Environment*. doi:10.1080/13549839.2012.716413

Crouch, P. (2012, October 23). Evolution or gentrification: do urban farms lead to higher rents? *Grist.*

DeSousa, C. A. (2004). The greening of brownfields in American cities. *Environmental Planning and Management*, *47*(4), 579–600. doi:10.1080/0964056042000243249

DETR. (1998). Sustainable Development: Opportunities for Change. Department of the Environment, Transport and the Regions. Consultation Paper on a Revised UK Strategy, DETR, London.

DETR. (2000). Our Towns and Cities: The Future. CM 4911, Department of the Environment Transport and the Regions. London: DETR.

Detroit Works Project. (2014). *Detroit future city: Detroit strategic framework plan.* Retrieved from http://detroitfuturecity.com/wp-content/uploads/2014/12/DFC_Full_2nd.pdf

Douglas, J. R. (1975). *Hydroponics: The Bengal System* (5th ed.). Madison, NY: Oxford University Press.

Draus, P. J., Roddy, J., & McDuffie, A. (2014). We don't have no neighbourhood: Advanced marginality and urban agriculture in Detroit. *Urban Studies (Edinburgh, Scotland)*, *51*(12), 2523–2538. doi:10.1177/0042098013506044

Dubbeling, M., Bracalenti, L., & Lagorio, L. (2009b). Participatory design of public spaces for urban agriculture, Rosario, Argentina. *Open House International*, *34*, 36–49.

Fitz Gerald, E., & Winston, N. (2005). Housing, equality and inequality. In M. Norris M & D. Redmond (Eds.), Housing Contemporary Ireland: Economy, Society, Space and Shelter (pp. 224-244). Institute of Public Administration.

Fuhrich, M., & Goderbauer, E. (2011). Urban restructuring: Making 'more' from 'less. In B. Müller (Ed.), *Ur-ban regional resilience: How do cities and regions deal with change* (pp. 49–58). Berlin: Springer.

Garvin, E. C., Cannuscio, C. C., & Branas, C. C. (2013). Greening vacant lots to reduce violent crime: A randomised controlled trial. *Injury Prevention*, *19*(3), 198–203. doi:10.1136/injuryprev-2012-040439 PMID:22871378

Gough, M. Z., & Accordino, J. (2013). Public Gardens as Sustainable Community Development Partners: Motivations, Perceived Benefits, and Challenges. *Urban Affairs Review*, *49*(6), 851–887. doi:10.1177/1078087413477634

Greenstein, R., & Sungu-Eryilmaz, Y. (2004). *Recycling the City: The Use and Reuse of Urban Land.* Cambridge, MA: Lincoln Inst of Land Policy.

Hall, P. (1998, May 28) Vacant looks. *Guardian.*

Hall, P. G. (1998). *Cities on Civilisation.* London: Wiedenfeld and Nicolson.

Hamilton, A. J., Burry, K., Mok, H. F., Barker, F., Grove, J. R., & Williamson, V. G. (2013). Give peas a chance? Urban agriculture in developing countries. A review. *Agronomy for Sustainable Development*, *34*(1), 45–73. doi:10.100713593-013-0155-8

Herzog, K. (2011, Sept 19). Milwaukee recognized for urban farms, aquaponics. *Journal Sentinel.*

Heynen, N., Perkins, H. A., & Roy, P. (2006). The political ecology of uneven urban green space. *Urban Affairs Review*, *42*(1), 3–25. doi:10.1177/1078087406290729

Hodgson, K., Caton Campbell, M., & Bailkey, M. (2011). *Urban agriculture: growing healthy, sustainable places.* Washington, DC: American Planning Association.

Hoffman, M., Westermann, J. R., Kowarik, I., & van der Meer, E. (2012). Perceptions of parks and urban derelict land by landscape planners and residents. *Urban Forestry & Urban Greening*, *11*(3), 303–312. doi:10.1016/j.ufug.2012.04.001

Hough, M., & Barrett, S. (1987). *People and City Landscapes: a study of people and open space in metropolitan areas of Ontario.* Toronto: Conservation Council of Ontario.

HUD. (2015). *HUD Aggregated USPS Administrative Data on Address Vacancies.* U.S. Department of Housing and Urban Development. Retrieved from http://www.huduser.org/portal/datasets/usps.html

Hui, S. C. M. (2011). *Green roof urban farming for buildings in high-density urban cities.* Invited paper for the Hainan China World Green Roof Conference 2011, Hainan, China.

Jakle, J. A., & Wilson, D. (1992). *Derelict Landscapes: The Wasting of America's Built Environment.* Lanham, MD: Rowman & Littlefield Publishers.

Kim, G., Miller, P. A., & Nowak, D. J. (2018). Urban vacant land typology: A tool for managing urban vacant land. *Sustainable Cities and Society*, *36*, 144–156. doi:10.1016/j.scs.2017.09.014

Kim, G., Miller, P. A., & Nowak, D. J. (2018). Urban vacant land typology: A tool for managing urban vacant land. *Sustainable Cities and Society*, *36*, 144–156. doi:10.1016/j.scs.2017.09.014

Kim, G., Miller, P. A., & Nowak, D. J. (2018). Urban vacant land typology: A tool for managing urban vacant land. *Sustainable Cities and Society*, *36*, 144–156. doi:10.1016/j.scs.2017.09.014

Kohler, N. (1999). The relevance of Green Building Challenge: An observer's perspective. *Building Research and Information, 27*(4/5), 309–320. doi:10.1080/096132199369426

Kohler, N., & Hassler, U. (2002). The building stock as a research project. *Building Research and Information, 30*(4), 226–236. doi:10.1080/09613210110102238

Kremer, P., Hamstead, Z. A., & McPhearson, T. (2013). A social-ecological assessment of vacant lots in New York City. *Landscape and Urban Planning, 120*, 218–233. doi:10.1016/j.landurbplan.2013.05.003

Kremer, P., Hamstead, Z. A., & McPherson, T. (2013). A social-ecological assessment of vacant lots in New York City. *Landscape and Urban Planning, 120*, 218–233. doi:10.1016/j.landurbplan.2013.05.003

Lawson, L. J. (2005). *City bountiful: a century of community gardening*. Berkeley: University of California Press.

Lovell, S. T. (2010). Multifunctional urban agriculture for sustainable land use planning in the United States. *Sustainability, 2*(8), 2499–2522. doi:10.3390u2082499

Mallach, A. (2012). *Laying the Groundwork for Change: Demolition, Urban Strategy, and Policy Reform*. Washington, DC: Brookings Institution.

Mathey, J., & Rink, D. (2010). Urban wastelands—A chance for biodiversity in cities? In *Urban Biodiversity and Design* (pp. 406–424). Oxford, UK: Blackwell. doi:10.1002/9781444318654.ch21

McGuire, V. C. (2007, Sept 2). Replacing neglect with peach trees. *The New York Times.*

McMillan, T. (2008, May 7). Urban farmers' crops go from vacant lot to market. *The New York Times.*

Metcalf, S. S., & Widener, M. J. (2011). Growing Buffalo's capacity for local food: A systems framework for sustainable agriculture. *Applied Geography (Sevenoaks, England), 31*(4), 1242–1251. doi:10.1016/j.apgeog.2011.01.008

Mok, H. F., Williamson, V. G., Grove, J. R., Burry, K., Barker, S. F., & Hamilton, A. J. (2014). Strawberry fields forever? Urban agriculture in developed countries: A review. *Agronomy for Sustainable Development, 34*(1), 21–43. doi:10.100713593-013-0156-7

Myers, D., & Wyatt, P. (2003). Rethinking urban capacity: identifying and appraising vacant buildings. In *Proceedings of the European Real Estate Society Conference*. ERES.

Myers, D., & Wyatt, P. (2004). Rethinking urban capacity: Identifying and appraising vacant buildings, *Building Research &amp. Information, 32*(4), 285–292. doi:10.1080/0961321042000221061

National Commission on Urban Problems. (1968). *Land Use in 106 Large Cities*. Washington, DC: Three Land Research Studies.

National Land Use Database (2000) *Previously Developed Land (PDL): Vacant Sites and Derelict Sites, and Other Previously Developed Land and Buildings that May be Available for Redevelopment, Data Specification, NLUD – a partnership Rethinking urban capacity.* Author.

Németh, J., & Langhorst, J. (2014). Rethinking Urban Transformation: Temporary Uses for Vacant Land. *Cities (London, England), 40*, 143–150. doi:10.1016/j.cities.2013.04.007

Newman, G. D., Bowman, A. O. M., Jung Lee, R., & Kim, B. (2016). A current inventory of vacant urban land in America. *Journal of Urban Design*, *21*(3), 302–319. doi:10.1080/13574809.2016.1167589

Newman, G. D., Bowman, A. O. M., Jung Lee, R., & Kim, B. (2016). A current inventory of vacant urban land in America. *Journal of Urban Design*, *21*(3), 302–319. doi:10.1080/13574809.2016.1167589

Newman, G. D., Bowman, A. O. M., Jung Lee, R., & Kim, B. (2016). A current inventory of vacant urban land in America. *Journal of Urban Design*, *21*(3), 302–319. doi:10.1080/13574809.2016.1167589

Norris, M., & Winston, N. (2009). Rising second home numbers in rural Ireland: Distribution, drivers and implications. *European Planning Studies*, *17*(9), 1303–1322. doi:10.1080/09654310903053448

Northam, R. M. (1971). Vacant Urban Land in the American City. *Land Economics*, *47*(4), 345–355. doi:10.2307/3145071

Nugent, R. (2000). The impact of urban agriculture on the household and local economies. In Growing cities. Growing Food, (pp 67–97). Feldafing: Dtsch Stift int Entw.

O' Bowman, A. M., & Pagano, M.A. (2004). Terra Incognita: Vacant Land and UrbanStrategies. Washington, DC: Georgetown University Press.

Peck, J., Theodore, N., & Brenner, N. (2009). Neoliberal urbanism: Models, moments, mutations. *SAIS Review (Paul H. Nitze School of Advanced International Studies)*, *29*(1), 49–66.

Pettit, H. (2014). Cities without Suburbs: A Census 2010 Perspective. *European Planning Studies*, *22*(9), 1989–1991. doi:10.1080/09654313.2014.924219

Rosol, M. (2005). Community gardens—A potential for stagnating and shrinking cities? Examples from Berlin. *Die Erde*, *136*(2), 165–178.

Rupprecht, C. D. D., & Byrne, J. A. (2014). Informal urban greenspace: A typology and trilingual systematic review of its role for urban residents and trends in the literature. *Urban Forestry & Urban Greening*, *13*(4), 597–611. doi:10.1016/j.ufug.2014.09.002

Schilling, J. & Logan, J. (2008). Greening the Rust Belt: A green infrastructure model for right-sizing America's shrinking cities. *Journal of the American Planning Association, 74*, 451-466.

Schmelzkopf, K. (1995). Urban community gardens as contested space. *Geographical Review*, *85*(3), 364–381. doi:10.2307/215279

Schutzbank, M. H. (2012). *Growing vegetables in Metro Vancouver: An urban farming census (Unpublished master's thesis).* Vancouver: University of British Columbia. Retrieved from https://circle.ubc.ca/handle/2429/43559

Smith, J. P., Li, X., & Turner, B. L. II. (2017). Lots for greening: Identification of metropolitan vacant land and its potential use for cooling and agriculture in Phoenix, AZ, USA. *Applied Geography (Sevenoaks, England)*, *85*, 139–151. doi:10.1016/j.apgeog.2017.06.005

Smith, J. P., Li, X., & Turner, B. L. II. (2017). Lots for greening: Identification of metropolitan vacant land and its potential use for cooling and agriculture in Phoenix, AZ, USA. *Applied Geography (Sevenoaks, England)*, *85*, 139–151. doi:10.1016/j.apgeog.2017.06.005

Smith, J. P., Li, X., & Turner, B. L. II. (2017). Lots for greening: Identification of metropolitan vacant land and its potential use for cooling and agriculture in Phoenix, AZ, USA. *Applied Geography (Sevenoaks, England)*, *85*, 139–151. doi:10.1016/j.apgeog.2017.06.005

Smith, J. P., Li, X., & Turner, B. L. II. (2017). Lots for greening: Identification of metropolitan vacant land and its potential use for cooling and agriculture in Phoenix, AZ, USA. *Applied Geography (Sevenoaks, England)*, *85*, 139–151. doi:10.1016/j.apgeog.2017.06.005

Sterpka, M. (2009, July 6). *Cleveland's for-profit urban gardens are growing*. Retrieved from https://www.cleveland.com/metro/2009/07/clevelands_forprofit_urban_gar.html

Stewart, W. P., Gobster, P. H., Rigolon, A., Strauser, J., Williams, D. A., & van Riper, C. J. (2019). Resident-led beautification of vacant lots that connects place to community. *Landscape and Urban Planning*, *185*, 200–209. doi:10.1016/j.landurbplan.2019.02.011

Stewart, W. P., Gobster, P. H., Rigolon, A., Strauser, J., Williams, D. A., & van Riper, C. J. (2019). Resident-led beautification of vacant lots that connects place to community. *Landscape and Urban Planning*, *185*, 200–209. doi:10.1016/j.landurbplan.2019.02.011

Stewart, W. P., Gobster, P. H., Rigolon, A., Strauser, J., Williams, D. A., & van Riper, C. J. (2019). Resident-led beautification of vacant lots that connects place to community. *Landscape and Urban Planning*, *185*, 200–209. doi:10.1016/j.landurbplan.2019.02.011

Stewart, W. P., Gobster, P. H., Rigolon, A., Strauser, J., Williams, D. A., & van Riper, C. J. (2019). Resident-led beautification of vacant lots that connects place to community. *Landscape and Urban Planning*, *185*, 200–209. doi:10.1016/j.landurbplan.2019.02.011

Syms, P. (1994). The funding of developments on derelict and contaminated sites. In R. Ball & A. Pratt (Eds.), *Industrial Property Policy and Economic Development* (pp. 63–82). London: Routledge.

Syms, P. (1998) *Re-using Brownfield Sites: The Problem of Releasing Brownfield Land for Development.* Planning and Property Paper No. PP6, Centre for Regional Economic and Social Research, Sheffield Hallam University.

Syms, P. (1999). Redeveloping brownfield land: The decision making process. *Journal of Property Investment & Finance*, *17*(5), 481–500. doi:10.1108/14635789910294903

Syms, P. (2001). *Releasing Brownfields*. London: RICS Foundation, Royal Institution of Chartered Surveyors.

Taylor, J. R., & Lovell, S. T. (2012). Mapping public and private spaces of urban agriculture in Chicago through the analysis of high-resolution aerial images in Google Earth. *Landscape and Urban Planning*, *108*(1), 57–70. doi:10.1016/j.landurbplan.2012.08.001

Tortorello, M. (2012, Oct 31). Urban gardens grow everything except gardeners. *The New York Times*.

United Nations. (2002). *Report of the World Summit on Sustainable Development*. Johannesburg, South Africa: United Nations.

Vandermeulen, V., Gellynck, X., Huylenbroeck, G. V., Orshoven, J. V., & Bomans, K. (2009). Farmland for tomorrow in densely populated areas. *Land Use Policy*, *26*(4), 859–868. doi:10.1016/j.landusepol.2008.10.014

Vaughan, K., Kaczynski, A., Wilhelm Stanis, S., Besenyi, G., Bergstrom, R., & Heinrich, K. (2013). Exploring the distribution of park availability, features, and quality across Kansas City, Missouri by income and race/ethnicity: An environmental justice investigation. *Annals of Behavioral Medicine*, *45*(S1), 28–38. doi:10.100712160-012-9425-y PMID:23334757

Voicu, I., & Been, V. (2008). The effect of community gardens on neighboring property values. *Real Estate Economics*, *2*(2), 241–283. doi:10.1111/j.1540-6229.2008.00213.x

Walker, R., & Walker, R. A. (1978). Two sources of uneven development under advanced capitalism: Spatial differentiation and capital mobility. *The Review of Radical Political Economics*, *10*(3), 28–37. doi:10.1177/048661347801000304

Wiechman, T. (2007). Between Spectacular Projects and Pragmatic Deconstruction. The Future of Shrinking Cities: Problems, Patterns, & Strategies of Urban Transformation in a Global Context, 5–16.

Williams, J. (2010, December 1). Some landowners using community gardens to get a tax break. *The Vancouver Sun*. Retrieved from http://thetrac.ca/resources/docuploads/2010/12/02/ Some%20landowners%20using%20community%20gardens%20to%20get%20a%20tax%20break.pdf

Chapter 4
Multi-Stakeholder and Multilevel Food Governance:
The Case of the Community of Portuguese-Speaking Countries

Sérgio Pedro
https://orcid.org/0000-0001-6790-538X
Centre for Social Studies, University of Coimbra, Portugal

ABSTRACT

The contemporary food system, in its global and local dimensions, is a central element of the debate on the sustainability of the planet, a debate that increasingly involves more stakeholders and areas of knowledge in the search for answers to the multiple questions related to the attainment of more sustainable patterns for food and agriculture. The present chapter analyses the participative multi-stakeholder and multilevel model of food governance of the Community of Portuguese Speaking Countries (CPLP), in which stakeholders from different societal and expertise sectors participate in equal manners in the process of co-construction of institutional, technical, and financing measures for the functioning of a given food system. The present chapter has the main goal of sharing and critically analysing the CPLP´s institutional context for the promotion of sustainable food systems as an example of an integrated methodological approach to support the creation of coordinated public policies and institutional conditions to implement a transition to more sustainable food systems and diets.

INTRODUCTION

The contemporary food system, in its global and local dimensions, is a central element of the debate on the sustainability of the planet and the future of life that inhabits it. It encompasses the entire range of actors and their interlinked value adding activities involved in the production, aggregation, processing, distribution, consumption, and disposal of food products that originate from agriculture, forestry or fisheries. Food systems are constructed upon the interconnection of economic, social and environmental dimensions, being a common thread linking all 17 Sustainable Development Goals (SDGs).

DOI: 10.4018/978-1-7998-2599-9.ch004

The multidisciplinary approach to the complexity of the current global food system challenges increasingly involves more stakeholders and areas of knowledge in the search for answers to the multiple questions related to food, coupled with the clear trend towards a global integration of food systems and economic consolidation and mergers between agribusiness corporations in the global food system (Feenstra, 2002; IPES Food, 2017). Other factors reveal the unsustainable status quo of the contemporary global food system. We highlight the contribution of the agricultural sector to the global greenhouse gas emissions (9%) (IPCC, 2015) and the US$3.5 trillion and US$500 billion per year that represent respectively, the global costs of malnutrition and obesity (WHO, 2018). Likewise, environmental degradation of soils, water quality, biodiversity loss (IPCC, 2015), the accentuation of chronic diseases related to poor diet and excessive ingestion of food chemicals (McKinsey Global Institute, 2014), and other negative impacts on agriculture and food systems reveal the challenges of food systems to achieving food and nutrition security, the implementation of the human right to adequate food and nutrition and, consequently, the Agenda 2030 objectives (Pedro, 2019).

These goals, as evidenced by recent macro research (Willett et al., 2019; IPCC, 2015), will not be achieved without the consideration of the natural resources of the planet and social impacts of the global food system as key parts of the sustainability equation. Thus, from the foregoing, the global food system needs an urgent transition towards promoting sustainable practices and creating regenerative, resilient and fair food systems, in order to protect the environment, promote biodiversity, traditional knowledge and cultures, and healthy diets (Pedro, 2018). This transition is as urgent as the social and economic impacts of unhealthy diets which are increasing and the impacts of climate change which are more evident, including the loss of traditional food systems and their associated cultural heritage (Willett et al., 2019).

The need for this transition is recognized in several global fora and covers a diverse thematic range of initiatives, which are increasingly linked to the dynamism and lessons learnt by governments, civil society, academia and the private sector in a relevant set of territories.

Although there are existing political commitments towards food governance, defined in this chapter as: the process that gathers different stakeholders in the design of public policies and institutional, technical, and financing measures for the functioning of the food system, there is still a lack of knowledge and concrete institutional and legal frameworks to guide the implementation of sustainable food policies in the national and local territories (Lever, Sonnino & Cheetham, 2019). With the aim of raising awareness of such frameworks, this chapter seeks to contribute to the conceptualization of multi-stakeholder and multilevel models of food governance as viable options. Thus, it is necessary to make a broad discussion about the goals, pros and cons, of the multi-stakeholder and multilevel models and the impacts of the resulting public policies for the transition to more sustainable food systems- which are food systems that deliver food security and nutrition for all, in such a way that the economic, social, and environmental bases to generate food security and nutrition for future generations are not compromised.

The objective of this paper is to critically examine the multi-stakeholder and multilevel institutional framework for the promotion of sustainable food systems adopted by the Community of Portuguese Speaking Countries (CPLP), therefore contributing to the integrated methodological approach to support the creation of coordinated public policies and institutional conditions to implement a transition to more sustainable food systems and diets, on local territories.

METHODOLOGY

Considering the multi-disciplinarity of the theme of this chapter an in-depth review of literature, encompassing a context analysis of the current global food system, the evolution of the food governance models, mechanisms of multi-stakeholder and multilevel models of food governance was carried out. Through the legal analysis of the CPLP´s legal framework the present chapter provides valuable in-depth analysis of the case study and contributes to the adapted transposition of such models to other countries and regions.

Moreover, before the interconnection of the presented institutional framework with other initiatives (United Nations and CPLP), the chapter analyses the relation of sustainable food systems public policies with several topics and global initiatives, such as:

- The **Sustainable Development Goals** (SDGs), particularly SDG 1 (eradicating poverty), 2 (eradicating hunger), 5 (gender equality), 6 (clean water and sanitation), 8 (decent work and economic growth), 10 (reduced inequalities), 11 (sustainable cities and communities), 12 (sustainable production and consumption), 13 (climate action), 15 (protect life on land), 16 (peace, justice and strong institutions) and 17 (partnerships for the goals);
- The **United Nations Decade of Family Farming**, adopted by the United Nations in 2018, particularly, the expected outcome of increasing sustainable food production and the resilient agricultural practices that contribute to the preservation of ecosystems and strengthen our ability to adapt to climate change, extreme weather phenomena, drought, flooding and other disasters, and gradually improve the quality of the soil.

EVOLUTION OF FOOD GOVERNANCE MODELS

Following the trend towards global integration and economic consolidation, the agri-food system is increasingly required to feed in larger quantities and more regularly throughout the year (Feenstra, 2002). Such requirements promote the concentration of production by a small number of producers capable of offering high production quantities and who specialize in a given product (Moital, Almeida, & Pinto-Correia, 2012). Such dynamics lead to the strengthening of a network of corporations whose power is unbalanced when compared with other actors in the private sector. A network that combines both corporations and multinational companies as well as small farmers, and which brings together the ability to influence the formulation of food policies of countries (IPES Food, 2017).

Moreover, the functioning of food systems equally depends upon associated regulations and governing institutions. The management of food systems can take place in a more or less participatory manner, involving a wide range of stakeholders in the food system concerned, such as representatives of producers, consumers, academia and the private sector and public officials seconded for this purpose or just involving only the latter, within a centralized cabinet logic (Pedro, 2019).

The evolution of food systems is a major factor in the equal evolution of food governance models, thus, the understanding and operationalization of the concept of food governance has been subject to changes over time.

Considering the necessary effort of planification during World War II, and after, for the reconstruction of the affected countries, up until the the fifties of the XXth century, food systems were subject to a planification focused in the specialization of crop production areas, either inland or in colonized territories (Hueston & McLeod, 2012), and the activity of policymaking concerning food and agriculture was mainly an issue restricted to public officials and other policy experts, focused on food production levels to ensure food security (Renting & Wiskerke, 2010).

The advances in the technologies associated with the processes of food production, processing and transportation in the second half of the XXth century represented an enlargement of the global food supply chain. The advent of the Green Revolution, beginning most markedly in the late 1960s, also revealed a structural reconfiguration of food systems implying new dimensions to be considered in food and agricultural policies. Elements such as environmental, ethical, political and social implications brought by the intensification of food production and dependent upon artificial inputs and intensive monoculture during the Green Revolution, are now part of the pillars of any food system and characterizing elements in its analysis. Presently, the added complexity of food systems that is evident in the XXIst century, associated with the geographical expansion and interdependence of food systems at the global level, implies a multi-dimensionality that requires a transversal set of analytical tools and the mobilization of a broad set of knowledge fields as well asy, new stakeholders and interests that take part in food and agriculture policy making. This transformation marks a change of political position of the State as a policy maker. The phenomenon of globalization, evidenced the decentralization of the position of the State in the governance and regulation of food systems (Arts, Leroy, & Tatenhove, 2006; Lang, Barling & Caraher, 2009) and the necessity of engaging in more participative models of food governance (Pedro, 2019). On the other side, the globalization of food system implies an added responsibility of the State to regulate food safety and quality (Havinga, Casey, & Waarden, 2015) in a context of global integration of food supply chains as well as creating and implementing a framework for the sustainability of the production, processing, transport and consumption of food. Notwithstanding, the regulatory activity is not anymore exclusive to the figure of the sovereign State, considering that the increasing influence of the private sector and growing intervention by civil society organizations resulted in the sharing of the policy making and intervention activity between public and private sector. In fact, from the 1990s onwards, private stakeholders took a leading role in global food safety regulation and the development of retail driven food safety self-regulation standards (Havinga, Casey, & Waarden, 2015; FAO, 2010). One example of the result of the self-regulatory activity by the private sector is the GLOBAL G.A.P (Global Good Agricultural Practices), a common standard for farm management practice created in the late 1990s by several European supermarket chains and their major suppliers. Such standards have become increasingly important as tools of chain coordination, as meta-management systems (Caswell, Brendahl & Hooker, 1998) and are not merely public goods to resolve market failures. They are also strategic instruments of market differentiation and market share and niche protection by food companies (Reardon et al., 1999).

A process that, considered jointly with the policymaking activity of the State, resulted in a policy landscape composed by several layers and levels, merging the local, national and international. This transformation of food governance since the 1990s, promoted by the added complexity associated to globalization of food chains, and the creation of self-regulatory standards by the private sector, introduced a new hybrid configuration of governance such as co-management, between civil society and state, state and private sector, civil society and the private sector (Lemos & Agrawal, 2006). Such hybridization of food governance has its foundational support in the interplay between different public and private stakeholders, including their relative interests and capacities, and their activities at different levels of the food system (Verbruggen & Havinga, 2017). Thus enabling an expansion of the issues addressed by food policies to not only the activities of production and consumption of food, but equally, to the social norms and cultures in which those activities are embedded, as well as the environment and natural resources which they depend upon to function. Such holistic viewpoint of the food system is attained when there is the recognition that the functioning of food systems depends upon the intervention of a broad set of stakeholders and expertise, such as food producers, distributors, sellers and consumers, health officers, teachers, environmental agencies and others. It is by analysing the process of historical evolution of food governance that we come to the analysis of the concepts of multi-stakeholder and multilevel governance and their proposal of adoption in the context of the management of food systems.

MULTI-STAKEHOLDER GOVERNANCE

In this section, we focus our analysis in two governance models that can occur or not at the same time. The first mention goes to the multi-stakeholder governance model, a participative governance model, in which stakeholders from different societal and expertise sectors such as civil society, academia, private sector and municipalities participate in equal manners in the process of co-construction of institutional, technical and financing measures for the functioning of a given food system. Such process is based on principles of transparency and broad participation to address the complexity of food systems. Concomitantly aiming to develop partnerships and strengthen networks and favourable partnerships between stakeholders. The participation of the wide range of stakeholders mentioned before has the added value of enabling the collection of different perspectives and possible solution proposals about the same food system during the decision-making process. In this regard, the situation of access to food markets could be better understood by civil society organizations who support disadvantaged social groups without access to food, through food aid programs, as well as the private sector as a relevant stakeholder in access and availability of food for consumption. It is also important to highlight the participation of municipalities and academia in the multi-stakeholder food governance process as, through their activity, they are, respectively, aware of the impact of the effective implementation of public policies at the local level and situating and conceptualizing the reality through oriented research to substantiate future public policy solutions.

The adoption of a consideration of the governance of a food system as being a shared responsibility between a wide range of knowledge and stakeholders creates the potential for a deeper assessment of the challenges and opportunities of that given food system. Participant stakeholders are equally responsible for the reality and are challenged to contribute with their perspectives and information concerning the status quo of the food system and the direct and indirect impacts of the implementation of policy measures in place. Thus, creating a "feedback loop mechanism" not only for allowing a better assessment of the impact of these policies, but also for the identification of possible review measures for better effectiveness.

In recent years the research conducted on food governance contributed to the identification of several impacts of this approach. Some of these researches include:

- Enhancing capacity for stakeholders to work within the complexity of food systems (Breeman, Dijkman, & Termeer, 2015);
    - Improving the evaluation of trade-offs in policy options, as drivers and outcomes will be reviewed and holistically assessed (HLPE, 2018);
    - Identifying synergies and leverage points for implementing context-specific solutions (Solon et al., 2019);

- Enhancing coordination of policy actions, institutional frameworks, and stakeholders, which strengthens overall food systems governance (HLPE, 2018);
- Supporting more efficient use of natural resources and lower environmental impacts, while simultaneously improving societal outcomes, such as human health and rural livelihoods (UN Environment Programme, 2016);
- Revealing underlying and root causes of unsustainable production and consumption patterns;
- Continuing systemic thinking and collaboration among food systems stakeholders; and
- Increasing capacity for the delivery of integrated sustainable food systems policies, and also for achieving a number of Sustainable Development Goals (HLPE, 2018).

The shape of multi-stakeholder processes depend upon the issues considered and the case by case reality, as well as the objectives, participants, timelines and degree of linkage into official decision-making. They take the form of dialogues, previous public consultations to legislation projects or monitoring processes to current legislation.

MULTILEVEL FOOD GOVERNANCE

Considering the integration of current food systems into a global food production, processing, transport and consumption system and also their environmental and social impact across state borders, the development of a multi-stakeholder governance model for food systems is one step in the path to encompass such complexity. The recognition that the globalization of food systems implies the need for transnational concerted efforts and coordination for the adoption of public policy measures that consider the macro and holistic view of food systems at the global level and their transnational impacts is another. In this context, and in order to confer stability to the decision-making process, a reflection on the institutionalization of multilevel decision-making processes and the development of public policies at the transnational level that engage food systems at the local level is pertinent. Multilevel food governance implies the concertation, based in the principle of subsidiarity of the supranational, national and local levels of decision making and their respective policy makers, in the effort to tackle the a-territorial challenges of contemporary food systems. Recognizing the limitations of the state-centric forms of government of policy issues with transnational implications such as food systems, the multilevel governance model was first analysed by Marks (1992) with the example of the European Community, and later expanded by Enderlein, Walti and Zurn (2010). Such model is based in the principle that all States are structured along shared multiple layers of government and public policy. Regardless of the issue, multilevel governance is the result of the interactions between institutions and organisations operating at different levels (Caponio & Correa, 2017). Its application to the context of food system governance is manifested through decision-making processes whose effects are manifested through the adoption of binding legal frameworks or guidelines at more than one level of decision (global, regional, national or local), ensuring the existence of a flux of information and representation from the local to the global, and vice versa.

The application of this governance model to the food and agriculture decision making at the global level is best represented by the United Nation´s Committee on World Food Security (CFS), self-defined as the "central United Nations political platform" for food security coordination at the global level between states (CFS, 2009). Its inception dates back to its founding in 1974, following the first World Food Conference as an intergovernmental committee within the United Nations Food and Agriculture

Organization (FAO). Being later reformed in 2009, following the Global Food Prices Crisis of 2007-2008, and the concomitant analysis that the fragmentation of global governance in food and agriculture, in its different and sometimes conflicting regimes, was a major obstacle, that needed to be overcome urgently (Schutter, 2014). The aim of the reform was to to transform the CFS into "the foremost inclusive international and intergovernmental platform for a broad range of committed stakeholders to work together in a coordinated manner and in support of country-led processes towards the elimination of hunger and ensuring food security and nutrition for all human beings" (CFS, 2009). Although the CFS decisions are non-binding, its function is to facilitate the policy coordination between various levels of governance (regional, national and local), creating a transnational coordination platform with a shared set of goals to be addressed by policy measures.

THE FOOD GOVERNANCE MODEL OF THE COMMUNITY OF PORTUGUESE SPEAKING COUNTRIES

Other examples at the regional scale can be provided as representations of multilevel and multi-stakeholder food governance models. This is the case of the Food and Nutrition Security Council of the Community of Portuguese Speaking Countries (CONSAN-CPLP) created in 2012 as a food governance supranational mechanism, having as legal attributions the policy making of food and agricultural issues of all the member countries of the international political community.

The CPLP is an intergovernmental organization created in 1996, currently with ten countries (Angola, Brazil, Cape Verde, Guinea-Bissau, Mozambique, Portugal, São Tomé and Príncipe, East-Timor and Equatorial Guinea) and ten associated observers (Czech Republic, Georgia, Hungary, Japan, Mauritius, Namibia, Senegal, Slovakia, Turkey and Uruguay) with a joint population of 250 million, (CPLP, 2011), across four continents having Portuguese as the official language, and operating as a multilateral fora for cooperation between Member States in several policy topics.

The Community is characterized by diversity, encompassing a broad range of situations of political, economic and social development, characterized by the scale of high (0.847) to low human national development index values (0.437) (UNDP, 2018), low (35.5) to high (54) income distribution inequality (World Bank, 2015) and with 11% of its population (258 million) undernourished (CPLP, 2015). The member countries share a common history of colonialism, internal political conflicts and unequal competition in the context of the liberalization of international trade. The agricultural sector of the majority of the member countries does not guarantee an adequate level of national food sufficiency. This historical context, results in a socio-economic situation of scarcity of adequate resources and public services at different levels in the CPLP countries, coupled with the incidence of adverse climate change impacts, particularly in the member countries of the African continent, as well as demographic and accelerated urbanization trends and the consequent concerns about the sustainability of natural resources and energetic sufficiency (Pedro, 2019).

The necessity for the creation of the CONSAN-CPLP, a mechanism that unites both the multi-stakeholder and multilevel models of food governance, emerged as a result ofthe official recognition that the economic and historical interrelation between member countries and within their territories and municipalities demand coordinated action. The creation of a common intergovernmental food and nutritional security agenda in coordination with the shared CFS agenda mentioned above (multilevel element) is as shown below (see Figure 1).

Figure 1. Diagram of the multilevel integration of the CONSAN-CPLP within the CFS, national and local decision-making levels.
Adapted from: (CPLP, 2011)

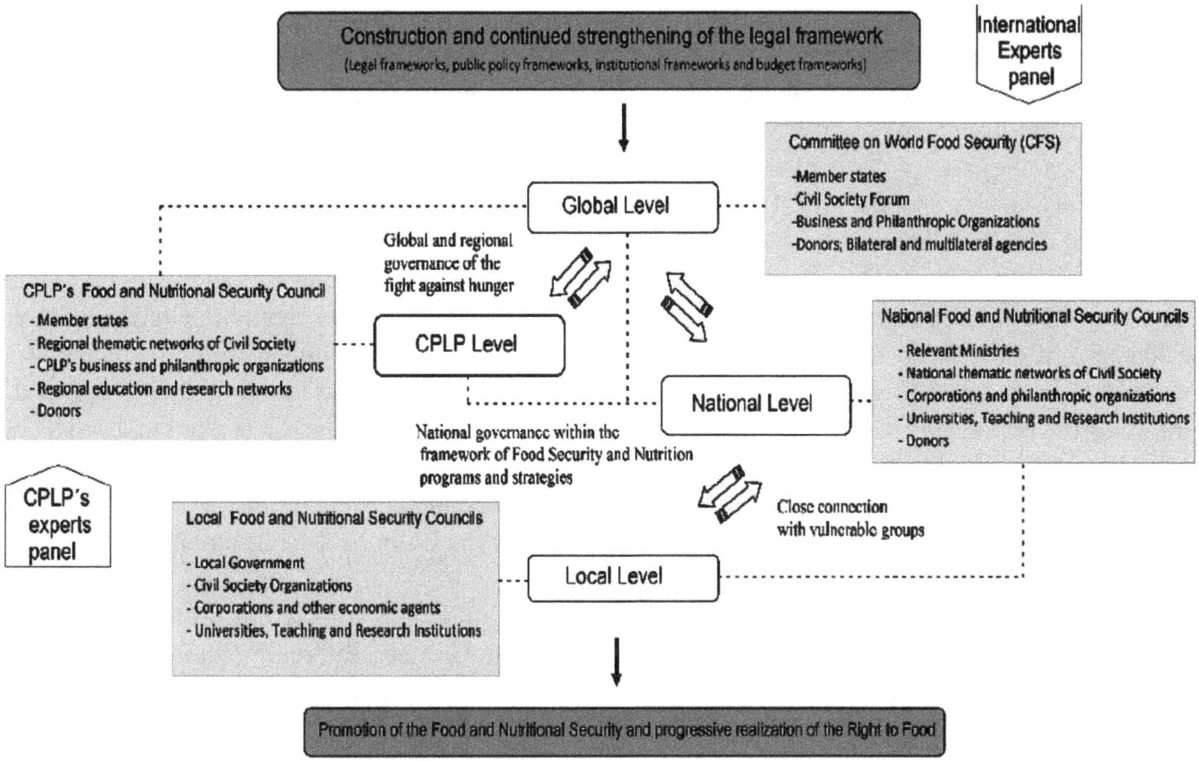

Considering the international development programs between member countries that encompassed political cooperation for the development of the food sector of the supported countries, the adoption of Food and Nutrition Security sectoral policies with individual priorities and, in many cases adopting a "top-down" reality was counterproductive. The lack of policy coordination verified in the member countries before the creation of the CONSAN-CPLP generated an inadequacy of the policies in place at the time to address the structural causes of hunger and poverty in the specific territorial contexts in which they were supposed to intervene (CPLP, 2011). With such a policy environment, the reform of the CFS in 2009 and its recognition as a reference of an existing international fora for coordination of food and agriculture policies was also an additional element that provided the necessary reasons for an institutional reorganization of the policy making process of the CPLP´s member countries. This led to the creation in 2012, of a regional food governance mechanism, with the political goal of eradicating hunger and poverty and progressively realizing the Human Right to Adequate Food and Nutrition at the national level.

Its statutes describe the CONSAN-CPLP as a multi-stakeholder platform with the task to coordinate policies and programs developed in the area of food and nutritional security and to advise the CPLP´s Conference of Heads of State and Governments (Article 2). In addition to its plenary meeting which involves the participation of government delegates (1 representative per country), the CONSAN-CPLP is equally composed of participants representing academia, private sector, parliamentarians, municipalities

(2 representatives each) and civil society (8 representatives) from all of the member countries. and the Presidency which is composed of the Permanent Technical Secretariat, Working Groups and a Panel of Technical Experts (Article 5).

Within these we highlight the role of the Working Groups through an in-depth reflection of specific areas considered relevant (Article 12) as it was the case with the development of the regional guidelines to support and promote family farming (2017) and regional guidelines to promote sustainable food systems and diets (ongoing).

The strategic long-term goals of the CONSAN-CPLP are framed by a Regional Strategy for Food and Nutrition Security enacted in 2011, which establishes the policy priorities and commitments to be developed in detail and operationalized by the decisions taken at the CONSAN-CPLP. Both the multi-stakeholder and multilevel elements of the CPLP´s case study are institutionalized in the foundational documents. The statutes of the CONSAN-CPLP and their provision for the creation of representation mechanisms of each sector (academia, civil society,...) present the commitment to multi-stakeholder governance. In the same manner, the Regional Strategy sets the background for its multilevel element, reinforcing the replicability of the CONSAN-CPLP´s model of food governance at the national and local levels, constituting a political commitment to be adopted by all member states. This commitment assumed by the CPLP´s member countries implies the creation of food councils equal to the CONSAN-CPLP at the national and local levels and that the decisions adopted by the CONSAN-CPLP are implemented at the national and local levels.

For this purpose, member states have adopted common guidelines for the construction and reinforcement of legal and institutional frameworks in their territories, and the reinforcement of the need for strong budgetary frameworks for food and agriculture. The regional strategy is based on three axes: 1) strengthening the governance and coordination for food security and nutrition sectors; 2) promoting access to food and improving the livelihoods of the most vulnerable groups and 3) increasing internal food availability based on small producers.

Following this supranational framework, CPLP´s member states have created, or are currently creating national food councils and strategies in line with the regional priorities established in the CONSAN-CPLP. It is in this context of multilevel governance that the CONSAN-CPLP fits into a framework that considers the CFS as an international food governance platform. At the national and local levels, the food and nutrition security councils of CPLP Member States establish a network of information flows and interaction between stakeholders, from local to international. In this way, the local stakeholders, through their participation in the local mechanisms of food governance, not only shape the local food systems but also contribute to the international debate in the global decision fora (CFS), bringing to it the specificities of the local and the impact at the local level of the application of the international legal framework on food and agriculture.

An example of the result of the functioning of the CONSAN-CPLP as a coordination mechanism is the creation of an Action Plan (the most recent refers to the timeframe of 2018-2020), which includes, among other political commitments, a program for the promotion of sustainable food systems and diets. This program provides for knowledge exchanges, the systematization of traditional food systems and their diets, and the development of public policy proposals, including the promotion of public food procurement based on sustainable diets and short food supply chains. Of particular interest to the present analysis are the CPLP´s guidelines for the support and promotion of family farming that are part of the Action Plan commitments for 2018-2020. The document produced under CONSAN´s Working Group on family farming and approved in 2017 details a set of common guidelines and priorities that aim to contribute to the support of family farmers through specific policy measures such as public procurement of local food produced by family farmers and the creation in every member country, a statue of a family farmer. With the legal scope, recognition, identification and promotion of family farming, the guidelines are the result of a multi-stakeholder process of decision for the consolidation of the access by family farmers to means of production, markets and guaranteed income, social protection, the valorisation of the preservation of biodiversity by family farming and recognition of the economic autonomy of rural women and younger farmers. The guidelines integrate the international level by articulating to international legal references and international commitments, closing the bridge between the generalized and sometimes vague commitments of international commitments and bringing them to a more local and concrete level. That is the case of the reference in the guidelines of the international voluntary guidelines to support the progressive realization of the right to adequate food and the international voluntary guidelines on the responsible governance of tenure of land, fisheries and forests, both developed by FAO. Going further, the CPLP´s guidelines for family farming, inspired by the text of the above-mentioned FAO´s guidelines bring their text to the reality of the community with more detailed commitments and food and agriculture policy measures. That is the case of the role that CPLP´s guidelines had in serving as the conceptual basis and legal framework for the creation of family farmer statutes and family farming councils in member countries. In addition to its ability to influence and shape the local context from the regional, the multilevel model also has the possibility to shape the international, in a bottom up flow logic. That was the case of the influence of the CPLP guidelines of family farming on the construction of the action plan of the United Nations Decade of Family Farming (2019-2028)(UN, 2018) as well as the intervention of the CPLP´s member countries in the political support for the approval in 2018 of the United Nations Declaration on the rights of peasants and other people working in rural areas.

CONCLUSION

The current global food system is a central element in the planet's environmental sustainability debate, and at the political level, a theme directly or indirectly related to 12 of the 17 Sustainable Development Goals. However, in the past and still in the present, food and agriculture policies were and are managed in a fragmented manner, handled by a multiplicity of institutions and public authorities at the regional, national and local levels most times, without any interaction among them and with different environmental, social and economic perspectives and results. This led to a profusion of sectoral policies, that only have the possibility to access a particular perspective of the food system, not considering the domino effect and articulation with other vectors at different geographical levels.

Recognizing the transnational character of the economic, environmental and political containments and impacts of food systems, the promotion of a territorialized view of the management of food systems based on local or national sovereignty represents a challenge for the attainment of the necessary coordination of food and agriculture policies for the realization of the SDGs and the promotion of more sustainable food systems.

With regards to the challenges presented above, a transition towards a path of multi-stakeholder and multilevel food governance is at the centre of debates in several geographical spheres, with some initiatives already implemented such as the CONSAN-CPLP, that can possible provide useful guidelines about multi-stakeholder and multilevel governance models as well as historical experience for the emergence of similar initiatives in the context of intergovernmental organizations such as the African Union or the Arab League.

Although there are political commitments, there is a lack of concrete institutional and legal guidelines for the implementation of sustainable food policies in the national and local territories that consider the transnational nature and interdependence of the global food system. The challenge is to guarantee a process coherently articulated with the main goals of the SDGs and with an efficient use of scarce resources that connects to transnational agendas while concretizing them considering the peculiarities of the territory of application. Thus, it is necessary to discuss broadly about the goals, benefits and impacts of the methodology and the impact of the public policies for the transition to more sustainable food systems, as well articulate how the management of food systems could be more inclusive and coordinated. We believe that, based on the literature analysed the present chapter sought to provide an example of food governance that integrates a multi-stakeholder and multilevel perspective of participatory governance of a food system, recognizing its natural interconnection within the local, national and international levels.

The development of food governance processes cannot however be taken for granted and are not a panacea, co-governance among food system actors is a work in progress and a shared learning opportunity for the success of food policies and the achievement of the 2030 Agenda. The model of food governance presented in this text represents solely an instrument for the institutionalization of a dialogue between actors that usually are not involved in the process of decision and monitoring of concerning the food system in which they live. The model presented provides a reflection of the process of mutation of food systems and addresses the negative impacts of the current global food system institutionally, giving voice to all its participants in governments, the private sector, academia, civil society, local and central public representatives in direct trans territorial communication.

REFERENCES

Arts, B., Leroy, P., & Tatenhove, J. (2006). Political Modernisation and Policy Arrangements: A Framework for Understanding Environmental Policy Change. *Public Organization Review. A Global Journal, 6*(2), 93–106.

Breeman, G., Dijkman, J., & Termeer, C. (2015). Enhancing food security through a multi-stakeholder process: The global agenda for sustainable livestock. *Food Security, 7*(2), 425–435. doi:10.100712571-015-0430-4

Caponio, T., & Correa, M. (2017). Theorising migration policy in multilevel states: The multilevel governance perspective. *Journal of Ethnic and Migration Studies, 44*(12), 1995–2010. doi:10.1080/1369183X.2017.1341705

Caswel, J., Bredahl, M., & Hooker, N. (1998). How Quality Management Metasystems Are Affecting the Food Industry. *Review of Agricultural Economics, 20*(2), 547-557.

CFS. (2009). *Reform of the committee on world food security. Final version.* Committee on world food security. Thirty-fifth Session. Agenda Item III. Retrieved from: http://www.fao.org/3/a-k7197e.pdf

CPLP. (2011). *Estratégia de Segurança Alimentar e Nutricional ESAN-CPLP. Parte 1: Enquadramento* [Food and Nutritional Security Strategy ESAN-CPLP. Part 1: Background]. Retrieved from: https://www.cplp.org/Admin/Public/Download.aspx?file=Files%2FFiler%2Fcplp%2FEstrategia%2FI-_Enquadramento_final.pdf

CPLP. (2015). *CPLP em números* [CPLP in numbers]. Retrieved from: https://www.CPLP.org/Default.aspx?ID=4447&Action=1&NewsId=3984&M=NewsV2&PID=10872

Enderlein, H., Walti, S., & Zurn, M. (Eds.). (2010). *Handbook on Multi-level Governance*. Cheltenham: Edward Elgar Publishing. doi:10.4337/9781849809047

FAO. (2010). *Private Food Safety Standards: Their Role in Food Safety Regulation and their Impact*. Retrieved from: http://www.fao.org/3/ap236e/ap236e.pdf

Feenstra, G. (2002). Creating space for sustainable food systems: Lessons from the field. *Agriculture and Human Values*, *19*(99).

Havinga, T., Waarden, F., & Casey, D. (Eds.). (2015). *The Changing Landscape of Food Governance—Public and Private Encounters*. Cheltenham, UK: Edward Elgar Publishing Limited.

HLPE. (2018). *Multi-stakeholder partnerships to finance and improve food security and nutrition in the framework of the 2030 Agenda*. A report by the High-Level Panel of Experts on Food Security and Nutrition of the Committee on World Food Security. Retrieved from: http://www.fao.org/3/CA0156EN/CA0156en.pdf

Hueston, W., & McLeod, A. (2012). Overview of the global food system: changes over time/space and lessons for future food safety. In *Institute of Medicine (US). Improving Food Safety Through a One Health Approach: Workshop Summary*. Washington, DC: National Academies Press (US). Retrieved from https://www.ncbi.nlm.nih.gov/books/NBK114491/

IPCC. (2015). *Agriculture, Forestry and Land Use. Fifth Assessment Report*. Retrieved from: https://www.ipcc.ch/site/assets/uploads/2018/02/ipcc_wg3_ar5_chapter11.pdf

IPES-Food. (2017). *Too big to feed: Exploring the impacts of mega-mergers, concentration, concentration of power in the agri-food sector*. Retrieved from: http://www.ipes-food.org/_img/upload/files/Concentration_FullReport.pdf

Lang, T., Barling, D., & Caraher, M. (2009). *Food Policy: Integrating Health, Environment and Society*. Oxford: Oxford University Press. doi:10.1093/acprof:oso/9780198567882.001.0001

Lever, J., Sonnino, R., & Cheetham, F. (2019). Reconfiguring local food governance in an age of austerity: Towards a place-based approach? *Journal of Rural Studies*, *69*, 97–105. doi:10.1016/j.jrurstud.2019.04.009

Marks, G. (1992). Structural Policy in the European Community. In A. Sbragia (Ed.), *Europolitics: Institutions and Policymaking in the "New" European Community*. Washington, DC: Brookings Institute.

McKinsey. (2014). *Overcoming obesity: An initial economic analysis*. Retrieved from: https://www.mckinsey.com/industries/healthcare-systems-and-services/our-insights/how-the-world-could-better-fight-obesity

Moital, F., Almeida, M., & Pinto-Correia, T. (2012). "Proposta de diversificação de actividades nas explorações agrícolas. In PROVE (Ed.), *Promover, Vender, Participar e Diversificar. Relatório final,* [Promote, Sell, Participate and Diversify. Final report]. ICAAM / University of Évora and Monte-ACE. Retrieved from: http://www.prove.com.pt/www/wst/files/I167-05XDRAFTXXRELATOXXRIOX27JAN.PDF

Pedro, S. (2018). Family Farming, the Environment and the Global Food Chain. In P. Cooney & W. Sacher (Eds.), *Environmental Impacts of Transnational Corporations in the Global South. Research in Political Economy* (Vol. 33). Emerald Publishing Limited. doi:10.1108/S0161-723020180000033008

Pedro, S. (2019). Food Governance: Multistakeholder Decision- Making Mechanisms in Food Systems. In W. Leal (Ed.), Encyclopaedia of the UN Sustainable Development Goals, (Vol 2: Zero Hunger). Springer Publishing

Reardon, T., Codron, J., Busch, L., Bingen, R., & Harris, C. (1999). Global Change. IN Agri-food Grades and Standards: Agribusiness Strategic Responses. In Developing Countries. International Food and Agribusiness Management Review, 2, 1-15.

Renting, H., & Wiskerke, J. (2010). *New Emerging Roles for Public Institutions and Civil Society in the Promotion of Sustainable Local Agro-Food Systems*. Paper presented at 9th European IFSA Symposium, Vienna, Austria.

Schutter, O. (2014). The Reform of the Committee on World Food Security: The Quest for Coherence in Global Governance. In *Rethinking Food Systems. Structural Challenges, New Strategies and the Law* (pp. 219–238). Dordrecht: Springer. doi:10.1007/978-94-007-7778-1_10

Solon, M., Mink, P., Dernini, S., Bortoletti, M., & Lomax, J. (2019). The One Planet Sustainable Food Systems Programme as a Multi-stakeholder Platform for a Systemic Approach. In B. Burlingame & S. Dernini (Eds.), *Sustainable diets: Linking Nutrition and Food Systems*. Wallingford: C.A.B. International. doi:10.1079/9781786392848.0178

UN. (2018). *United Nations Decade of Family Farming (2019–2028). Resolution adopted by the General Assembly on 20 December 2017*. Retrieved from: https://undocs.org/A/RES/72/239

UN Environment Programme (2016). *Collaborative framework for food systems transformation: A multi-stakeholder pathway for sustainable food systems*. Author.

Verbruggen, P., & Havinga, T. (Eds.). (2017). *Hybridization of food governance: Trends, types and results*. Cheltenham: Edward Elgar. doi:10.4337/9781785361708

WHO. (2018). *2018 Global Nutrition Report*. Retrieved from: https://www.who.int/nutrition/globalnutritionreport/en/

Willett, W., Rockström, J., Loken, B., Springmann, M., Lang, T., Vermeulen, S., . . . Majele Sibanda, L. (2019). *Food in the Anthropocene: the EAT–Lancet Commission on healthy diets from sustainable food systems.* Retrieved from: https://www.thelancet.com/journals/lancet/article/PIIS0140-6736(18)31788-4/fulltext

Chapter 5
The Price Insurance Demand of Rice Producers in the Vietnamese Mekong Delta

Huynh Viet Khai
https://orcid.org/0000-0001-8969-5387
Can Tho University, Vietnam

ABSTRACT

Using the dichotomous choice contingent valuation method, this chapter helps shed light on the potential for marker-based insurance schemes in Vietnam by empirically exploring the demand for minimum price insurance among rice households. The study showed that the majority of rice farmers accepted the guaranteed price of VND 4,500 per kg, and their accepted insurance fee was about 13% of the guaranteed price and 30% of the break-even price. Farmers growing rice under a monoculture system were less likely to pay for the proposed insurance service, while those with access to any formal credits were more likely to pay for it.

INTRODUCTION

Since the 1980s, a growing process of globalization has witnessed the proliferation of markets for many financial instruments to manage risks, such as futures, options, swaps, etc. This development has also generated new ways to help farmers hedge against unforeseen price declines, based on the use of such market instruments, either directly by farmers, or via marketing and financial intermediaries. Recently, these market-based insurance schemes are also being piloted in developing countries. For instance, a recent initiative of the International Task Force (ITF) on Commodity Risk management, has proposed using market-based derivative instruments to provide price insurance for internationally traded commodities (ITF, 1999), while other proposals have suggested using market-based weather insurance to cover yield or crop income risks (Skees et al., 2006). Varangis et al. (2002) have suggested using combinations of the above instruments to manage agricultural market risks in developing countries.

DOI: 10.4018/978-1-7998-2599-9.ch005

The institution of any policy or instrument designed to ensure producers against the risk of an unexpected price decline must be evaluated based on benefit-cost analysis. While the costs can be estimated relatively easily, the benefits are often more difficult to ascertain, as the underlying demand for such policies by the affected groups must be assessed. Yet, this information is important to help governments decide on the usefulness and modalities of such market-based price insurance schemes in assisting smallholders to deal with commodity price risks.

However, the above proposals are related to insurance services or supply sides. They did not consider the demand of consumers who benefit from these services. In other words, what yield or price insurance for products would they desire to obtain and how much would they be willing to pay for it? These important questions should be answered if we want to encourage these kinds of services to operate well in the developing world. This chapter tried to determine and analyze the willingness to pay (WTP) of rice farmers in Vietnam for commodity price insurance by using the Dichotomous Choice Contingent Valuation Method (DC CVM). We begin by describing our data and empirical methodology. We then proceed by analyzing the perception of the service of price insurance among rice farmers in the sample, the estimated WTP and overall demand for rice price insurance. Some important conclusions are withdrawn in the final part.

METHODOLOGY

The indirect and direct methods have popularly been applied to calculate the WTP to keep households away from the uncertainties of the market price. One weakness of an indirect method is the use of exogenous time-series information on shocks of prices and yields from household surveys that are not aimed at exploring issues of vulnerability and insurance. Alternatively, the direct method engages household interviews designed particularly to estimate the vulnerability and demand for insurance by households. Thus, this study applies the direct approach (Alexander et al., 2007).

The direct or "contingent valuation" (CV) methods are based on direct questioning of agents (producers, households, etc.) on how much they are willing to pay for avoiding an undesirable event, or for having available a possibly welfare improving instrument such as a given amount of an insurance contract.

The major problems with this approach largely have to do with the specification of the "scenario" or the "benchmark" against which the agent is supposed to compare the current situation, and express a monetary value for what it is worth to him/her to move to the new situation, or avoid a bad one. It is not always easy to specify this scenario appropriately, especially if it involves a rather improbable event, and this lies at the heart of most criticisms of this approach. However, in the case of well-specified risks, such as price or yield variations, it is likely that farm households are familiar not only with their normal values, but also with their variability over time, and hence the above criticism may not be valid.

The basic theory of the CV approach has been known for some time, and a comprehensive survey can be found in Hanemann and Kanninen (1998). The idea favored by current CV practice is to ask each respondent a closed question, namely whether they would accept to pay a given amount to obtain a given change in their status quo. Hence the answers obtained are of the "Yes" or "No" type, necessitating a theory of how to translate these discrete responses into meaningful WTP estimates. Following Hanemann and Kanninen (1998), we suppose that a respondent is asked to consider the change from the status quo q^0 to q^1, where q^1 refers to the value of a yet non-existent good, such as an insurance contract, and presumably, the latter choice is preferable to the former. Denote the indirect utility of respondents as

$v(p,q,y,s,\varepsilon)$, where p is a vector of prices for all the market goods currently available, y is the respondent's income, s is a vector of respondent characteristics, and ε is the stochastic component of utility. Then if the respondent is asked whether he would be willing to pay an amount A to obtain q^1, his answer would be "Yes" if the following condition holds:

$$Pr\left\{response\,is\,"Yes"\right\} = Pr\left\{v\left(p,q^1,y-A,s,\varepsilon\right) \geq v\left(p,q^0,y,s,\varepsilon\right)\right\} \tag{1}$$

where Pr denotes the probability. If we denote by B the maximum WTP for the change from q^0 to q^1, then B is defined implicitly by the condition:

$$v\left(p,q^1,y-B,s,\varepsilon\right) = v\left(p,q^0,y,s,\varepsilon\right) \tag{2}$$

This implies that B is a function of all the same variables that enter the function v(.). Hence condition (1) can be written equivalently as:

$$Pr\left\{response\,is\,"Yes"\right\} = Pr\left\{B\left(p,q^0,y,s,\varepsilon\right) \geq A\right\} \tag{3}$$

As B is a random variable, let $G_B(.)$ be the cumulative distribution function (*cdf*) of B. Then (3) translates into the following:

$$Pr\left\{response\ is\,"Yes"\right\} = 1 - G_B\left(A\right) \tag{4}$$

When $G=\Phi$, namely the standard normal *cdf*, and when B has a mean equal to μ and variance equal to σ^2 then one has a logit model:

$$Pr\left\{response\,is\,"Yes"\right\} = \Phi\left(\frac{\mu - A}{\sigma}\right) \tag{5}$$

Hence if we estimate a logit model of the type:

$$Pr\left\{response\,is\,"Yes"\right\} = \Phi\left(\sum_{i=1}^{n}\alpha_i X_i - \beta A\right) \tag{6}$$

The no stochastic part of the WTP can be derived by the simple formula:

$$E\left(B\right) = \frac{\sum_{i=1}^{n}\alpha_i X_i}{\beta} \tag{7}$$

DATA COLLECTION

Primary data were collected during the field survey in Can Tho city, one of the biggest rice producers in the Mekong River Delta. We interviewed 364 rice farmers from February to March 2010 to get detailed household-level information related to WTP for price insurance, production costs, income, and social and economic characteristics.

We first explained and described how the price insurance contract would work, and then asked whether farmers were interested in the insurance. If they said "yes", we asked whether they were willing to pay a minimum price contract with the prices of VND 2,500 per kg, VND 3500 per kg, and VND 4,500 per kg in the next harvest season. If the next season's rice price is higher than the contracted minimum rice price, farmers would sell the rice at the market price. Conversely, if the next season's rice price is lower than the minimum price signed in the contract. The hypothetical insurance contracts are supposed to be structured at the time of the survey, and the contract would take effect at the time of the new crop harvested in the next marketing year.

The dichotomous choice contingent valuation method was applied in the study. Each farmer was interviewed whether he/she would be willing to pay a certain amount for each of these contracts; the answer was yes or no. Four different bid values (insurance fees) were chosen for each contract. The contract with a minimum price of VND 2,500 per kg had bid values of VND 35, 75, 125 and 300 per kg; The contract with a minimum price of VND 3,500 per kg had bid values of VND 75, 125, 300 and 625 per kg; The contract with a minimum price of VND 4,500 per had bid values of VND 125, 300, 625 and 925 per kg. Each farmer was randomly interviewed whether he/she would be willing to pay one of these bid values and answered whether he/she accepted only one bid value of each contract.

RESULTS AND DISCUSSIONS

Since price insurance is rather new and not familiar with Vietnamese farmers, the consumers' behavior or interest of this service could be investigated to operate or promote this service to farmers in the future. This part tries to answer the following questions; Do farmers have an interest in the price insurance service? What are the characteristics of the farmers who favor this service? Logit function with the dependent variable of accepting to buy price insurance is applied to investigate which farmers are interested in the price insurance service.

Table 1 shows the expression of interest by households in price insurance for rice production. About 46% of farmers in the sample indicated that they are interested in the insurance service. Most of them have an average age of 49 years, and the education level of 6 years. Almost all farmers produce rice in medium-scale (from 0.5 ha to less than 1 ha), constituting to about 43% of total farmers in the sample. Their average total income is VND 14 million.

Table 2 shows the results of Logit selection regression concerning interest in minimum price insurance by rice farmers. We utilize the following independent variables in the model. Firstly, the variables of household characteristics are education, age, gender. Secondly, the variables of farming scale, income structure (a proxy for the degree of risk aversion), and farming conditions are large size, medium size, per capita income, share rice in total income, access to credit, training attendance, the occurrence of serious diseases during the study time and rice monoculture. Finally, the variable of recent rice price is a proxy of current market conditions.

Table 1. Descriptive statistics of some important factors related to price insurance interest

Variables	Description	Mean	S.D.
Accept	(1= accepting to buy price insurance, 0 = otherwise)	0.456	0.499
Large size	(1 = farm size equal or more than 1ha, 0 = otherwise)	0.247	0.432
Medium size	(1= farm size from 0.5ha to less than 1ha, 0 = otherwise)	0.431	0.496
Rice price	(The market price of rice production in VND/kg)	4,100.820	301.097
Incpercap	(Per capita household income in VND 1,000)	14,623.920	59,379.650
Riceincrate	(Share of rice income in total household income)	0.489	0.296
Mono	(1 = rice monoculture, 0 = otherwise)	0.588	0.493
Diseases	(1 = diseases happening during the study year, 0 = otherwise)	0.412	0.493
Training	(1= respondents attending trainings, 0 = otherwise)	0.404	0.491
Gender	(1= male respondent, 0 = female respondent)	0.926	0.262
Edu	(Education level in years)	6.176	2.870
Age	(The age of respondents in years)	48.599	10.986
Credit	(1 = Access formal credit, 0 = otherwise)	0.305	0.461

Table 2. Logit selection regression concerning interest in minimum price insurance by rice farmers[ψ]

Variables	Logit Function		Marginal Effect		95% Conf. Interval	
	Coefficient	S.E.	dy/dx	S. E.	Lower	Upper
Large size	1.398***	0.448	0.335***	0.097	0.144	0.525
Medium size	0.906***	0.311	0.221***	0.074	0.077	0.366
Rice price	-0.319	1.398	-0.079	0.345	-0.755	0.598
Incpercap	0.187	0.210	0.046	0.052	-0.055	0.148
Riceincrate	-0.327	0.588	-0.081	0.145	-0.365	0.204
Mono	0.084	0.234	0.021	0.058	-0.092	0.134
Diseases	0.112	0.235	0.028	0.058	-0.086	0.141
Training	0.568**	0.244	0.140**	0.060	0.023	0.257
Gender	1.285**	0.551	0.272***	0.090	0.097	0.447
Edu	0.016	0.043	0.004	0.010	-0.017	0.025
Age	-0.016	0.011	-0.004	0.003	-0.009	0.001
Credit	0.447*	0.250	0.111*	0.062	-0.011	0.232
Constant	-0.829	11.732				
Pseudo R^2			0.104			
Observations (N)			364			

Notes: (1) ***, **, * indicate statistical significance at the 0.01, 0.05 and 0.1 level respectively
(2) ψ Dependent variable is a dummy equal to one if the answer is "yes" to the interest question.

There are some variables that have a significant impact on the desirability of insurance service, including *large size farms, medium size farms, training, gender, and access to credit.* The results reveal that the large-scale (more than 1 ha) and medium-scale farmers (from 0.5 ha to less than 1 ha) demand price insurance 33.5% and 22.1% higher than the small-scale ones (less than 0.5 ha), respectively. Farmers who attended short training are more interested in the proposed price insurance than those who did not participate in short training, while those who have access formal credit also pay higher attention than those who have never borrowed money from banks. However, some social-characteristics of farmers such as the education level, age, and per capita income have no significant relationship with farmers' interest in the proposed insurance service.

Table 3. Willingness to pay for price insurance service at different guaranteed prices

Guaranteed Price (VND/kg)	Willingness to Pay	
	Number	**Percent (%)**
2,500	17	10.24
3,500	29	17.47
4,500	101	60.84
Other	19	11.45
Total	**166**	**100.00**

Table 3 shows rice farmers' willingness to pay for price insurance service at different guaranteed prices. About 60.84% of the farmers were willing to pay for VND 4,500 per kg because they found this price profitable, while fewer farmers were willing to pay VND 2,500 and VND 3,500 per kg. Thus, this study only estimates the insurance demand at the price of VND 4,500 per kg.

The demand for guaranteed insurance of VND 4,500 per kg was estimated only from the households who state that they were interested in the price insurance. With the price of VND 4,500 per kg, farmers were asked on their preference for the insurance fees of VND 125, 300, 625, and 925 per kg using the dichotomous choice contingent valuation method. Figure 1 shows the mean WTP for the guaranteed price of VND 4,500 per kg. Over 80% of farmers accept to buy insurance at the lowest bid of VND 125 per kg, while around 20% of them demand the insurance service at the highest fee of VND 925 per kg.

Table 4 shows the results of the two logit regressions to estimate the willingness to pay for price insurance service. Model 1 includes only the bid value, while Model 2 includes the other variables such as farmers' characteristics and factors related to minimum price insurance. The coefficients of the bid value in the two models are statistically significant and negative as expected, revealing the existence of the WTP for the guaranteed price of VND 4,500 per kg. The results also show that rice monoculture (*Mono*) has a negatively impact on the probability of saying 'yes' for the proposed insurance service, while access to credit (*Credit*) and medium-size positively influence the probability of accepting the proposed insurance service. The predictive power is quite high, with more than 78 percent in Model 1 and 80 percent in Model 2. From the findings in Table 4, the medium and mean of WTP for the guaranteed price of VND 4,500 per kg are calculated by using equation (7) for the parametric method (utilizing the directly estimated values of the coefficients), which are presented in Table 5.

Figure 1. The demand for insurance service at the insurance price of VND 4,500

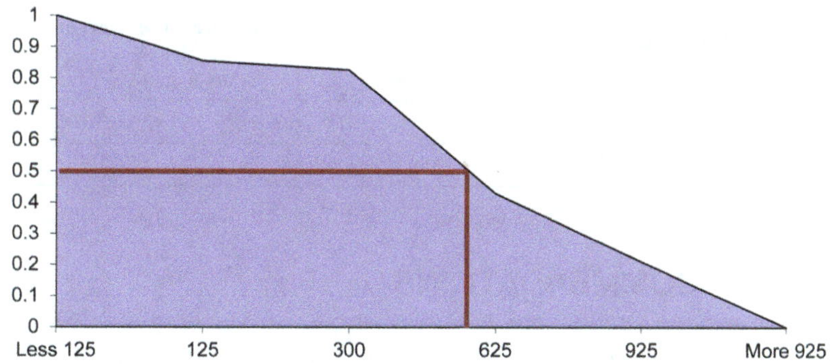

Table 4. The results of logit regression for estimating the willingness to pay for price insurance service

Variables	Model 1		Model 2			
	Coef.	**S.E.**	**Coef.**	**S.E.**	**dy/dx**	**S. E.**
Bid	-0.0042***	0.0006	-0.0048***	0.0007	-0.0011***	0.0002
Large size			0.6303	0.8615	0.1372	0.1785
Medium size			1.2104*	0.6534	0.2658*	0.1361
Rice price			-2.7294	3.8142	-0.6176	0.8646
Incpercap			-0.2508	0.4400	-0.0567	0.0994
Riceincrate			0.2263	1.1439	0.0512	0.2591
Mono			-0.9342**	0.4625	-0.2025**	0.0944
Diseases			0.5232	0.4472	0.1170	0.0983
Edu			-0.1109	0.0758	-0.0251	0.0171
Age			-0.0015	0.0213	-0.0003	0.0048
Training			0.0455	0.4709	0.0103	0.1066
Credit			0.7610*	0.4589	0.1648*	0.0941
Constant	2.4739***	0.3749	27.8130	31.6339		
Pseudo R-squared	0.2524		0.3364			
Correct prediction (%)	78.313		80.120			
Observation	166		166			

Notes: ***, **, * indicate statistical significance at the 0.01, 0.05 and 0.1 level respectively

Table 5 provides the estimates of WTP for the guaranteed price of VND 4,500 per kg, the medium WTP was estimated at around VND 475 per kg. This accounted for 23.8% of the break-even price and 10.6% of this guaranteed price. The mean WTP was calculated around VND 596 per kg, accounted for 29.8% of break-even price and 14.5% of the current market price.

Table 5. Rice farmers' willingness to pay for price insurance service

	Premium (VND/kg)	The Percent of Break-Even Price (VND 2,000/kg)	The Percent of Market Price (VND 4,100/kg)	The Percent of Guarantee Price (VND 4,500/kg)
Medium WTP	475	23.8	11.6	10.6
Mean WTP	596	29.8	14.5	13.2

CONCLUSION AND RECOMMENDATIONS

The study applied the dichotomous choice contingent valuation method to estimate the demand for price insurance service by rice farmers in Vietnam. This study concludes that farmers prefer the guaranteed price of VND 4,500 per kg and we estimated the price insurance demand based on the guaranteed price of VND 4,500 per kg. The study also investigated farmers who cultivated rice over 0.5ha, were used to attend short training or access any formal credits were more interested in the price insurance service. The insurance fees that farmers have the willingness to pay for insurance were about 13% of the guaranteed price and equal to about 30% of the break-even price of rice production.

Since less than 50% of interviewed farmers are interested in the price insurance service. The reasons may be that this service is new and farmers are not confident of it. Therefore, to make this service more acceptable among farmers, the government should provide financial assistance to pilot small-scale service operations. The target group is mainly farmers with medium and large scale (over 0.5ha) or those who access to formal credit. The acceptable rate of insurance fee is about 14.5% of the current market price. In the long run, the information of this service could be advertised on TV or through short training often given by extension services. This price insurance service could be expanded when Vietnamese farmers are familiar and understand clearly the operation of this service. However, the benefit of service suppliers is not included in the study because the study only focuses on the analysis of price insurance demand and does not involve the supply side of this service, it is recommended that an in-depth study on supply-side will be needed to develop this service more sustainably.

ACKNOWLEDGMENT

We would like to express our gratitude to Prof. Mitsuyasu Yabe, Kyushu University for his invaluable comments, which led to a significant improvement of the manuscript. Also, the staff in the Department of Environmental and Resource Economics, College of Economics, Can Tho University for assisting in data collection. This chapter is based on an article originally published in the Asian-African Journal of Economics and Econometrics (AAJEE) in 2011. This material is reprinted with the permission of AAJEE.

REFERENCES

Alexander, S., Panayotis, K., & Luc, C. (2007). The Stated Benefits From Commodity Price and Weather. In C. Luc & S. Alexander (Eds.), *Rural Household Vulnerability and Insurance Against Commodity Risks: Evidence From the United Republic of Tanzania*. Food and Agriculture Organization of the United Nations.

Hanemann, W. M., & Kanninen, B. (1998). *The Statistical Analysis of Discrete-Response Data*. Working Paper No.798, Department of Agricultural and Resource Economics and Policy, University of California.

International Task Force on Commodity Risk Management in Developing Countries (ITF). (1999). Dealing with commodity price volatility in developing countries: A proposal for a market based approach. Discussion paper for the roundtable on commodity risk management in developing countries, World Bank.

Skees, J. R., Hartell, J., & Hoa, J. (2006). Agricultural Commodity Markets and Trade: New Approaches to Analyzing Market Structure and Instability. In A. Sarris & D. Hallam (Eds.), *Weather and Index-Based Insurance for Developing Countries: Experience and Possibilities*. Edward Elgar and Food and Agriculture Organization of the United Nations.

Varangis, P., Larson, D., & Anderson, J. (2002). *Agricultural Markets and Risks: Management of the Latter Not the Former*. World Bank, Policy Research Working Paper No. 2793.

Section 3

Agricultural Resource Management and Management of Food Resources

Chapter 6
Multiple Scenarios-Based Impact Analysis of Predicted Land-Use Change on Ecosystem Services Value:
Evidence From Nigeria

Aisha Olushola Arowolo
Department of Agricultural Economics and Farm Management, Federal University of Agriculture, Abeokuta, Nigeria

Raheem Olatunji Aminu
Department of Agricultural Economics and Farm Management, Federal University of Agriculture, Abeokuta, Nigeria

Shakirat Bolatito Ibrahim
Department of Agricultural Economics and Farm Management, Federal University of Agriculture, Abeokuta, Nigeria

Abdus Samie
Institute of Agricultural and Resource Economics, University of Agriculture, Faisalabad, Pakistan

Funminiyi Peter Oyawole
Department of Agricultural Economics and Farm Management, Federal University of Agriculture (FUNAAB), Abeokuta, Nigeria

ABSTRACT

This chapter investigates how land-use/land-cover (LULC) changes under different scenarios will affect ecosystem services provisions in Nigeria using multiple data sources. The Markov and dynamics of land system models were integrated to predict future LULC changes while the value transfer methodology was adopted to evaluate the economic value of ecosystem services. The results revealed varying patterns and trends of LULC change under the baseline, forest protection priority, and sustainable economic growth scenarios. Based on the predicted LULC change, the total ecosystem services value in Nigeria will decline under the baseline and forest protection priority scenarios but increase in the sustainable economic growth scenario. The sustainable economic growth scenario showed major positive impacts on the ecosystem service functions of recreation, climate regulation, soil formation, and erosion control. This study concludes that the sustainable economic growth scenario is the best to ensure expected production while safeguarding the environment in Nigeria.

DOI: 10.4018/978-1-7998-2599-9.ch006

INTRODUCTION

The wide range of goods (such as food, fodder, fuelwood, timber, game animals, and pharmaceutical products) and services (such as purification of air and water, regulation of climate, regeneration of soil fertility) provided by ecosystems are critical to the existence and welfare of humanity (Costanza et al., 1997; MEA, 2005). In recent years, driven by the growing needs arising from the expanding human population, rapid urban growth and economic development, humans have extremely and extensively modified the ecosystems more than ever before (Erle & Pontius, 2007; Wang et al., 2018). The natural covers, particularly forests, savannahs and grasslands have substantially been replaced by agriculture, and nearly 40% of the earth's ice-free land surface is presently being used to cultivate crops or as pastures (Foley et al., 2005).

Among all human actions, land-use/land-cover (LULC) change has emerged as one of the most pressing issues increasingly drawing the public and scientific attention as it is recognized as an underlying driver of global environmental and climate change (Islam et al., 2018; Halmy et al., 2015; Kim et al., 2014) even though it occurs locally (Sleeter et al., 2013). LULC change is also a significant force that impacts biodiversity through the loss, degradation, and fragmentation of habitat (Baan et al., 2012; Newbold et al., 2015; Elmqvist et al., 2016), which causes decline in ecosystem integrity as well as genetic losses that may impede future scientific advances in agriculture and pharmaceutics (de Sherbinin, 2002; Antwi-Agyakwa, 2014). It affects air quality and increases the risk of infectious diseases (Foley et al., 2005), threatens food security (Hettig et al., 2016), impacts on water resources quality (Houet et al., 2010; Uriarte *et al.*, 2011; Singh et al., 2013), and also determine, in part, the vulnerability of places and people to climatic, economic and socio-political perturbations (Lambin & Geist, 2006). Due to LULC change, about 60% of various ecosystem services have been degraded in the past 50 years alone (Wang et al., 2018).

Over the past decades, research has made considerable advancement in modelling LULC changes (NRC, 2014; Azadi et al., 2017), evaluating the values of ecosystem goods and services (De Groot et al., 2012; Costanza et al., 2014), and examining their variation response to LULC dynamics (Newbold et al., 2015). These studies highlight the profound influences of land changes, particularly from the natural ecosystems to artificial landscapes on the provision of ecosystem services. For example, land changes to agriculture and urban use are noted to negatively affect the provision of other crucial ecosystem services such as nutrient cycling, climate regulation (Peng et al., 2006; Li et al., 2007), erosion control and genetic resources (Portela & Rademacher, 2001), disturbance regulation (Zhao et al., 2004; Wang et al., 2006), soil fertility (Schroter et al., 2005; Collard & Zammit, 2006), recreation opportunities (Nahuelhual et al., 2014) and water regulation (Schroter et al., 2005; Fiquepron et al., 2013), which are mainly provided by the natural ecosystems.

Measuring the economic value of ecosystem services and quantifying their variations in response to land-use changes is an attractive land-use decision support tool (Förster et al., 2015). The valuation methods of ecosystem services exist in two major categories. The first category includes the primary data-based valuation approaches, which are original valuation approaches that rely on either direct and indirect market price information on ecosystem services or creation of hypothetical markets to elicit ecosystem services values (Pascual et al., 2010). The primary data-based approaches are data-driven and complex. Thus, they applied more often in small-scale studies, where the aim is to focus on the valuation of a smaller number of ecosystem services types (Wang et al., 2018). The second category is the proxy-based approaches that do not necessitate the use of primary data from within the study region.

They rely on "benefit transfer" otherwise known as "value transfer" where a prior estimated ecosystem services values from original (primary) valuation studies in one or more location are transferred to other areas assumed to have similar demographics, economic and ecological conditions (Plummer, 2009; Richardson et al., 2015; Yi et al., 2017). The proxy-based approaches often make use of the land-cover data such that the monetary values for the different ecosystem services are transferred by land-cover class. A major limitation of these approaches is that they assume homogeneity of ecosystems services value within a particular LULC type and do not consider possible variations over space and time (Plummer, 2009; Eigenbrod et al., 2010; Song et al., 2015; Wang et al., 2018). However, they are used more often due to their minimal cost and simplicity of operation, the widespread availability of LULC data, and their effectiveness in modelling the trade-offs among multiple ecosystem services due to land-use changes (Wang et al., 2018).

Modelling in land change science involves the use of artificial representations of the interactions within the land-use system to improve our understanding of its dynamics and potential future development (Verburg et al., 2006). LULC change models are useful in identifying the drivers of change and understanding their spatiotemporal inter-relationships, predicting future scenarios of change regarding location, and quantity of change as well as when the change will happen (Veldkamp & Lambin, 2001). The models have provided the capability to propose and evaluate land-use policies and offer the possibility of visualizing alternative future scenarios (Chaudhuri & Clarke, 2013). Over the years, a range of land change modelling approaches has been developed, and many authors have categorized these models into different groups from various viewpoints. A recent classification by the US National Research Council (NRC, 2014), noted to be widely accepted by international researchers (Camacho Olmedo et al., 2018), grouped land change modelling approaches into six main categories. They include (1) machine learning and statistical, (2) cellular, (3) sector-based economic, (4) spatially disaggregated economic, (5) agent-based, and (6) hybrid/integrated approaches.

Land change is the result of multiple interactions between the human-environment systems operating across a range of spatial scales (Lambin et al., 2003) and no single model is capable of considering all of the processes of LULC change at different scales (Verburg et al., 2008). Hence, hybrid models have been advocated recently for studying land-use/land-cover changes (O'Sullivan et al., 2016). These integrated modelling techniques combine multiple approaches into a single model or modelling framework to represent various aspects of land change patterns and processes (NRC, 2014). Land change modelling through this novel approach of hybridization take advantage of the strength of the individual models and minimize some of their inherent limitations.

In this case study, an integrated approach was applied to simulate the future land-use change in Nigeria in a spatially explicit manner. The integrated approach coupled the Markov Chains (MC) model, the Dynamics of Land System (DLS) model and the logit regression model to simulate land-use change under three scenarios (i.e., baseline, forest protection priority and sustainable economic growth scenarios). The coupled model involves two procedures for the simulation of future land-use in Nigeria. First, the MC model was applied to obtain the future land-use demand based on historical land-use conversion rates and assumptions. Second, the predicted trends and quantity of land use changes are spatially allocated to specific locations by the DLS model, depending on the factors that influence the spatial pattern of the distribution of distinctive land-use types, which are determined via the logit regression model.

The present study is focused on Nigeria, a country that is home to Africa's largest population and has undergone significant LULC changes over the past four decades. Notably, agriculture has traded places with the natural ecosystems with the transition from savanna and forests to agriculture constituting the most substantial LULC change in terms of area (World Bank, 1998; Abubakar, 2015; Arowolo & Deng, 2018). While numerous studies have provided a thorough insight of past and present LULC change at varying scales (Abubakar, 2015; AC-Chukwuocha, 2015; Jibril and Liman, 2014; Ejaro & Abdullahi, 2013; Oyinloye & Oloukoi, 2012; Njoku et al., 2010; Abbas, 2009; World Bank, 1998), the potential future change, particularly in a spatially explicit and scenario-based manner, are still poorly understood most especially at the national level. Besides, beyond analyzing the dynamics of LULC in Nigeria, quantitative assessments to reveal the impact of the past/present LULC changes on ecosystem services value are scanty (Ayanlade, 2012; Ayanlade & Proske, 2016) and how probable future LULC change may likely impact the provisioning of ecosystem services is lacking. Thus, in this study, we aim to investigate how land-use changes under different scenarios will affect the provisioning of ecosystem services in Nigeria by combing an integrated land-use change simulation model and a proxy-based ecosystem services valuation model.

STUDY AREA

The study area, Nigeria, is located in the western part of Africa between latitudes of 4°16′13.50″ - 13°53′31.24″ N and longitudes of 2°40′6.35″ - 14°40′35.09″ E (Figure 1). It comprises of 36 administrative states and a Federal Capital Territory situated in Abuja. The total area coverage of the country is about 937, 052.155 km^2 (NPC, 2010). Nigeria shares land border in the west with the Republic of Benin, in the north with the Niger Republic, in the east with Chad and the Republic of Cameroon, while the shores of the Atlantic Ocean forms the southern limit (Figure 1). Nigeria is characterized by two main seasons: the wet season that lasts from mid-March to November in the south and from May to October in the north, and the dry season, which occupies the rest of the year (Oyenuga, 1967).

Nigeria experiences a tropical climate that varies with elevation from south to north. The southern part has a tropical rainforest climate with abundant rainfall usually above 2,000 mm annually, and a mean temperature of about 27°C (Oginni & Adebamowo, 2013). The western to central Nigeria has a tropical savannah climate, otherwise known as the tropical wet and dry climate. Temperatures in these regions are above 18 °C throughout the year, and annual rainfall is about 1,500 mm (Eludoyin et al., 2013). The Sahel or tropical dry climate predominates in northern Nigeria where rainfall is less than 600 mm per annum (Abiodun et al., 2011) and temperatures climb as high as 40 °C during the dry season. A cool mountain climate is found in areas well over 1,520 meters (4,987 ft) above sea level (Ileoje, 2001).

Nigeria is the most populous African country ranking as the seventh most populous country in the world, with a total estimated population of over 200 million people in 2015 (United Nations, 2019). Attributable to the population pressure and associated demand to feed the largely growing population, previous studies have reported significant LULC change in Nigeria, most substantially into agricultural land, which expanded on both forests and non-forests vegetation (Abubakar, 2015; Abbas, 2009; World Bank, 1998). Similarly, Arowolo et al. (2018) reported that conversions to cultivated land dominate the LULC change processes during the years 2000 and 2010. In the study, cultivated land was estimated to increase from 244, 826.19 km^2 in 2000 to 291,718.68 sq. km in 2010, invading the surrounding ecological lands (i.e., forests, grassland, shrubland, wetland and water areas), which in total decreased by 6.83%.

Figure 1. Study area: geographic location and land-use/land-cover of Nigeria

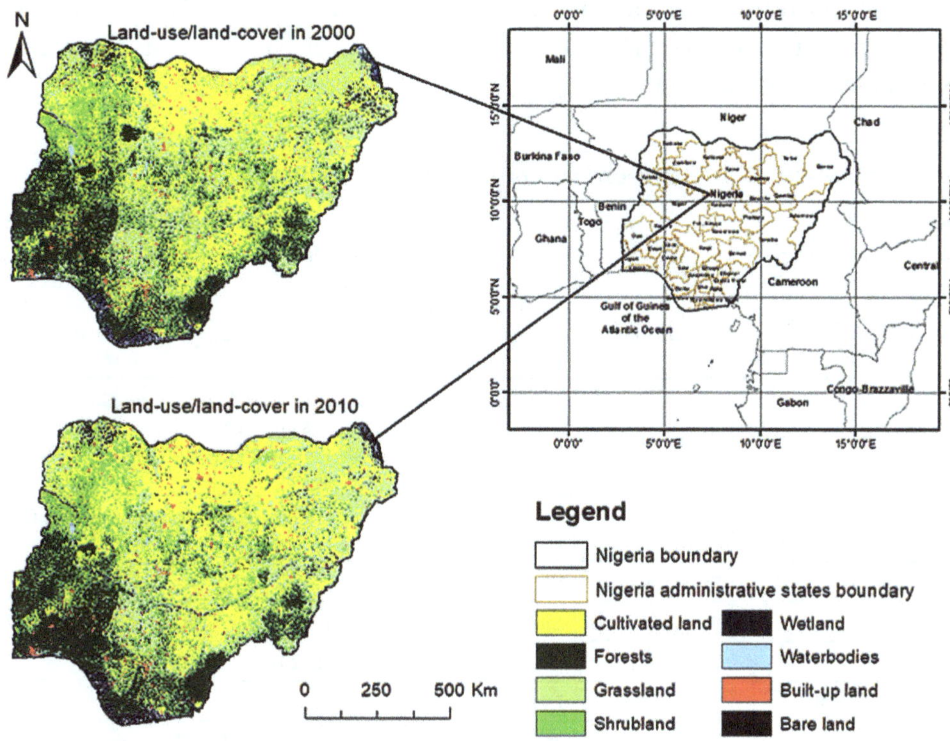

DATA COLLECTION AND PROCESSING

The datasets required for this study came from various sources. We obtained the land data of Nigeria for the years 2000 and 2010 (Figure 1) from the global land-cover maps (GlobeLand30) database of the National Geomatics Centre of China (NGCC, 2014). GlobeLand30 with a resolution of 30m were generated based on high resolution remotely sensed data imageries (Chen et al., 2015) and overall accuracy of 79.6% in 2000 and 83.5% in 2010 were reported (Zhang et al., 2015; Zhang et al., 2016). The GlobeLand30 datasets consist of 10 major classes, but only 8 land-use/land-cover types are found in Nigeria (Figure 1).

Other spatial data sourced include the landform data provided by the Joint Research Centre (JRC) of the European Commission (Meybeck et al., 2001), Agro-ecological Zones (AEZs) data obtained from the Global Trade and Analysis Project (GTAP) database (Ramankutty et al., 2007), elevation and slope data extracted from the processed 90 m resolution Shuttle Radar Topographic Mission (SRTM) Digital Elevation Model (DEM) (Jarvis et al., 2008), soil properties data (soil pH and soil depth) obtained from the Africa Soil Information Service (AfSIS) Sentinel Site database (Hengl et al., 2015), road network data extracted from the 2010 global roads data released by NASA Socioeconomic Data and Applications Center (SEDAC) (CIESIN/ITOS, 2013), data on water areas acquired from the database of DIVA-GIS (2016) and main cities data prepared based on topographic maps and Google Earth images.

In addition to the spatial data, statistical data (including temperature, rainfall, population and GDP) were also utilized. Temperature and rainfall data for the period of 2000 to 2010 were collected from the

Nigeria Meteorological Agency (NIMET, 2015). State-wise population and Gross Domestic Product (GDP) data were obtained from the National Bureau of Statistics, Nigeria (NBS, 2016) and Canback Global Income Distribution Database (C-GIDD, 2016) respectively. Table 1 lists the sources and preprocessing procedures of the data in this study.

Table 1. Sources and preprocessing procedures of the data in this study

Data Group	Sub-group	Typology/ Resolution	Description/Processing/Sources
Land dataset	Land-use/land-cover	Raster, 30m	Land-use/land-cover maps for the years 2000 and 2010 with eight land-use classes in Nigeria (Figure 1), further description of the dataset can be found in Arowolo et al. (2018).
Geophysical	Landform	Raster, 1km	Topographic structure classified based on relief roughness and elevation (Meybeck et al., 2001). The landform has 12 classes, which were reduced to 5 (Plains, Lowlands, Plateaus, Hills and Mountains) for use in logistic modelling.
	Agro-ecological Zones (AEZs)	Feature	Demarcation of the study area based on the moisture regime (Ramankutty et al., 2007).
	Elevation	Raster, 90m	Elevation extracted from DEM (Jarvis et al., 2008).
	Slope	Raster, 90m	Slope gradient derived from DEM (Jarvis et al., 2008).
	Soil PH	Raster, 250m	pH value of soil (Hengl et al., 2015).
	Soil Depth	Raster, 250m	Depth of soil (Hengl et al., 2015).
Climate	Temperature	Tabulated	Monthly temperature and rainfall data from 43 weather stations (NIMET, 2015). Surfaces were created using the spline interpolation algorithm (Hutchinson, 2006), the optimal method for spatial interpolation of temperature and rainfall in Nigeria (Arowolo et al., 2017).
	Rainfall	Tabulated	
Socio-economic	Population	Tabulated	State-wise population for the years 2000 and 2010 (NBS, 2016), Converted into raster in ArcGIS.
	Gross Domestic Product (GDP)	Tabulated	State-wise GDP for years 2000 and 2010 (C-GIDD, 2016) Converted into raster in ArcGIS
Accessibility	Roads	Feature	GIS layer for 2010 global roads CIESIN/ITOS, 2013. The Euclidean distances from each pixel to the closest road were calculated using the spatial analyst extension tool in ArcGIS
	Water	Feature	GIS layer for water generated from the digital chart of the world DIVA-GIS, 2016. The Euclidean distances from each pixel to the nearest water source were calculated using the spatial analyst extension tool in ArcGIS
	Cities	Feature	GIS layer for cities generated based on the GPS coordinates Google Earth. The Euclidean distances from each pixel to the closest road were calculated using the spatial analyst extension tool in ArcGIS

Scenario-Based Land-Use Change Simulation in Nigeria

Land-Use Change Scenarios Design

In this study, we designed three scenarios for simulating the LULC change dynamics in Nigeria. They include the baseline, forest protection priority and sustainable economic growth scenarios. The baseline scenario represents a baseline scenario without any comprehensive plan or policy intervention. In this scenario, future land demand was estimated based on historical conversion trends of land-use/land cover. The forest protection priority scenario assumed a strict implementation of spatial policies about forests aimed at conserving the remaining forests ecosystem. Thus, the area of forests was designated as a restricted region wherein change is not allowed. The sustainable economic growth scenario is designed to create an environment that enables the co-existence of growth and development on an enduring and sustainable basis in Nigeria as envisioned in Nigeria's vision 20:2020 (Eneh, 2011). The aim of the vision was for Nigeria to be among the top 20 strong economies of the world by the year 2020 (Bature, 2013). The GDP is planned to grow at an estimated annual average of about 13.8% between 2010 and 2020, to be mainly contributed by the industry, manufacturing and services sectors while the agricultural sector continues to grow based on high productivity. Thus, based on Nigeria's vision 20:2020, we constructed the sustainable economic growth scenario and assumed that several useful measures such as the use of modern farm inputs (fertilizer and improved seeds), farmers' access to credit, provision of irrigation facilities etc., would be taken to increase agricultural productivity, thus, reducing the need for more of the cultivated land-use type. Further, forests, wetland and water bodies are designated as important ecosystems that are crucial in maintaining a sustainable environment and thus are not allowed to change into any other land-use categories. We also hypothesize rehabilitation of bare lands into forests under this scenario.

Markov Chains Model

The use of Markov chains analyses for land-use change modelling was first proposed by Burnham (1973) and has been applied in many studies to model LULC change (e.g., Kumar et al., 2014; Fan et al., 2008; Iacono et al., 2015; Dadhich & Hanaoka, 2010; Ma et al., 2012). A Markovian analysis is based on the core principle of a continuation of historical development and uses a change matrix of land-use constructed over a historical period to derive the probability of change of one land-use category into another in some observed period under the assumption of a constant rate and amount of change. These probabilities are then used to calculate the future land areas of different land types in a non-spatial manner (NRC, 2014) as shown below:

$$A_{t+1} = A_t P_{ij} \tag{1}$$

where A_{t+1} and A_t represents the states of land-use at a future point t+1 and a historical period t, respectively. P_{ij} denotes the probability of land use i shifting to land use j and can be expressed as follows:

$$P_{ij} = \begin{vmatrix} P_{11} & P_{12} & \cdots & P_{1n} \\ \vdots & \vdots & \vdots & \vdots \\ P_{m1} & P_{m2} & \cdots & P_{mn} \end{vmatrix} \left(0 \le P_{ij} \le 1; \sum_{j=1}^{n} P_{ij} = 1 \right) \tag{2}$$

where m and n are the number of land-use types.

Based on LULC change from 2000 to 2010, the Markov model was used to predict the land requirements for the different land-use types in 2030 under the three future scenarios discussed above. The status of land-use in 2030 under the baseline scenario was predicted based on the land change matrix from

2000 to 2010, and following the stated assumptions under the forest protection priority and sustainable economic growth scenarios, the historical land change matrix was modified to predict the future land demand for these scenarios.

Markov models are trendy for land-use change modelling because of their simplicity of implementation (van Schrojenstein Lantman et al., 2011; NRC, 2014). However, the MC model suffers from the ability to predict the locations and pathways of LULC change; i.e., it is not spatially explicit. To overcome this deficiency, several authors have integrated the Markovian concepts with other simulation rules and concepts (Guan et al., 2011; Hu et al., 2013; Sinha & Kumar, 2013; Keshtkar & Voigt, 2016) in which case the Markovian model is used to estimate the quantities of land change in the future while another type of cellular model is used to simulate the spatial patterns. In this study, we used the Dynamics of Land System (DLS) model, a cellular model, to allocate the predicted land requirements from the Markov chains model to geographical space.

Dynamics of Land System (DLS) Model

The DLS model framework was developed by Deng et al. (2008). The DLS model is capable of integrating multiple sources of data for simulating the spatiotemporal patterns of all kinds of land-use. Studies in different regions of the world have proven the DLS model to be robust in simulating the dynamics of land-use changes (Deng et al., 2010; Yuan et al., 2013; Ke et al., 2014; Zhen et al., 2014; Li et al., 2015; Hasan et al., 2017; Samie et al., 2017; Najmuddin et al., 2017; Jiang et al., 2015). The model integrates two distinctive features: Firstly, it combines a dual approach to attain a balance: a scenario analysis of land demand at a regional level and spatial disaggregation of land uses at a detailed pixel level. Secondly, it considers the interactions between land-uses influencing factors and the interactions between the nearby pixels for these influencing factors. Besides, the DLS model allows for the setting of simulations restrictions. Here, the model sets areas where a particular land-use type cannot appear as restricted regions (restriction areas may include various protected areas such as nature reserves, forest parks, water sources) and removes these regions so that they are not inputted into the model.

Three main modules make up the DLS model. Firstly, a spatial regression analysis module that establishes the relationship between the different land-uses and influencing factor. Secondly, a scenario analysis non-spatial land demand module, which identifies the regional change in demand for the different land-uses for a given period. Thirdly, a spatially explicit allocation module that allocates land-use changes from a regional level into individual grid cells based on a predefined transformation rule.

The results of land-use change simulation models are commonly validated through a comparison of actual and simulated land use pattern. Thus, to validate the DLS model used in this study, the year 2000 was taken as a base year and the land-cover map and other driving factors data for the period were used as model input, and land-cover map for 2010 was simulated. After that, we assessed the agreement between the actual and simulated maps using Kappa coefficient of agreement (Equation 3), the most extensively used index (Vliet, 2009; Parsa & Salehi, 2016) for agreement assessment between actual map and LULC model outputs.

$$Kappa = \frac{P_o - P_c}{P_p - P_c} \tag{3}$$

where, P_o is the percent correct for the model's output, P_c is the expected percent correct due to chance and P_p is the percent correct for a perfect classification (i.e., 100%).

The contingency matrix from the comparison of the actual and simulated land-use maps of 2010 is presented in Table 2. The table indicated that the overall agreement between the actual and simulated land use pattern is 86.62%. The corresponding Kappa coefficient was 0.802 whereas a value above 0.6 is regarded as substantial (Viera & Garrett, 2005; Rwanga & Ndambuki, 2017) and good/satisfactory (Piyathamrongchai & Batty, 2007; Saraux et al., 2013) agreement between the predicted and actual land-use patterns. This indicates that the simulation result was good, and the DLS model was suitable for simulating the spatial pattern of land-use in Nigeria.

Table 2. The contingency matrix of actual and simulated land-use in the year 2010

	LULC	Simulated Land Use							
		Cultivated Land	**Forests**	**Savannahs**	**Wetland**	**Water Bodies**	**Built-up Land**	**Bare Land**	**Total**
Actual land use	Cultivated land	258924	8626	22956	81	110	407	793	291897
	Forests	12093	173408	18278	1340	324	541	384	206368
	Savannahs	17633	23034	347018	496	393	902	4580	394056
	Wetland	1070	738	442	12389	453	12	58	15162
	Water bodies	208	172	139	794	6922	18	127	8380
	Built-up land	1009	240	766	18	7	6586	19	8645
	Bare land	968	174	4483	53	183	93	5202	11156
	Total	291905	206392	394082	15171	8392	8559	11163	935664
	Expected agreement 32.39%								
	Overall agreement 86.62%								
	Kappa 0.802								

The significant influencing factors of the distribution pattern of each LULC type incorporated into the spatial regression analysis module of the DLS model were identified using the binary logistic regression model (Equation 4).

$$\ln\left(\frac{p_q}{1-p_q}\right) = \beta_0 + \beta_1 X_{1,q} + \beta_2 X_{2,q} + \ldots + \beta_k X_{k,q} \tag{4}$$

Where, p_q is the probability of the occurrence of a particular LULC type in a grid cell q, X's are the LULC pattern influencing factors (Table 3) and β's are the coefficients to be estimated through logistic regression using the actual LULC pattern as dependent variable, which takes a value of 1 if a specific LULC type exists in a given pixel q and 0 if otherwise.

Given that the multicollinearity diagnostics indicate that the model predictors (influencing factors) are not highly correlated, we selected all the 13 predictor variables to conduct the binary logistic regression analysis, and coefficient estimates of the significant influencing factors of the spatial distribution of each LULC type are presented in Table 4. Specifically, the significant driving factors differ for each LULC types. For example, all the factors significantly influence the distribution pattern of cultivated land, forests and savannahs, whereas, the distribution pattern of wetland, water bodies, built-up land and bare land spatial were significantly explained by a smaller number of driving factors. For each land-use type, we selected those specific significant driving factors for land-use change simulation, based on which the spatially explicit allocation module in the DLS model can disaggregate the regional land change demands into individual grids under different scenarios.

The Relative Operating Characteristics (ROC) method that evaluates the predicted probabilities by comparing them with the observed values (Verburg et al., 2002) was used to test the accuracy of the logit regression model. The result showed that the selected driving factors could reasonably explain the spatial patterns of the LULC types. The area under the ROC's curve (AUC) was greater than 0.70 (Table 4) for all the seven land-use types whereas a value of above 0.5 has been indicated to be acceptable (Kindu et al., 2015).

Table 3. List of LULC pattern influencing factors used in binary logistic regression modelling

Factor Category	Sub Factors	Description
Geo-physical factors	Landform: Plains Lowlands Plateaus Hills Mountains	Topographic structure Reference category is plains
	AEZs: Arid zone Dry semi-arid zone Moist semi-arid zone Sub-humid zone Humid zone Highly humid zone	Agro-Ecological Zones (AEZs) Reference category is highly humid zone
	Elevation	Digital Elevation Model (DEM)
	Slope	Slope gradient derived from DEM
	Soil PH	PH values of soil.
	Soil depth	Depth of soil (cm)
Climatic factors	Mean temp	Average mean temperature from 2000 to 2010 (° C)
	Mean rain	Average mean rainfall from 2000 to 2010 (mm)
Socio-economic factors	POPDEN	Population density change from 2000 to 2010 (Persons/km^2)
	GDP/capita	GDP per capita change from 2000 to 2010 (US$)
	City distance	Euclidean distance to nearest major city (km)
Accessibility factors	Road distance	Euclidean distance to nearest major road (km)
	Water distance	Euclidean distance to nearest water source (km)

Table 4. Coefficients of significant influencing factors of the spatial distribution patterns of LULC in Nigeria in 2010

Drivers	Cultivated Land	Forests	Savannahs	Wetland	Water Bodies	Built-up Land	Bare Land
Lowlands	−0.3511***	0.5264***	−0.0566***	0.1404*	−0.4529***	-	−0.1896***
Plateaus	−0.8894***	0.5979***	0.3753***	0.6299***	0.4229***	-	-
Hills	−1.4599***	0.6314***	0.3867***	-	−0.8772***	−0.4446***	−0.6213***
Mountains	−2.1658***	1.2464***	0.4674***	-	2.3566***	−1.0143***	-
Arid zone	1.2275***	−5.4565***	3.1704***	3.7508***	−4.7363***	1.1069***	5.3550***
Dry semi-arid zone	2.3431***	−4.9675***	2.0867***	6.5500***	−1.6170***	-	4.4575***
Moist semi-arid zone	2.9387***	−3.9933***	1.4483***	3.0493***	-	0.2726**	3.1401***
Sub-humid zone	2.2615***	−2.3452***	1.2098***	-	-	0.4159***	2.3274***
Humid zone	0.8952***	−1.4186***	1.2059***	−2.8778***	−0.1573*	0.7433***	0.7999***
Elevation	−0.0016***	−.0044***	0.0036***	−0.0361***	−0.0027***	-	0.0020***
Slope	−0.1629***	0.1253***	−0.0714***	−1.7513***	−1.9370***	−0.0340*	−0.0943***
Soil depth	0.0176***	0.0093***	−0.0249***	0.0060***	-	0.0069***	−0.0133***
Soil PH	0.0927***	−0.2786***	0.0468***	0.0761***	0.0294***	0.2197***	-
Mean temp	−0.6778***	−1.0650***	1.2072***	−0.3801***	-	−0.3972***	1.1909***
Mean rain	0.0003***	−0.0031***	0.0021***	−0.0006***	-	0.0005***	0.0027***
POPDEN	0.0001***	0.0001***	−0.0010***	0.0006***	0.0001***	0.0010***	0.0009***
GDP/capita	−0.0005***	0.0003***	−.0001***	0.0003***	-	0.0003***	−0.0002***
City distance	−0.0013***	0.0063***	−0.0020***	0.0087***	−0.0016***	−0.0188***	−0.0116***
Road distance	−0.0107***	0.0009***	0.0060***	0.0283***	0.0042***	−0.0959***	0.0136***
Water distance	0.0004***	0.0148***	−0.0068***	−0.1148***	−0.0494***	−0.00569***	−0.0104***
Constant	8.4923	49.0862	−36.9950	4.5769	−1.8902	−7.8353	−41.1186
Area under ROC	0.7530	0.8818	0.7144	0.9731	0.8519	0.8425	0.7844

Valuation of Ecosystem Services

In this study, we determined the potential impacts of land-use changes on the provisioning of ES under the three developed scenarios via a proxy-based ES valuation approach proposed by Costanza et al., (1997). Costanza et al. (1997) estimated the global economic value of 17 ecosystem services per unit area provided by 16 main biomes (i.e., land cover such as croplands, forest, and wetlands). Afterwards, the estimates were revised (Costanza et al., 2014) based on a more extensive database of more than 300 case studies all over the world (De Groot et al., 2012). Many studies have applied both valuation coefficients (Costanza *et al.*, 2014; Costanza et al., 1997) to quantify the effects of LULC change on ecosystem services value in different regions of the world (Kreuter et al., 2001; Zhao et al., 2004; Wang et al., 2006; Li et al., 2007; Hu et al., 2008; Polasky et al., 2011; Zang et al., 2011; Li et al., 2014; Long et al., 2014; Song & Deng, 2017; Yi et al., 2017) through a proxy method that matches the land-use type to equivalent biomes. To attach value coefficients to the 8 LULC types in Nigeria (Figure 1), they were compared with Costanza (Costanza et al., 2014) classified biomes. Cropland was used as the nearest equivalent biome for cultivated land, tropical forest for forests, grass/rangelands for grasslands and shrublands (which were lumped together and hereafter referred to as savannahs), urban for built-up land and desert/tundra/ice and rock for bare land, and the value coefficients for each LULC type were transferred accordingly. The total ESV by adding up the values of all types of ecosystem services provided by per unit area of each land use:

$$ESV_t = \sum_{k=1}^{n} \left(A_i \times VC_i\right) \tag{5}$$

Where; ESV_t is the estimated total ESV (US$), A_i and VC_i are the area (ha) and ecosystem service value coefficients (US$/ha/year) of the *i*th type of LULC.

The percentage change in ESV during the monitoring period was estimated using the following equation:

$$ESV_{t,pc} = \frac{ESV_{t,end\ year} - ESV_{t,start\ year}}{ESV_{t,start\ year}} \times 100\% \tag{6}$$

In this expression, $ESV_{t,cr}$ is the percentage change rate of the total ESV from the start year to end year, $ESV_{t,star\ tyear}$ is the total ESV at the beginning of the monitoring period, $ESV_{t,end\ year}$ is the total ESV at the end of the monitoring period. In addition to quantifying the impact of LULC change on the total *ESV*, we also estimated the effect of such changes on individual ecosystem functions within the study area. The value of the different ecosystem functions was calculated using the following equation:

$$ESV_f = \sum_{k=1}^{n} \left(A_i \times VC_{fi}\right) \tag{7}$$

where; ESV_f is the calculated ecosystem service value of function 'f', A_i is the area (ha) of the LULC type '*i*'. VC_{fi} is the value coefficient of function 'f' (US$ ha-1 year-1) for LULC type '*i*'.

Predicted Land-Use/Land-Cover (LULC) Change From 2010 to 2030 Under Different Scenarios

The land-use data of the year 2010 was taken as the base year and LULC change pattern from 2010 to 2030 under three different scenarios were simulated with the DLS model. Cultivated land, forests, savannahs, wetland, water area, built-up land, and bare land were considered for the land-use change simulation analysis in Nigeria. The predicted area and the changing trend of each land-use type under the three scenarios are presented in Table 6. In 2010, the savannahs accounted for the lion's share (42.11%) of the total land area in Nigeria, followed by cultivated land (31.18%) and forests (22.08%), while the combined areas of wetland, water bodies, built-up land and bare land was only 4.73% of the country's total land area. The land-use change simulation results revealed the existence of competition among the seven land-use types under the three scenarios. Cultivated land and built-up land will keep increasing under all three scenarios. However, cultivated land under the sustainable economic growth scenario

Table 5. Equivalent ecosystem services value coefficients (Costanza et al., 2014) of the LULC types in Nigeria in US$/ha/year

Ecosystem Service Categories/Functions	Cultivated Land	Forests	Savannahs	Wetland	Water Bodies	Built-up Land	Bare Land
Provisioning							
Food production	2,323	200	1,192	952	106	0	0
Raw materials	219	84	54	416	0	0	0
Regulating							
Gas regulation	0	12	9	0	0	0	0
Climate regulation	411	2,044	40	200	0	905	0
Disturbance regulation	0	66	0	4,596	0	0	0
Water regulation	0	8	3	1,789	7,514	16	0
Water supply	400	27	60	959	1,808	0	0
Waste treatment	397	120	75	111,345	918	0	0
Supporting							
Erosion control	107	337	44	3,507	0	0	0
Soil formation	532	14	2	0	0	0	0
Nutrient cycling	0	3	0	577	0	0	0
Pollination	22	30	35	0	0	0	0
Biological control	33	11	31	303	0	0	0
Habitat/refugia	0	39	1,214	12,452	0	0	0
Genetic resources	1,042	1,517	1,214	243	0	0	0
Recreation and culture							
Recreation	82	867	26	2,199	2,166	5,740	0
Cultural	0	2	167	636	0	0	0
Total Ecosystem Value	**5,568**	**5,381**	**4,166**	**140,174**	**12,512**	**6,661**	**0**

foresees a steady and lower rate of expansion through strategies that will help improve agricultural productivity and consequently reduce the need for more of this land type. In this scenario, cultivated land will increase by only 3.28% of its cover that existed in 2010. This relative change is about 14.35% and 9.95% lesser compared to that of the baseline and forest protection priority scenarios, respectively. On the contrary, the sustainable economic growth scenario forecasted a higher expansion of built-up land compared to the other two scenarios and the increase is higher by about 13.44% under the baseline scenario and 15.07% under the forest protection priority scenario.

As regards forests, the baseline scenario presented a 14.51% decrease relative to 2010 but the sustainable economic growth scenario will have an optimum impact on the forests land-use type in Nigeria with a considerable gain in forest cover by 2030 by about 16077 km^2 (Table 6). This represents an increase of about 7.78% of its cover that existed in 2010 and a recovery of nearly half of the area that projected to decline under the baseline scenario. The forest cover gain under the sustainable economic growth scenario will be achieved mainly through a decreased pressure from cultivated land growth, protection of the remaining forests, and hypothesized plan of afforesting bare lands into forests.

In all the three scenarios, savannahs will decrease but at the highest rate under the forest protection priority scenario (Table 6). This is because, given the assumption of strict implementation of spatial policies to prevent the forests from converting to other land-use types, the pressure of cultivated land expansion on the savannahs will tend to increase more resulting into its higher rate of decrease in this scenario as against the baseline scenario. However, a decreased pressure from cultivated land expansion through measures that will increase agricultural productivity under the sustainable economic growth scenario will make savannahs decrease at a lower rate as compared to that foreseen the forest protection priority scenario. Table 6 also reveals that wetland and water areas will reduce under the baseline and forest protection priority scenarios, most likely due to the pressure of cultivated land growth. This decrease could also be due to climate change factors, as it was reported in other studies elsewhere (Savage (Savage et al., 2009). Comparison between the projections for the baseline and sustainable economic growth scenarios clearly demonstrates the importance of proper land-use planning and other regulations in protecting the natural lands and sustainable management of the ecosystems. For instance, under the sustainable economic growth scenario, the wetland and water area remain stable from 2010 to 2030, forest cover increased and cropland increased at a lower rate, but bare land and savannahs have larger declines than under baseline scenario.

Table 6. Predicted area and changing trend of land-uses under the different scenarios (in sq. km)

Scenario	Year	Cultivated Land	Forests	Savannah	Wetland	Water Bodies	Built-up Land	Bare Land
Baseline	2010	291719	206609	394061	15231	8317	8639	11147
	2030	343123	176623	374733	14134	6574	10522	10015
	Change by 2030 (%)	+17.62	−14.51	−4.90	−7.21	−20.95	+21.79	−10.15
Forest Protection Priority	2010	291719	206609	394061	15231	8317	8639	11147
	2030	330291	206609	357809	14095	6740	10381	9799
	Change by 2030 (%)	+13.22	0.00	−9.20	−7.46	−18.96	+20.16	−12.09
Sustainable Economic Growth	2010	291719	206609	394061	15231	8317	8639	11147
	2030	301274	222686	371153	15231	8317	11683	5380
	Change by 2030 (%)	+3.28	+7.78	−5.81	0.00	0.00	+35.23	−51.73

Ecosystem Services Values (ESV) and Their Changes Under Different Scenarios

The estimated total ESV of Nigeria by 2030 were about US$ 655.53, US$ 657.17 and US$ 673.86 billion under baseline, forest protection priority and sustainable economic growth scenarios, respectively (Table 7). Of this value, wetlands contributed the highest, making up 30.22%, 30.06% and 31.68%, respectively under baseline, forest protection priority and sustainable economic growth scenarios. Due to the predicted land-use/land-cover dynamics, the total ESV of Nigeria would decrease under the baseline and forest protection priority scenarios but increase under the sustainable economic growth scenario. The ESV loss under forest protection priority scenario was about US$ 10.27 billion, which is about 1.54% of the estimated 2010 total ecosystem service value. The total ecosystem service value was further reduced under the baseline scenario by an amount of US$ 11.91 billion, which is about 1.784% of the estimated 2010 total ecosystem service value and about 1.2 times higher than the loss estimated under forest protection priority scenario. The estimated loss of total ecosystem service value in these scenarios is an indicator of the potential scarcity of ecosystem services in Nigeria. On the contrary, the sustainable economic growth scenario produced gains in ecosystem services value, wherein the total ecosystem service value increased by about US$ 6.43 billion.

As per each LULC types, the estimated ecosystem services value and their changes differed under the considered scenarios. For instance, under the baseline scenario, the ecosystem service value of cultivated land and built-up land increased, and those of the other LULC types decreased. Cultivated land accounted for about US$ 191.02 billion (29.14%) of the total ecosystem service value, which is about US$ 28.59 billion higher than the estimated value for 2010. However, the ecosystem service value of built-up land showed a minimal increase by approximately US$ 1.25 million (Table 7) and its contribution to the estimated total ecosystem service value for the year 2030 is 1.07%. The respective contribution of forests and savannahs were about US$ 95.04 billion (14.5%) and US$ 156.11 billion (23.81) of the total ecosystem service value, while the losses were about US$ 16.14 and US$ 8.05 billion of the 2010 estimated value, respectively. The ecosystem services value of wetland showed an enormous decline by an amount of US$15.38 billion, whereas a reduction in the ecosystem services value of the water body was about US$ 2.18 billion.

Under the forest protection priority scenario, estimated ecosystem service value of savannahs, wetland and water bodies decreased to US$ 149.06 billion, US$ 197.58 billion, and US$ 8.43 billion from the estimated value of US$ 164.17 billion, US$ 213.50 billion and US$ 10.41 billion in 2010, respectively. On the other hand, the ecosystem service value of cultivated land increased from about US$ 162.43 billion to US$ 184.01 billion.

The loss and gain of ecosystem services from the savannahs and cultivated land ecosystems were more balanced under sustainable economic growth scenario while the improved situation of the forest ecosystem has also positively affected the changes of ecosystem service value. For instance, the ecosystem service value of forests improved by about US$ 8.65 billion, while the total ecosystem services value in Nigeria grew by approximately US$ 6.43 million (Table 7).

The study also examined the impact of the probable future LULC change on the individual ecosystem functions, and the results are presented in Table 8. The food production of the service category of provisioning, climate regulation and waste treatment of the service category of regulating, habitat/refugia and genetic resources of the service category of supporting were the top five service functions with the highest economic values, contributing about US$ 530.93, 530.63 and 544.21 billion under baseline, forest protection priority and sustainable economic growth scenarios, respectively (Table 8).

Of the 17 ecosystem functions, 12 will decline under the baseline scenario while 11 will decrease under the forest protection priority scenario by 2030 (Table 8). In these two scenarios, the value of the ecosystem function of water regulation will decline more rapidly than any other service function at a rate of 16.53% in the baseline and 15.06% under the forest protection priority scenario. The next four most affected service functions include gas regulation (−8.86%), disturbance regulation (−8.40%), nutrient cycling (−7.69%) and climate regulation (−6.94%) under the baseline scenario; and cultural (−8.93%), habitat/refugia (−8.60%), nutrient cycling (−6.97%) and disturbance regulation (−6.24%) under the forest protection priority scenario. However, the sustainable economic growth scenario showed significant improvement of changes with only six (6) service functions showing a decreasing trend by 2030. Among the ecosystem functions that showed a positive change in economic value in this scenario, recreation recorded the highest rate of increase (10.05%), followed by climate regulation (6.79%), soil formation (3.29%) and erosion control (3.16%).

Table 7. Ecosystem Service Value (ESV) by LULC type and the change (in US$ billion) in Nigeria under the different scenarios

		LULC Type							Total
		[a]CL	FS	SH	WL	WB	BU	BL	
Base Year 2010	ESV	162.43	111.18	164.17	213.50	10.41	5.75	0.00	667.44
Baseline Scenario	ESV	191.02	95.04	156.11	198.12	8.23	7.01	0.00	655.53
	% of Total	29.14	14.50	23.81	30.22	1.25	1.07	0.00	100
	ESV Change	28.59	−16.14	−8.05	−15.38	−2.18	1.25	0.00	−11.91
	ESV Change Rate By 2030								−1.784%
Forest Protection Priority Scenario	ESV	184.01	111.18	149.06	197.58	8.43	6.91	0.00	657.17
	% of Total	28.00	16.92	22.68	30.06	1.28	1.05	0.00	100
	ESV Change	21.58	0.00	−15.10	−15.93	−1.97	1.16	0.00	−10.27
	ESV Change Rate By 2030								−1.538%
Sustainable Economic Growth Scenario	ESV	167.72	119.83	154.62	213.51	10.41	7.78	0.00	673.86
	% of Total	24.89	17.78	22.95	31.68	1.54	1.15	0.00	24.89
	ESV Change	5.29	8.65	−9.54	0.00	0.00	2.03	0.00	6.43
	ESV Change Rate By 2030								0.967%

[a]CL, FS, SH, WL, WB, BU and BL refers to cultivated land, forests, savannah, wetland, water bodies, built-up land and bare land respectively.

CONCLUSION

Increasing human activities worldwide have significantly altered the natural ecosystems and consequently, the services they provide. This is no exception in Nigeria, where land-use/land-cover has undergone a series of dramatic changes over the years as evident from previous studies and substantiated by the current study. However, estimating the impact of such changes on a wide range of ecosystem services is seldom attempted. Thus, using the value transfer methodology, we evaluated changes in the value of ecosystem services in response to predicted land-use/land-cover dynamics in Nigeria.

According to the predicted future change in land-use/land-cover, the total ecosystem service value in Nigeria will decrease by 2030 under the baseline and forest protection priority scenarios by about 1.78% and 1.54% respectively. In the baseline scenario, 12 of the ecosystem service functions will decline while 11 will decrease under the forest protection priority scenario by 2030. The impact will be severe on water regulation (−16.53%), gas regulation (−8.86%), disturbance regulation (−8.40%), nutrient cycling (−7.69%) and climate regulation (−6.94%) in the baseline scenario, while the service functions of water regulation (−15.06%), cultural (−8.93%), habitat/refugia (−8.60%), nutrient cycling

Table 8. Estimated value of ecosystem functions (ESV_f) and the change under different scenarios in Nigeria using coefficients of Costanza et al. (2014) (in US$ billion)

Ecosystem Service Category/ Function								Change Under Different Scenarios						
	2010	Baseline		Forest Protection Priority		Sustainable Economic Growth		Baseline		Forest Protection Priority		Sustainable Economic Growth	
	ESV_f	ESV_f	%	ESV_f	%	ESV_f	%	ESV_f	%	ESV_f	%	ESV_f	%
Provisioning													
Food production	120.409	129.310	19.726	124.965	19.016	120.207	17.838	8.901	7.392	4.556	3.784	−0.202	−0.168
Raw materials	10.886	11.608	1.771	11.491	1.749	11.105	1.648	0.723	6.637	0.606	5.564	0.219	2.016
Regulating													
Gas regulation	0.603	0.549	0.084	0.570	0.087	0.601	0.089	−0.053	−8.858	−0.033	−5.414	−0.001	−0.220
Climate regulation	56.883	52.936	8.075	58.466	8.897	60.744	9.014	−3.948	−6.940	1.583	2.782	3.861	6.787
Disturbance regulation	8.364	7.662	1.169	7.842	1.193	8.470	1.257	−0.702	−8.397	−0.522	−6.245	0.106	1.269
Water regulation	9.271	7.739	1.181	7.875	1.198	9.282	1.377	−1.533	−16.530	−1.396	−15.059	0.011	0.117
Water supply	17.555	18.992	2.897	18.494	2.814	17.841	2.648	1.437	8.183	0.939	5.346	0.286	1.629
Waste treatment	187.374	176.528	26.929	175.842	26.757	187.773	27.865	−10.846	−5.789	−11.532	−6.155	0.398	0.213
Supporting													
Erosion control	17.160	16.229	2.476	17.016	2.589	17.702	2.627	−0.931	−5.426	−0.143	−0.836	0.543	3.162
Soil formation	15.888	18.573	2.833	17.942	2.730	16.411	2.435	2.686	16.904	2.054	12.931	0.523	3.294
Nutrient cycling	0.941	0.869	0.132	0.875	0.133	0.946	0.140	−0.072	−7.687	−0.066	−6.970	0.005	0.513
Pollination	2.641	2.596	0.396	2.599	0.396	2.630	0.390	−0.045	−1.691	−0.042	−1.576	−0.011	−0.418
Biological control	2.873	2.916	0.445	2.854	0.434	2.851	0.423	0.043	1.506	−0.019	−0.659	−0.022	−0.765
Habitat/refugia	67.611	63.781	9.730	61.795	9.403	64.893	9.630	−3.830	−5.665	−5.816	−8.602	−2.718	−4.021
Genetic resources	109.949	108.377	16.533	109.558	16.671	110.597	16.412	−1.572	−1.430	−0.391	−0.355	0.648	0.589
Recreation and culture													
Recreation	31.439	29.672	4.526	32.071	4.880	34.599	5.134	−1.767	−5.621	0.632	2.010	3.159	10.049
Cultural	7.591	7.192	1.097	6.913	1.052	7.212	1.070	−0.399	−5.251	−0.678	−8.928	-0.379	−4.997
Total	667.438	655.529	100	657.170	100	673.863	100	−11.909	−1.784	−10.268	−1.538	6.425	0.963

(−6.97%) and disturbance regulation (−6.24%) will decline more rapidly under the forest protection priority scenario. In contrast to the other two scenarios, the sustainable economic growth scenario will record a gain in the total ecosystem service value by about 0.97% and will mainly positively impact the service functions of recreation (+10.05%), climate regulation (+6.79%), soil formation (+3.29%) and erosion control (+3.16%). Improvement in ecosystem functions of climate regulation and erosion control is of particular importance in Nigeria, a country that is exceedingly vulnerable to climate change and incessantly impacted by natural disasters such as flooding.

The predicted decline in the total ecosystem services value by 2030 under the baseline and forest protection priority scenarios indicates that massive land changes from the natural ecosystems into agricultural uses will have a severe impact on the sustainable provisioning of ecosystem services in Nigeria. Therefore, we advocate that the sustainable economic growth scenario that includes policies on multiple land-use types and with a positive impact on ecosystem services value is the ideal scenario for the sustainability of the study area. This scenario can be a reference for policymakers to ensure a sustainable land-use management. The study recommends that strategies that will help improve agricultural productivity, which will minimize the opening up of new lands and conserve the natural ecosystems should be a priority in future land-use management in Nigeria.

REFERENCES

Abbas, I. I. (2009). An overview of land cover changes in Nigeria, 1975–2005. *Journal of Geography and Regional Planning*, *2*, 62–65.

Abiodun, B. J., Salami, A. T. & Tadross, M. (2011). *Climate change scenarios for Nigeria: understanding biophysical impacts.* A report published by Building Nigeria's Responses to Climate Change (BNRCC).

Abubakar, K. (2015). *Land use land cover changes and their implications in Nigeria using geospatial technique* (MSc. Dissertation). Ahmadu Bello University, Zaria, Nigeria.

AC-Chukwuocha, N. B. (2015). Analysis of structural changes in land use and land cover of Otamiri watershed in Owerri, south east Nigeria. *Journal of Environment and Earth Science*, *5*, 51–59.

Antwi-Agyakwa, K. T. (2014). *Assessing the Effect of Land Use-Land Cover Change on Weija Catchment* (MSc Thesis). Kwame Nkrumah University of Science and Technology (KNUST), Kumasi, Ghana.

Arowolo, A. O., Bhowmik, A. K., Qi, W., & Deng, X. (2017). Comparison of spatial interpolation techniques to generate high-resolution climate surfaces for Nigeria. *International Journal of Climatology*, *37*, 179–192. doi:10.1002/joc.4990

Arowolo, A. O., & Deng, X. (2018). Land-use/land-cover change and statistical modelling of cultivated land change drivers in Nigeria. *Regional Environmental Change*, *18*(1), 247–259. doi:10.100710113-017-1186-5

Arowolo, A. O., Deng, X., Olatunji, O. A., & Obayelu, A. E. (2018). Assessing changes in the value of ecosystem services in response to land-use/land-cover dynamics in Nigeria. *The Science of the Total Environment*, *636*, 597–609. doi:10.1016/j.scitotenv.2018.04.277 PMID:29723833

Ayanlade, A. (2012). Evaluating Environmental Change Impacts on Ecological Services in the Niger Delta of Nigeria. *Ife Research Publications in Geography*, *11*, 111–125.

Ayanlade, A., & Proske, U. (2016). Assessing wetland degradation and loss of ecosystem services in the Niger Delta, Nigeria. *Marine and Freshwater Research*, *67*, 828–836. doi:10.1071/MF15066

Azadi, H., Barati, A. A., Rađaani, P., Taheri, F., Gebrehiwot, K., Witlox, F., & Lebailly, P. (2017). Evolution of land use-change modeling: Routes of different schools of knowledge. *Landscape and Ecological Engineering*, *13*(2), 319–332. doi:10.100711355-016-0311-9

Baan, L., Alkemade, R., & Koellner, T. (2012). Land use impacts on biodiversity in LCA: A global approach. *The International Journal of Life Cycle Assessment*, *18*(6), 1216–1230. doi:10.100711367-012-0412-0

Bature, B. N. (2013). The G20 and the Nigerian Vision 20:2020. *IOSR Journal Of Humanities And Social Science*, *15*, 53-60.

Burnham, B. O. (1973). Markov intertemporal land use simulation model. *Journal of Agricultural and Applied Economics*, *5*(1), 253–258. doi:10.1017/S0081305200011110

C-GIDD. (2016). *GDP and income distribution data*. Available at https://www.cgidd.com/

Camacho Olmedo, M. T., Paegelow, M., Mas, J. F., & Escobar, F. (2018). Geomatic Approaches for Modeling Land Change Scenarios. An Introduction. In M. T. Camacho Olmedo, M. Paegelow, J.F., Mas, & F. Escobar (Eds.), Geomatic Approaches for Modeling Land Change Scenarios. Lecture Notes in Geoinformation and Cartography. Cham: Springer.

Chaudhuri, G., & Clarke, K. C. (2013). The SLEUTH Land Use Change Model: A Review. *The International Journal of Environmental Resources Research*, *1*, 88–104.

Chen, J., Chen, J., Liao, A., Cao, X., Chen, L., Chen, X., ... Mills, J. (2015). Global land cover mapping at 30 m resolution: A POK-based operational approach. *ISPRS Journal of Photogrammetry and Remote Sensing*, *103*, 7–27. doi:10.1016/j.isprsjprs.2014.09.002

CIESIN/ITOS. (2013). Global roads open access dataset, Version 1 (gROADSv1). Palisades, NY: NASA Socioeconomic Data and Applications Center (SEDAC).

Collard, S. J., & Zammit, C. (2006). Effects of land-use intensification on soil carbon and ecosystem services in Brigalow (Acacia harpophylla) landscapes of Southeast Queensland, Australia. *Agriculture, Ecosystems & Environment*, *117*(2-3), 185–194. doi:10.1016/j.agee.2006.04.004

Costanza, R., d'Arge, R., de Groot, R., Farber, S., Grasso, M., Hannon, B., ... van den Belt, M. (1997). The value of the world's ecosystem services and natural capital. *Nature*, *387*(6630), 253–260. doi:10.1038/387253a0

Costanza, R., de Groot, R., Sutton, P., van der Ploeg, S., Anderson, S. J., Kubiszewski, I., ... Turner, R. K. (2014). Changes in the global value of ecosystem services. *Global Environmental Change*, *26*, 152–158. doi:10.1016/j.gloenvcha.2014.04.002

Dadhich, P. N., & Hanaoka, S. (2010). Remote sensing, GIS and Markov's method for land use change detection and prediction of Jaipur district. *Journal of Genomics*, *4*, 9–15.

De Groot, R., Brander, L., van der Ploeg, S., Costanza, R., Bernard, F., Braat, L., . . . McVittie, A. (2012). Global estimates of the value of ecosystems and their services in monetary units. Ecosystem Services, 1, 50-61.

de Sherbinin, A. (2002). *Land-Use and Land-Cover Change, A CIESIN Thematic Guide*. Palisades, NY: Center for International Earth Science Information Network of Columbia University.

Deng, X., Jiang, Q. O., Zhan, J., He, S. & Lin, Y. (2010). Simulation on the dynamics of forest area changes in Northeast China. *J. Geogr. Sci., 20*, 495-509.

Deng, X., Su, H., & Zhan, J. (2008). Integration of Multiple Data Sources to Simulate the Dynamics of Land Systems. *Sensors (Basel)*, *8*(2), 620–634. doi:10.33908020620 PMID:27879726

DIVA-GIS. (2016). *Free spatial data*. Author.

Eigenbrod, F., Armsworth, P. R., Anderson, B. J., Heinemeyer, A., Gillings, S., Roy, D. B., ... Gaston, K. J. (2010). Error propagation associated with benefits transfer-based mapping of ecosystem services. *Biological Conservation*, *143*(11), 2487–2493. doi:10.1016/j.biocon.2010.06.015

Ejaro, S. P., & Abdullahi, U. (2013). Spatiotemporal analyses of land use and land cover changes in Suleja local government area, Niger state, Nigeria. *Journal of Environment and Earth Science*, *3*, 72–83.

Elmqvist, T., Zipperer, W., & Güneralp, B. (2016). Urbanization, habitat loss, biodiversity decline: solution pathways to break the cycle. In K. Seta, W. Solecki, & C. Griffith (Eds.), *Routledge handbook of urbanization and global environmental change*. London, UK: Routledge.

Eludoyin, O. M., Adelekan, I. O., Webster, R. E., & Eludoyin, A. O. (2013). Air temperature, relative humidity, climate regionalizationand thermal comfort of Nigeria. *International Journal of Climatology*, *34*(6), 2000–2018. doi:10.1002/joc.3817

Eneh, O. C. (2011). Nigeria's Vision 20:2020-Issues, Challenges and Implications for Development Management. *Asian Journal of Rural Development*, *1*(1), 21–40. doi:10.3923/ajrd.2011.21.40

Erle, E., & Pontius, R. (2007). Land-use and land-cover change. In C. J. Cleveland (Ed.), *Encyclopaedia of Earth*. Washington, DC: Environmental Information Coalition, National Council for Science and the Environment.

Fan, F., Wang, Y., & Wang, Z. (2008). Temporal and spatial change detecting (1998-2003) and predicting of land use and land cover in core corridor of Pearl River Delta (China) by using TM and ETM + images. *Environmental Monitoring and Assessment*, *137*(1-3), 127–147. doi:10.100710661-007-9734-y PMID:17564805

Fiquepron, J., Garcia, S., & Stenger, A. (2013). Land use impact on water quality: Valuing forest services in terms of the water supply sector. *Journal of Environmental Management*, *126*, 113–121. doi:10.1016/j.jenvman.2013.04.002 PMID:23681358

Foley, J., DeFries, R., Asner, G., Barford, C., Bonan, G., Carpenter, S., ... Snyder, P. (2005). Global consequences of land use. *Science*, *309*(5734), 570–574. doi:10.1126cience.1111772 PMID:16040698

Förster, J., Barkmann, J., Fricke, R., Hotes, S., Kleyer, M., Kobbe, S., ... Wittmer, H. (2015). Assessing ecosystem services for informing land-use decisions: A problem-oriented approach. *Ecology and Society*, *20*(3), 31. doi:10.5751/ES-07804-200331

Guan, D., Li, H., Inohae, T., Su, W., Nagaie, T., & Hokao, K. (2011). Modeling urban land use change by the integration of cellular automaton and Markov model. *Ecological Modelling*, *222*(20-22), 3761–3772. doi:10.1016/j.ecolmodel.2011.09.009

Halmy, M., Gessler, P., Hicke, J., & Salem, B. (2015). Land use/land cover change detection and prediction in the north-western coastal desert of Egypt using Markov-CA. *Applied Geography (Sevenoaks, England)*, *63*, 101–112. doi:10.1016/j.apgeog.2015.06.015

Hasan, S. S., Deng, X., Li, Z., & Chen, D. (2017). Projections of future land use in Bangladesh under the background of baseline, ecological protection and economic development. *Sustainability*, *9*(4), 505. doi:10.3390u9040505

Hengl, T., Heuvelink, G. B. M., Kempen, B., Leenaars, J. G. B., Walsh, M. G., Shepherd, K. D., ... Tondoh, J. E. (2015). *Mapping soil properties of Africa at 250 m resolution: random forests significantly improve current predictions*. Academic Press.

Hettig, E., Lay, J., & Sipangule, K. (2016). Drivers of Households' Land-Use Decisions: A Critical Review of Micro-Level Studies in Tropical Regions. *Land (Basel)*, *5*(4), 32. doi:10.3390/land5040032

Houet, T., Loveland, T., Hubert-Moy, L., Gaucherel, C., Napton, D., Barnes, C., & Sayler, K. (2010). Exploring subtle land use and land cover changes: A framework for future landscape studies. *Landscape Ecology*, *25*(2), 249–266. doi:10.100710980-009-9362-8

Hu, H., Liu, W., & Cao, M. (2008). Impact of land use and land cover changes on ecosystem services in Menglun, Xishuangbanna, Southwest China. *Environmental Monitoring and Assessment*, *146*(1-3), 147–156. doi:10.100710661-007-0067-7 PMID:18157650

Hu, Y., Zheng, Y., & Zheng, X. (2013). Simulation of land-use scenarios for Beijing using CLUE-S and Markov composite models. *Chinese Geographical Science*, *23*(1), 92–100. doi:10.100711769-013-0594-9

Hutchinson, M. F. (2006). *Anusplin version 4.36 user guide*. Canberra, Australia: Centre for Resource and Environmental Studies.

Iacono, M., Levinson, D., El-Geneidy, A., & Wasfi, R. (2015). A Markov chain model of land use change. *TeMA Journal of Land Use, Mobility and Environment*, *8*, 263–276.

Ileoje, N. P. (2001). *A New Geography of Nigeria* (New Revised Edition). Ibadan, Nigeria: Longman Publishers.

Islam, K., Jashimuddin, M., Nath, B., & Nath, T. (2018). Land use classification and change detection by using multi-temporal remotely sensed imagery: The case of Chunati wildlife sanctuary, Bangladesh. *The Egyptian Journal of Remote Sensing and Space Sciences*, *21*(1), 37–47. doi:10.1016/j.ejrs.2016.12.005

Jarvis, A., Reuter, H. I., Nelson, A., & Guevara, E. (2008). *Hole-filled seamless SRTM data V4. International Centre for Tropical Agriculture*. CIAT.

Jiang, Q., Ma, E., Zhan, J., & Shi, N. (2015). Seasonal and interannual variation in energy balance in the semiarid grassland area of China. *Advances in Meteorology, 2015*, 2015. doi:10.1155/2015/120620

Jibril, M. S., & Liman, H. M. (2014). Land use and land cover change detection in Ilorin, Nigeria using satellite remote sensing. *Journal of Natural Sciences Research, 4*, 123–129.

Ke, X., Ma, E., & Yuan, Y. (2014). Scenario simulation of the influence of land use change on the regional temperature in a rapidly urbanizing region: A case study in southern-Jiangsu, China. *Advances in Meteorology, 2014*, 2014. doi:10.1155/2014/159724

Keshtkar, H., & Voigt, W. (2016). A spatiotemporal analysis of landscape change using an integrated Markov chain and cellular autoamta models. *Modeling Earth Systems and Environment, §§§*, 2.

Kim, I., Le, Q., Park, S., Tenhumen, J., & Koellner, T. (2014). Driving Forces in Archetypical Land-Use Changes in a Mountainous Watershed in East Asia. *Land (Basel), 3*(3), 957–980. doi:10.3390/land3030957

Kreuter, U. P., Harris, H. G., Matlock, M. D., & Lacey, R. E. (2001). Change in ecosystem service values in the San Antonio area, Texas. *Ecological Economics, 39*(3), 333–346. doi:10.1016/S0921-8009(01)00250-6

Kumar, S., Radhakrishnan, N., & Mathew, S. (2014). Land use change modeling using a Markov model and remote sensing. *Geomatics, Natural Hazards & Risk, 5*(2), 145–156. doi:10.1080/19475705.2013.795502

Lambin, E., & Geist, H. (Eds.). (2006). *Land-use and land-cover change – local processes and global impacts*. Berlin: Springer-Verlag. doi:10.1007/3-540-32202-7

Lambin, E., Geist, H., & Lepers, E. (2003). Dynamics of land-use and land-cover change in tropical regions. *Annual Review of Environment and Resources, 28*(1), 205–241. doi:10.1146/annurev.energy.28.050302.105459

Li, F., Ye, Y. P., Song, B. W., Wang, R. S., & Tao, Y. (2014). Assessing the changes in land use and ecosystem services in Changzhou municipality, People's Republic of China, 1991–2006. *Ecological Indicators, 42*, 95–103. doi:10.1016/j.ecolind.2013.11.012

Li, R. Q., Dong, M., Cui, J. Y., Zhang, L. L., Cui, Q. G., & He, W. M. (2007). Quantification of the impact of land-use changes on ecosystem services: A case study in Pingbian County, China. *Environmental Monitoring and Assessment, 128*(1-3), 503–510. doi:10.100710661-006-9344-0 PMID:17278007

Li, Z., Deng, X., Wu, F., & Hasan, S. S. (2015). Scenario analysis for water resources in response to land use change in the middle and upper reaches of the Heihe River Basin. *Sustainability, 7*(3), 3086–3108. doi:10.3390u7033086

Long, H. L., Liu, Y. Q., Hou, X. G., Li, T. T., & Li, Y. R. (2014). Effects of land use transitions due to rapid urbanization on ecosystem services: Implications for urban planning in the new developing area of China. *Habitat International, 44*, 536–544. doi:10.1016/j.habitatint.2014.10.011

Ma, C., Zhang, G. Y., Zhang, X. C., Zhao, Y. J., & Li, H. Y. (2012). Application of Markov model in wetland change dynamics in Tianjin Coastal Area, China. *Procedia Environmental Sciences*, *13*, 252–262. doi:10.1016/j.proenv.2012.01.024

MEA. (2005). *Ecosystems and human well-being: Biodiversity synthesis*. Washington, DC: World Resources Institute, Island Press.

Meybeck, M., Green, P., & Vorosmarty, C. J. (2001). A new typology for mountains and other relief classes: An application to global continental water resources and population distribution. *Mountain Research and Development*, *21*(1), 34–45. doi:10.1659/0276-4741(2001)021[0034:ANTFMA]2.0.CO;2

Nahuelhual, L., Carmona, A., Aguayo, M., & Echeverria, C. (2014). Land use and ecosystem services provision: A case study of recreation and ecotourism opportunities in southern Chile. *Landscape Ecology*, *29*(2), 329–344. doi:10.100710980-013-9958-x

Najmuddin, O., Deng, X., & Siqi, J. (2017). Scenario analysis of land use change in Kabul River Basin–A river basin with rapid socio-economic changes in Afghanistan. *Physics and Chemistry of the Earth Parts A/B/C*, *101*, 121–136. doi:10.1016/j.pce.2017.06.002

NBS. (2016). *National population estimates*. National Bureau of Statistics, FCT.

NGCC. (2014). 30 meter global land cover dataset (GlobeLand30). http://www. globallandcover.com/GLC30Download/index.aspx, Accessed date: 18 February 2016.

Newbold, T., Hudson, L., Hill, S., Contu, S., Lysenko, I., Senior, R., ... Purvis, A. (2015). Global effects of land use on local terrestrial biodiversity. *Nature*, *520*(7545), 45–50. doi:10.1038/nature14324 PMID:25832402

NIMET. (2015). Abuja, Nigeria: Nigerian Meteorological Agency.

Njoku, J. D., Ebe, T. E., & Pat-Mbano, E. C. (2010). Detection and mapping of land use and land cover classes of a developing city in south-eastern region of Nigeria using multi-band digital remotely-sensed data. *African Research Review*, *4*(3), 315–330. doi:10.4314/afrrev.v4i3.60272

NRC. (2014). *Advancing Land Change Modeling: Opportunities and Research Requirements*. Washington, DC: The National Academic Press.

O'Sullivan, D., Evans, T., Manson, S., Metcalf, S., Ligmann-Zielinska, A., & Bone, C. (2016). Strategic directions for agent-based modeling: Avoiding the YAAWN syndrome. *Journal of Land Use Science*, *11*(2), 177–187. doi:10.1080/1747423X.2015.1030463 PMID:27158257

Oginni, A., & Adebamowo, M. (2013). An evaluation of the socio-cultural effects of climate change on vulnerable Africa: Making a case for urgent action towards adaptation in Nigeria. *Br. J. Arts Soc. Sci.*, *13*, 141–149.

Oyenuga, V. A. (1967). Agriculture in Nigeria. Rome, Italy: FAO.

Oyinloye, R. O., & Oloukoi, J. (2012). Spatio-temporal assessment and mapping of the landuse landcover dynamics in the central forest belt of southwestern Nigeria Research. *Journal of Environment and Earth Science*, *4*, 720–730.

Parsa, V. A., & Salehi, E. (2016). Spatio-temporal analysis and simulation pattern of land use/cover changes, case study: Naghadeh, Iran. *Journal of Urban Management*, *5*(2), 43–51. doi:10.1016/j.jum.2016.11.001

Pascual, U., Muradian, R., Brander, L., Gómez-Baggethun, E., Martín-López, B., Verma, M., ... Polasky, S. (2010). The Economics of Valuing Ecosystem Services and Biodiversity. In P. Kumar (Ed.), *The Economics of Ecosystems and Biodiversity: The Ecological and Economic Foundations*. London: Earthscan.

Peng, J., Wang, Y. L., Wu, J. S., Yue, J., Zhang, Y. A., & Li, W. F. (2006). Ecological effects associated with land-use change in China's southwest agricultural landscape. *International Journal of Sustainable Development and World Ecology*, *13*(4), 315–325. doi:10.1080/13504500609469683

Piyathamrongchai, K., & Batty, M. (2007). Integrating Cellular Automata and Regional Dynamics Using Gis. In E. Koomen, J. Stillwell, A. Bakema, & H. Scholten (Eds.), *Modelling Land-Use Change: Progress and Applications*. Dordrecht: Springer.

Plummer, M. L. (2009). Assessing benefit transfer for the valuation of ecosystem services. *Frontiers in Ecology and the Environment*, *7*(1), 38–45. doi:10.1890/080091

Polasky, S., Nelson, E., Pennington, D., & Johnson, K. A. (2011). The impact of land-use change on ecosystem services, biodiversity and returns to landowners: A case study in the State of Minnesota. *Environmental and Resource Economics*, *48*(2), 219–242. doi:10.100710640-010-9407-0

Portela, R., & Rademacher, I. (2001). A dynamic model of patterns of deforestation and their effect on the ability of the Brazilian Amazonia to provide ecosystem services. *Ecological Modelling*, *143*(1-2), 115–146. doi:10.1016/S0304-3800(01)00359-3

Ramankutty, N., Hertel, T., Lee, H. & Rose, S. (2007). *Global spatial data of 18 Agro-ecological Zones (AEZs)*. Academic Press.

Richardson, L., Loomis, J., Kroeger, T., & Casey, F. (2015). The role of benefit transfer in ecosystem service valuation. *Ecological Economics*, *115*, 51–58. doi:10.1016/j.ecolecon.2014.02.018

Rwanga, S. S., & Ndambuki, J. M. (2017). Accuracy Assessment of Land Use/Land Cover Classification Using Remote Sensing and GIS. *International Journal of Geosciences*, *8*(04), 611–622. doi:10.4236/ijg.2017.84033

Samie, A., Deng, X., Jia, S., & Chen, D. (2017). Scenario-based simulation on dynamics of land-use-land-cover change in Punjab Province, Pakistan. *Sustainability*, *9*(8), 1285. doi:10.3390u9081285

Saraux, A., Tobón, G. J., Benhamou, M., Devauchelle-Pensec, V., Dougados, M., Mariette, X., ... Schaeverbeke, T. (2013). Potential classification criteria for rheumatoid arthritis after two years: Results from a French multicenter cohort. *Arthritis Care and Research*, *65*(8), 1227–1234. doi:10.1002/acr.21982 PMID:23401491

Savage, M., Dougherty, B., Hamza, M., Butterfield, R., & Bharwani, S. (2009). Socioeconomic Impacts of Climate Change in Afghanistan. Department of International Development & Stockholm Environment Institute.

Schroter, D., Cramer, W., Leemans, R., Prentice, I. C., Araujo, M. B., Arnell, N. W., ... Zierl, B. (2005). Ecosystem service supply and vulnerability to global change in Europe. *Science*, *310*(5752), 1333–1337. doi:10.1126cience.1115233 PMID:16254151

Singh, S., Srivastava, P., Pandey, A., & Gautam, S. (2013). Integrated assessment of groundwater influenced by a confluence river system: Concurrence with remote sensing and geochemical modelling. *Water Resources Management*, *27*(12), 4291–4313. doi:10.100711269-013-0408-y

Sinha, P., & Kumar, L. (2013). Markov land cover change modeling using pairs of time-series satellite images. *Photogrammetric Engineering and Remote Sensing*, *79*(11), e1051. doi:10.14358/PERS.79.11.1037

Sleeter, B. M., Sohl, T. L., Loveland, T. R., Auch, R. F., Acevedo, W., Drummond, M. A., ... Stehman, S. V. (2013). Land-cover change in the conterminous United States from 1973 to 2000. *Global Environmental Change*, *23*(4), 733–748. doi:10.1016/j.gloenvcha.2013.03.006

Song, W., Deng, X., Yuan, Y., Wang, Z., & Li, Z. (2015). Impacts of land-use change on valued ecosystem service in rapidly urbanized North China Plain. *Ecological Modelling*, *318*, 245–253. doi:10.1016/j.ecolmodel.2015.01.029

Song, W., & Deng, X. Z. (2017). Land-use/land-cover change and ecosystem service provision in China. *The Science of the Total Environment*, *576*, 705–719. doi:10.1016/j.scitotenv.2016.07.078 PMID:27810757

United Nations. (2019). *World population prospects 2019*. New York: United Nations.

Uriarte, M., Yackulic, C., Lim, Y., & Arce-Nazario, J. (2011). Influence of land use on water quality in a tropical landscape: A multi-scale analysis. *Landscape Ecology*, *26*(8), 1151–1164. doi:10.100710980-011-9642-y PMID:26146455

van Schrojenstein Lantman, J., Verburg, P. H., Bregt, A., & Geertman, S. (2011). Core principles and concepts in land-use modelling: A literature review. In E. Koomen & J. Borsboom-Van Beurden (Eds.), Land-Use Modelling in Planning Practice. Springer Science+Business Media B.V. doi:10.1007/978-94-007-1822-7_3

Veldkamp, A., & Lambin, E. F. (2001). Editorial: Predicting land-use change. *Agriculture, Ecosystems & Environment*, *85*(1-3), 1–6. doi:10.1016/S0167-8809(01)00199-2

Verburg, P. H., Eickhout, B., & Meijl, H. V. (2008). A multi-scale, multi-model approach for analyzing the future dynamics of European land use. *The Annals of Regional Science*, *42*(1), 57–77. doi:10.100700168-007-0136-4

Verburg, P. H., Kok, K., Pontius, R. G., & Veldkamp, A. (2006). Modeling land-use and land-cover change. In E. F. Lambin & H. J. Geist (Eds.), *Land-use and land-cover change: local processes and global impacts*. Berlin: Springer. doi:10.1007/3-540-32202-7_5

Viera, A. J., & Garrett, J. M. (2005). Understanding interobserver agreement: The kappa statistic. *Family Medicine*, *37*, 360–363. PMID:15883903

Vliet, J. V. (2009). Assessing the Accuracy of Changesin Spatial Explicit Land Use Change Models. In *Proceedings of the 12th AGILE International Conference on Geographic Information Science, 2009 Germany*. Leibniz Universität Hannover.

Wang, Y., Li, X., Zhang, Q., Lia, J., & Zhou, X. (2018). Projections of future land use changes: Multiple scenarios-based impacts analysis on ecosystem services for Wuhan city, China. *Ecological Indicators, 94*, 430–445. doi:10.1016/j.ecolind.2018.06.047

Wang, Z., Zhang, B., Zhang, S., Li, X., Liu, D., Song, K., ... Duan, H. (2006). Changes of land use and of ecosystem service values in Sanjiang Plain, Northeast China. *Environmental Monitoring and Assessment, 112*(1-3), 69–91. doi:10.100710661-006-0312-5 PMID:16404535

World Bank. (1998). *The assessment of vegetation and land use changes in Nigeria between 1976/78 and 1993/95*. Washington, DC: World Bank.

Yi, H., Güneralp, B., Filippi, A. M., Kreuter, U. P., & Güneralp, İ. (2017). Impacts of land change on ecosystem services in the San Antonio River Basin, Texas, from 1984 to 2010. *Ecological Economics, 135*, 125–135. doi:10.1016/j.ecolecon.2016.11.019

Yuan, Y., Zhao, T., Wang, W., Chen, S., & Wu, F. (2013). Projection of the spatially explicit land use/cover changes in China, 2010–2100. *Advances in Meteorology, 2013*, 9. doi:10.1155/2013/908307

Zang, S. Y., Wu, C. S., Liu, H., & Na, X. D. (2011). Impact of urbanization on natural ecosystem service values: A comparative study. *Environmental Monitoring and Assessment, 179*(1-4), 575–588. doi:10.100710661-010-1764-1 PMID:22013594

Zhang, W., Chen, J., Liao, A., Han, G., Chen, X., Chen, L., ... Zhang, J. (2016). Geospatial knowledge-based verification and improvement of GlobeLand30. *Science China. Earth Sciences, 59*(9), 1709–1719. doi:10.100711430-016-5318-4

Zhang, Y., Zhao, L., Liu, J., Liu, Y., & Li, C. (2015). The impact of land cover change on ecosystem service values in urban agglomerations along the coast of the Bohai Rim, China. *Sustainability, 7*(8), 10365–10387. doi:10.3390u70810365

Zhao, B., Kreuter, U. L. B., Ma, Z., Chen, J., & Nakagoshi, N. (2004). An ecosystem service value assessment of land-use change on Chongming Island, China. *Land Use Policy, 21*(2), 139–148. doi:10.1016/j.landusepol.2003.10.003

Zhen, L., Deng, X., Wei, Y., Jiang, Q., Lin, Y., Helming, K., ... Hu, J. (2014). Future land use and food security scenarios for the Guyuan district of remote western China. *Iforest-biogeosciences and Forestry, 7*(6), 372–384. doi:10.3832/ifor1170-007

Chapter 7
Implications of Land Grabbing on Sustainable Food Systems

Agatha Osivweneta Ogbe
Federal University of Agriculture, Abeokuta, Nigeria

Sarah Edore Edewor
Federal University of Agriculture, Abeokuta, Nigeria

ABSTRACT

Land is one of the most valuable assets required for agricultural production. In Africa, smallholder agricultural producers are faced with a lot of challenges that have highly impacted on productivity and sustainable food systems. The global demand for agricultural land for food and bio-fuel production has increasingly led to the emergence of land grabbing after the 2007-08 food price crisis. The rural poor are the victim of land grabbing as they are faced with declining farmlands, low income generation, and loss of livelihood activities. These have affected the food security status of the rural poor as farmlands are taken from them. The proponents of land grabbing revealed that developing countries are expected to benefit from investments inflow on grabbed land, development of infrastructure, increased income generation, and job creation. They argue that investment in agriculture is necessary to stimulate agricultural production; however, this situation has brought negative effects as most investors failed to keep their end of the transaction on land acquisition deals.

INTRODUCTION

Land is one of the most valuable assets required for agricultural production. In Africa, smallholder agricultural producers are faced with a lot of challenges which highly impact their productivity and ultimately their ability to achieve food security. Among the Sustainable Development Goals proposed by the United Nations, Goal 2 is targeted at "Ending hunger, achieve food security and improved nutrition and promote sustainable agriculture". Achieving this goal involves the production and distribution/disposal (through sales) of food crops.

DOI: 10.4018/978-1-7998-2599-9.ch007

Smallholder farmers grow more that 70% of world's food (Elver, 2015) and have access to small farmland holdings of at most 2 hectares which could be through land acquisition (purchase, lease or inherited). Land acquisition refers to the process of securing a piece of land for public or private use. It could be legally or illegally through willful or forceful acquisition of other peoples' land (Yusuf et al., 2014).

Any purchase of large-scale land larger than 200 hectares or twice the median which is equivalent to ten times the size of an average small farm is referred to as land grabbing (Oxfam international, 2012). A land acquisition is considered as a land grab when at least one of these happens: there is a violation of human rights (right to local population food production, right to own land, or there is violation of the equal rights of women land ownership) (Lee, 2015); when contracts are unclear and transparent with binding commitments on employment and benefit-sharing; when informed consent are disregarded; or meaningful participation of local communities are ignored.

The World Bank estimated (World Bank, 2010) that in 2009, forty-five million hectares of farmland in developing countries have been subject of transactions or negotiations. These large-scale land purchase or acquisition as it is sometimes called accentuates the rapid increase in yield that they can produce and the additional employment they can provide. However, the benefits of this additional agricultural production are not often felt locally (Anseeuw et al., 2012a; D'Odorico & Rulli, 2013). This is because the loss of access to land can ultimately spell significant dietary, social, cultural and economic consequences for rural communities in the targeted areas (Borras et al., 2011; De Schutter, 2011). Loss of farm land to large scale acquisition can lead to partial or total loss of investment due to insufficient compensations by the individuals or organisations purchasing the land. It may ultimately lead to outmigration.

The world's poorest countries experience most land grab as these countries do not have formal land right and also do not have governments who are willing and able to advocate for the local populations at the expense of losing contract from these land grabbing actors. With land grabbing, local populations are excluded from the use of large parcels of land for production purposes from which household income is obtained through sales of produce from the land (Graham, 2010).

In most cases, local people are usually the first victims of the impacts of land grabbing as they are faced with at least one of the following consequences: smallholder farming decline (negative influence on agricultural jobs); affects women in particular (their ability to acquire land as a production resource); difficulties accessing land by domestic farmers as a result of rising land prices; deterioration of local population communities' or grabbed country's food security; contempt for ownership of and usage rights to land, which can lead to conflict and/or endanger vulnerable populations.

On this premise, this chapter seeks to determine the implication of land grabbing on sustainable food systems. It will specifically seek to determine if land grabbing is occurring, identify the drivers of land grabbing, the extent of land grabbing and the impact of land grabbing on food systems.

CONCEPT OF LAND GRABBING

Several terms are used interchangeably to define demand for large scale land acquisition globally. Terms such as large-scale agriculture investment, global land grabbing and international agriculture investment have often been used to define land grabbing. "Land grabbing is the control (whether through ownership, lease, concession, contracts, quotas, or general power) of larger than locally-typical amounts of land by any persons or entities (public or private, foreign or domestic) via any means (legal or illegal) for purposes of speculation, extraction, resource control or commodification at the expense of peasant farmers, agroecology, land stewardship, food sovereignty and human rights." (European Coordination Via Campesina, 2016).

McElvenney (2012) in his definition of this concept viewed land grabbing using time period as a key element behind land acquisition. He perceived land grabbing as a term that describes land purchase or a long lease of land by foreign and domestic investors typically characterised by a period of between 50-99 years. Similarly, German *et al.* (2011) and Cotula et al. (2009a) stated that global land grabbing is often linked with corporate interests or foreign governments. These entities operate single handed to secure agricultural land on a large scale to combat food and bio-fuel insecurities in their home countries. This was further reiterated by Merian research (2009) and Sindayigaya (2012) that land grabbing refers to loss of land by rural population as a result of land scale acquisition by foreign businesses for commercial agricultural production and exports.

Another definition by Kachingwe (2012) involves the size of the land acquired. Land grabbing is believed to be occurring if the area of land secured runs in excess of thousands of hectares. In most cases, the method of acquisition does not take into account the local land users entitlements either through mutual consultation, informed consent or adequate compensation for their land based livelihoods loss. Large scale land acquisition becomes a problem because the deals are often not transparent and they lead to displacement and loss of livelihood.

In land transactions, the actors are the governments and the state agencies. In most cases in developing countries, the government facilitate, promote and make decisions on land deals often times without the involvement of the local land users and holders who mostly end up been displaced. In order to promote land deals at times, the government utilise tax payer funds and public policies and provide some kinds of enticement to investors through tariff waivers and tax holidays (ActionAid, 2014).

Land deals are based on contracts which find its origin in national and international law. These contracts are binding and offer a level of protection to the parties purchasing the land. These contracts define the costs, risks and the benefits (Cotula et al., 2011). The recent surge in land deals can be attributed to the global food and oil crisis in 2007 and 2008 which lead to the acquisition of farmlands in developing countries by some food importing countries so as to guarantee their populations' food security (Grain, 2008). This act have been perpetrated by countries such as China, South Korea, Japan, Saudi Arabia and the Gulf States and the focus has been on low and lower middle income countries in Africa such as Tanzania, Sudan, Nigeria, Sierra Leone, Laos and Madagascar amongst others (Cotula et al., 2009a; Kaloustian et al., 2009).

Anseew et al. (2012) stated that of the 1217 commercial large-scale acquisition globally, 83.2 million hectares of the land deals are in developing countries. It is disheartening to note that little of the outputs from these lands go into the domestic markets. In some cases, the lands are left idle to appreciate in value and sold off at a profit or they are cultivated with crops for exports or with crops to be used as biofuels (Geary, 2012).

Simply put, land grabbing refers to the acquisition of poor rural farmers' agricultural land by governments, individual(s) and institution(s) through outright sale or lease agreements to the point where the interest of the original owners of the land are threatened. One of the striking features of land grabbing is that land deals are usually not made in the interest of the local landowners as the size of the land is well above the average land holding of local owners in the region.

Land Grabbing: Evidence from Literature

Land grabbing can take the form of legal or illegal, forceful or willing acquisition of land belonging to other people or group of persons for private or public use. Land grabbing directly result from the aftermath of the worldwide food price spike of 2007-08 which affected the prices of internationally traded staple foods (maize, rice, soy and wheat). The concern was that food insecure nations are appropriating land overseas with the goal of securing food supply and by so doing, displacing people in the host countries who are generally food insecure from their land themselves (Edelman et al., 2015). Land grabbing have been viewed since 2008 from various dimension: 'green grabbing', 'water grabbing', 'biofuels and biomass', 'financialisation of agriculture', 'seizing of land used by peasant', indigenous communities and pastoralists for industrial, mining and urbanisation projects (Ghosh, 2010; Borras et al., 2010; Mehta et al., 2012).

Land grabbing involved both domestic (members of the community or country) and foreign (these investors act in the interest of a foreign government) actors (C2A Notes, 2010). Deducing from this statement, land grabbing by foreigners is described as the act of acquiring portions of land by individuals who are not members of the community in which the land is grabbed. In most cases these lands are acquired through political and economic means especially in areas where the right of the people over the land resources are not usually protected (C2A Notes, 2010) from Table 1. Following Land Matrix reports, the total number of land deals throughout the world since 2000 is 85.5 million hectares of which 43.5 million was determined by contract (Rafiee & Stenberg, 2018), from which Africa accounts for 42% of global land deals showing that the Africa region is heavily targeted (Nolte et al., 2016).

The acquisition of developing countries farmlands by other countries seeking to ensure food supply for themselves has increased the pressure on natural resources and water scarcity among others. Land acquisition to some extent has the potential to inject investment into agriculture and rural poor areas in developing countries, however, concerns have been raised as to the impact of this to the rural poor who are at the verge of losing their access to use and control the land they depend on for their livelihood (von Braun & Meinzen-Dick, 2009). These land deals deny poor rural communities' access to their farmland and this impact on their income generation from Table 2. The major foreign actors involved in land grabbing are the rich capital countries (Gulf State) as well as the countries with large population (China, South Korea, India) that are faced with food security concerns and are seeking opportunities for food production overseas as a result of land abundance and lower production cost in developing countries. Large scale land acquisition does not just occur in Africa but also in Southeast Asia, Latin America and some powerful big countries in the world from Tables 3 to 5.

Nigeria is not left out from the occurrence of large scale land acquisition, as this is evidence from the presence of big investment companies operating in Nigeria. In Nigeria, some reported cases of large scale land acquisitions by companies include Wilmer PZ, Lafarge Africa and Real oil Plantation around Ekong Anaku (Ojo & Offiong, 2018), acquisition of land for the establishment of multi-national oil companies in Niger Delta region; development of commercial farms in Kwara State (Ariyo & Mortimore, 2011); establishment of Banana plantation at Uekan in Ogoni, Niger Delta region by Union De Iniciativa S.A. De C.V (a Mexican Agricultural Firm) and construction of ultra-modern sugar factory in Ododwu Ibaji in Kogi State (Aruwa, 2011).

Table 1. Number and magnitude of land deals in Africa

Country	Number of Deals	Magnitude (1000ha)	
		Min	Max
Ethiopia	26	2.892	3.524
Madagascar	24	2.745	
Sudan	20	3.171	4.899
Tanzania	15	1.717	11.000
Mali	13	2.417	2.419
Mozambique	10	10.305	
Uganda	7	1.874	1.904
DR Congo	6	11.048	
Nigeria	6	821	
Zambia	6	2.245	
Ghana	5	89	
Malawi	5	307	
Senegal	5	510	
Kenya	4	135	150
Liberia	4	421	
The Republic of Congo	3	10.240	
Angola	3	223	
Cameroon	3	30	
Egypt	3	54	
Zimbabwe	2	101	
Algeria	1	2	
Libya	1	35	40
Morocco	1	21	
Mauritania	1	15	
Namibia	1		
Niger	1		
Zanzibar	1		
Total	177	51.415	63.111

Source: (Friis & Reenberg, 2010)

Drivers and Forms of Land Grabbing

Over the years, land grabbing has been attributed to be perpetrated by different actors such as foreign nations, investors or corporate entities, domestic, state and non-state, natural and legal persons with the required capital to make large-scale investments in agricultural farmlands both domestically and abroad in the event to make up for deficiency faced in food supply or speculation on the price of cultivatable lands (Lorenzo et al., 2009; Deininger & Byerlee, 2011; Anseeuw et al., 2012b). The incident of 2007-

Table 2. Summary findings of land grabbed by countries

Country	Total Lost Income ($)	Total People Affected	% of Population
Nigeria	331,781,421	153,439	0.10
Sierra Leone	501,467,190	610,031	10.40
Liberia	225,161,293	478,476	11.98
Ghana	332,672,327	206,456	0.85
Cameroon	203,675,121	90,845	0.46
Morocco	926,336,692	201,836	0.63
Mozambique	2,443,013,473	2,710,813	11.59
Congo	13,127,064	4,136	0.10
DR Congo	105,572,483	319,605	0.48
Madagascar	158,298,340	165,997	0.80
Tanzania	305,055,452	215,955	0.48
Uganda	19,237,881	15,379	0.05
Ethiopia	809,980,299	785,701	0.95

Source: (Davis et al., 2014)

08 global food crisis sparked fear among net food importing countries due mainly to the security of food supplies of their country. In order to guarantee food security, net food importing countries invest in suitable agricultural lands outside it national borders. Apart from the fear sparked by the 2007-08 global food price crisis, arid countries as well as oil-rich countries in the Gulf States (Saudi Arabia and United Arab Emirates) invested in suitable agricultural land (farmlands) outsider their national border due to their scare land, poor climatic condition and poor soil (Haralambous et al., 2009). The food or agricultural produce produced by these countries or their agents on farmlands elsewhere is specifically meant for export or repatriation back to the home countries (Friis & Reenberg, 2010).

It is also important to note that the need to meet food demand by food importing nations during the 2007-08 food price crises was not the only driver of land grabbing, however, the desire by nations to attain a reliable energy supply and the production of bio-fuel has made some countries invest in large scale acquisition of land. In most cases, poor developing countries have been the destination of large land acquisition due to easy and cheap land access. The demand for biofuel is predominantly for the European and Western markets (Kosciejew, 2011), following the European Union legislation of member nations to use 20% of clean energy obtained mainly from bio-fuel by 2020 (Schaffnit, 2012) and the US domestic energy policy for the use of 36 billion gallons of renewable biofuel by 2022 (Anseeuw et al., 2012b; Arezki et al., 2011; Brüntrup, 2011; Cotula et al., 2009b; Deininger & Byerlee 2011; Lonza et al., 2011; Mittal, 2009). Achieving this target require the cultivation of large expanse of land for bio-fuel stock production. The major crops grown for the bio-fuel stock production are jatropha, palm oil, cassava and sugar cane. The biggest biofuel production investor is the United Kingdom followed by the United States, India, Norway and Germany respectively.

Institutional factor (especially government legislation) has also contributed to large land acquisition across the global. Some government policies initiated with respect to land acquisition, land transfer and land use have been misused and abused by some official which has led to some illegal land negotiations not favourable for ordinary citizens who are the traditional land owners.

Table 3. Table of media reports on overseas land investment to secure food supplies (government sector to government sector) 2006-09

Region	Target Country	Investors Country	Nature of Deal	Status of Deal	Date Announced/ Signed	Source	Article Date
Africa	Kenya	Qatar	40,000 ha leased for fruit and vegetable cultivation in exchange for funding US$2.3 billion port	Signed	Nov-08	https://www.nation.co.ke/News/-/1056/513528/-/view/printVersion/-/3wecp8z/-/index.html; www.arabianbusiness.com/543415?tmpl=print&page=	Jan-09
	Malawi	Djibouti	Unknown area of farmland leased	Signed	Apr-09	http://www.coastweek.com/3216-15.htm	Apr-09
	Mali	Libya	100,000 ha secured for rice	Signed	n.a	http://www.lemonde.fr/planete/article/2009/04/15/securite-alimentaire-2-5-au-mali-les-nouvelles-mises-enculture-beneficient-surtout-aux-investisseurslibyens_1180879_3244.html#ens_id=1178742	Apr-09
	Sudan	Egypt	Land secured to grow 2 million tons of wheat annually	Signed	n.a	https://economictimes.indiatimes.com/articleshow/msid-3092250,prtpage-1.cms	Jun-08
	Sudan	Saudi Arabia (Hail Agricultural Development Co)	9,200-10,117 ha leased for wheat, vegetables, and animal feed; 60% paid by Saudi government	Signed	Feb-09	http://articles.latimes.com/2008/sep/28/world/fgfood28?s=o&n=o&sessid=0e7c6eb6e53f67fceb588fa4fae5d361ef214fa3&uuid=14b1c8e3172d15ada60954a128f2a6d1d3afe0fc&pg=0&pgtp=article&eagi=&cat=finance+%26+insurance&pe_id=4321477&page_type=article&exci=2008%7C09%7C28%7Cworld%7Cfg-food28; https://www.reuters.com/article/marketsNews/idUSLK10422520090320?sp=true	Sep-08; Mar-09
	Sudan	Jordan	25,000ha secured for livestock and crops	Signed	n.a	https://www.gulftimes.com/site/topics/printArticle.asp?cu_no=2&item_no=240524&version=1&template_id=48&parent_id=28	Sep-08
Asia	Cambodia	Kuwait	Land leased for rice	Signed	Aug-08	https://uk.reuters.com/article/oilRpt/idUKBKK33108620080822; http://www.atimes.com/atimes/Southeast_Asia/JI26Ae01.html	Aug-08
	Philippines	Bahrain	10,000ha secured for agro-fishery	Signed	Feb-09	http://english.bna.bh/?ID=76356; http://www.manilastandardtoday.com/?page=business5_mar30_2009	Feb-09
	Pakistan	UAE (Abraaj Capital)	324,000ha purchased	Under implementation	n.a	http://www.ft.com/cms/s/c6536028-1f9b-11dd-9216-000077b07658,dwp_uuid=fc3334c0-2f7a-11da-8b51-00000e2511c8,print=yes.html; http://viewswire.eiu.com/index.asp?layout=VWArticleVW3&article_id=83461793&text=food%20outsourcing&rf=0	May-08
	Indonesia	Saudi Arabia (Bin Laden Group	500,000 ha secured in US$4.3 billion investment for rice; put on hold by Bin Laden Group	Discontinued	n.a	http://www.thejakartaglobe.com/business/article/11843.Html	Mar-09

Source: (von Braun & Meinzen-Dick, 2009) IFPRI Policy Brief 13

Table 4. Media reports on overseas land investment to secure food supplies (private sector to government sector) 2006-09

Region	Target Country	Investors Country	Nature of Deal	Status of Deal	Date Announced/ Signed	Source
Africa	Angola	Lonrho (UK)	25,000 ha leased for rice. Lonrho is negotiating for a further 125,000 ha in Mali and Malawi	Signed	n.a	http://www.ft.com/cms/s/0/63460024-e342-11dd-a5cf-0000779fd2ac.html
	Nigeria	Trans4mation Agric-tech Ltd (UK)	10,000 ha secured	Signed	n.a	http://greenbio.checkbiotech.org/news/firm_plans_rice_cassava_production
	Ethiopia	Unknown private investors (Saudi Arabia)	Land leased in exchange for US$100 million investment	Signed	Mar-09	https://af.reuters.com/article/investingNews/idAFJOE5350BS20090406?pageNumber=1&virtualBrandChannel=0
	Sudan	Jarch Capital (USA)	400,000 ha in Southern Sudan signed with local army commander	Signed	n.a	https://www.reuters.com/article/marketsNews/idUSLK10422520090320?sp=true
	Madagascar	Daewoo (South Korea)	1.3 million ha secured for maize	Discontinued	Nov-08	https://www.reuters.com/article/marketsNews/idUSLJ18138720090320?sp=true
	DR Congo	Agriculture South Africa (South Africa)	10 million ha offered to farmers' union	Unknown	n.a	https://af.reuters.com/article/investingNews/idAFJOE53E0GL20090415; http://www.witness.co.za/index.php?showcontent&global[_id]=21838

Source: (von Braun & Meinzen-Dick, 2009) IFPRI Policy Brief 13

Table 5. Table of media reports on overseas land investment to secure food supplies (private sector to private sector) 2006-09

Region	Target Country	Investors Country	Nature of Deal	Status of Deal	Date Announced/ Signed	Source
Africa	East Africa Agri-business (Ethiopia)	Dubai World Trading Company (UAE)	5,000 ha secured in joint venture for tea	Signed	Apr-09	http://www.ethiopianreview.com/news/2009/04/ethiopian-firm-signs-300-million-tea-deal-witth-dubaicompany/

Source: (von Braun & Meinzen-Dick, 2009) IFPRI Policy Brief 13

Land grabbing is of two main forms: Domestic land grab and International or Transnational land grab. The domestic land grab refers to all land deals that are perpetrated by local citizens, companies as well as national governments (Levien, 2011). The international or transnational land grab is perpetrated by foreign citizens, governments, institutions and corporation entities (Amanor, 2012).

Extent of Land Grabbing

The implications of land grabbing are felt in different sectors of host countries. It ranges from the effect on local communities where these lands are grabbed to their agricultural activities and the loss of livelihood. These implications could be positive or negative with local people usually ending as the first victims of land grabbing.

In most cases, employment creation have been the main reason for large land deals according to the investors (Cotula et al., 2009a; Fisseha, 2011; Richards, 2013). In some cases, the nature of the employment created has been associated with poor working conditions and low wage (Fisseha, 2011). Similarly, limited jobs would exist due to the highly mechanized nature of commercial agriculture (Kachika, 2010; Hilhorst et al., 2011) and due to the nature of the crops (maize and soya) often grown. This therefore becomes problematic as it affects the livelihood of those in the community who are highly dependent on the promise of employment.

Since commercial agriculture is highly capital intensive and dependent on skilled labour for operating machines, the rural dwellers are more deprived which results into increased poverty. An example of this case scenario occurred in Indonesia and Cambodia where migrants from surrounding areas were selected over the local people (Practvuthy, 2011). Theting and Brekke (2010) also reported about some Eastern African countries (Kenya, Tanzania and Mozambique) that large scale agricultural land grab investors failed to fulfill their promise of job creation to the host communities.

On the positive side (win-win deal) of land grabbing, it has been reported that large scale land acquisition has the potentials to increase food and agro-fuel production investments flow into rural communities of developing countries presenting growth opportunities for rural livelihood promotion (Ojo & Offiong, 2010; Twene, 2016). The investments have the possible ability of boosting the agricultural sector, stimulation of rural economies through processing industries development within the rural communities, creation of employment, low cost of production, increased returns for farmers (who are engaged in contract farming for these big investors), provision of infrastructure (roads, schools, health centres, water services) as well as livelihood diversification (Haralambous et al., 2009).

Infrastructure development becomes important because it can sustain daily activities such as crop production, quality of life as well as the economy of the rural areas (Halseth & Ryser, 2006). Access to basic infrastructure such as rural electrification, feeder roads and other transportation infrastructure, education (access to school), health facilities (primary health centers or clinics), storage and processing facilities and access to irrigation amongst others are lacking in some rural communities. It is believed that increase in agricultural investments will be beneficial to rural areas in developing countries (Liu et al., 2013) due to the fact that they often bring with them some form of infrastructural improvement.

In some case, these foreign companies or projects provides smallholder food producers with advanced technologies, access to markets, access to information and knowledge and access to financing and lower borrowing costs (Thorpe, 2013). Waterhouse et al. (2010) reported that large scale agricultural investments resulted in the development of some social and related infrastructure, for instance in Zambia, new buildings were constructed. Despite the positive outcomes associated with large scale agricultural land acquisition, the negative impact outweighs it (Polack et al., 2013) as investors fail in keeping to their end of the transaction on land acquisition deals.

The implications of land grabbing could further be summed up in two headings below:

Economic Effects of Large-Scale Land Acquisitions

It has been argued that large scale land acquisition will promote the livelihood of the community in which it is taking place. But according to varying views, this has not been the case as the negative impacts has led to the displacement of the agricultural land owners or users with a resultant loss of their incomes as a result of premature sales of products or total loss of crops. In Nigeria as it is in most African countries, the agriculture sector employs a large proportion of the employed.

With farming as the main occupation of most people employed in the agriculture sector, land thus becomes a crucial resource as an input for food production and for commercial purposes hence the dependence (Cotula et al., 2008). Most indigenous people rely on agricultural production and non-timber forest products for their household income (Practvuthy, 2011). This was why Gobena (2010) stated that any loss to their land holdings will have a resultant effect on their income which will negatively affect their welfare.

Social Effects of Large-Scale Land Acquisitions

The basic necessity of life is food and this needs to be satisfied before any other developmental issue (Kuwornu et al., 2013). A decline in food production will affect the food security of most net consumers of food and since most small holder farmers are net consumers of what they produce, any associated loss will result in food insecurity and welfare loss. Veldés et al. (2010) and Cruz (2010) reported that majority (over 80%) of the worlds' smallholder farmers are food insecure and largely dependent on land as their primary source of livelihoods.

Hence the reason why Guerena et al. (2014), stated that investments in agriculture becomes problematic when it affects smallholder agriculture's access to land. Large bodies of literature Wahl *et al.* (2009), Huggins (2012), and De Zoysa (2013) are concerned about farmers and the national food security as a result of large land acquisitions. The numbers of the poor and vulnerable in developing countries keep increasing globally as farmers face issues of climate change, price volatility, competing demands for land (Lundius, 2009). For all indications, Geary (2012) stated that about two third of large-scale agricultural land deals are mostly in countries suffering from a serious case of hunger and poverty who are still recipient of food aid from World Food Program.

Another social effect of land grabbing is social conflict. In Sub-Saharan Africa, land is becoming more controversial and a source of conflict traditionally, its access has been considered relatively democratic (Yamano & Deininger, 2005). Factors such as agricultural commercialization, pressure on population and urbanization have been attributed to causing an increase in land conflicts coupled with the current African land tenure systems which is incapable of resolving these conflicts (Cotula et al., 2004; van Donge, 1999).

Large scale acquisition of agricultural lands such as that of the Dominion Farms in Taraba state in Nigeria for commercial agriculture might likely result in or cause nonviolent or violent social conflicts in form of riots, protests, or small scale rebellion in response to the land grabbed (Meinzen-Dick & Markelova, 2009; Oakland Institute, 2013).

Land Grabbing and Sustainable Food System

Agricultural production by smallholder producer is the backbone of Africa continent. In Africa, access to landholdings and its resources is crucial to livelihood, as smallholder producers heavily depend on it for food production (Rafiee & Stenberg, 2018). Food is one of the essential needs of people and a basic human right. The world poorest and most food insecure people have three characteristics: they live in rural areas; rely on agricultural to survive, and do not all own the land cultivated. The rural areas of most developing countries have large number of small farmers and people who depend on agriculture for their livelihoods. The process of large-scale land acquisition due to land grabbing displaces these farmers from their livelihood activities which is food production.

Land is not only important for agricultural production but also a social and cultural identity for people. Most land deals in host countries are geared towards exporting produces to the investors' countries leaving the host countries to suffer from food insecurity (Ojo & Offiong, 2010) with small scale farmers worldwide depending on land access for their food security (Daniel & Mittal, 2009). Big investment on grabbed land induce food insecurity issues as high-quality land are diverted from local production (food and livestock grazing) which form the bases of rural livelihood income generation activities (Action Aid International, 2008). Not only that grabbed land means loss of livelihood to famers, it also implied that their land assets are transferred and transformed from small farm holdings to large industrial farm meant to produced mainly for international markets (GRAIN, 2008)

Large scale acquisition of agricultural land though may bring more capital, new technologies and best agricultural practices, it seriously affect food system as a large number of small farms will be submerged as the intimate link of farmers with food production system will be disturbed. Displacement of small scale farmers will accelerate the number of food insecure individual in developing countries. Land grabbing have rapidly expanded in Europe's agriculture especially in the eastern countries (Kay, 2016). This occurrence has an adverse effect on the livelihood of small scale farmers as well as other agricultural workers (Borras et al., 2013) by marginalization of many small holder farmers.

One of the aftermaths of large-scale acquisitions is that rural farm households are being displaced and this results to a decline in their standard of living in terms of loss of land holdings and livelihood. The size of land grabbed by investors (government or private individuals or groups) have an adverse effect on the welfare of household due to inadequate compensations received by households from these investors due to high cost of living, couple with the realization of the fact that households cultivating these landholding loss their livelihood and unable to get another employment opportunities (GRAIN, 2015; Adepoju et al., 2018). Following a report by Makutsa (2010) addressing the effect of land grabbing on livelihood in Kenya, indicated that Tana delta (home to many land grab cases) in Kenya will experience severe food deficit if proposed investment on grabbed land for agriculture take off in that region.

CONCLUSION

This chapter focuses on the implications of land grabbing on sustainable food system by identifying the drivers of large scale land acquisition, determining the extent of occurrence and access the impact of large scale land acquisition on food system. It was observed that there is great occurrence of large scale land acquisition and this is predominant in developing countries with Sub-Saharan African been heavily targeted. The argument from proponent of land grabbing has it that developing countries are expected to benefits from the inflow of investments on grabbed land. However, this situation has brought negative effects as most investors failed in keeping to their end of the transaction on land acquisition deals. With the foregoing, it is necessary to develop an appropriate reform or code of conduct or policy framework in land governance in developing countries to manage or regulate agricultural land grabbing and address situations leading to possible eviction of local farmers' from their land holdings. The reform or code of conduct will be expected to determine the extent and methods of land acquisition, land use as well as ownership arrangement which will serve as a roadmap for all land deals for a sustained food system. This will enhance the proper utilisation of land resource by grabbed countries for agricultural purposes and also contribute to household food production, consumption and income creation from final disposal of produces through sales by host countries.

RECOMMENDATIONS

Basing on the findings of this chapter, the following recommendations are prominent:

1. A well-defined policy framework for large scale land acquisition which is sufficiently free and inform the consent of affected parties (investors and rural land owners) in any land deal should be developed. This should state clearly the compensation package for local land owners to avoid conflict.
2. Sustainable strategies should be put in place to safe guard smallholder land ownership with titles which essentially will promote land management practices and avoid situation where there is conflict as to the control and use of the land.
3. There should be proactive large scale land acquisition strategies that protect the rights of smallholder farmers by ensuring that investors do not hold on to land they do not utilize for many years. This mechanism should be put in place to allocate investors only the land they can utilize within a reasonable span of time and to withdraw land from investors who do not comply with agreed land development plans.
4. Agricultural land investments strategies should carefully evaluate what they are doing, there should be carefully evaluation of the negative and positive effects to smallholder livelihoods and agricultural production before land is granted to investors. A strong watch dog instrument should be developed as this is fundamental in ensuring that investments by investors deliver the expected outcome to majority of smallholder farmers and nation at large.

REFERENCES

C2A Notes. (2010). *Land grabbing, a threat to food security*. Agriculture and food in question, Issue No. 3.

Action Aid International. (2008). *Food, Farmers and Fuel: Balancing Global Grain and Energy policies with Sustainable Land Use*. Durban, South Africa: Action Aid International Secretariat.

ActionAid. (2014). *The Great Land Heist: How the world is paving the way for corporate land grabs*. Retrieved from www.actionaid.org/publications/great-land-heist

Adepoju, A. O., Ewolor, S., & Obayelu, O. A. (2018). Welfare implications of Domestic land grabs among rural households in Delta State, Nigeria. *Rural Sustainability Research*, *39*(334), 19–25. doi:10.2478/plua-2018-0003

Amanor, S. K. (2012). Global Resource Grabs. Agribusiness Concentration and Smallholder; Two West African Case Studies. *The Journal of Peasant Studies*, *39*(3-4), 731–749. doi:10.1080/03066150.2012.676543

Anseeuw, W., Boche, M., Breu, T., Giger, M., Lay, J., & Messerli, P. (2012b). *Transnational Land Deals for Agriculture in the Global South: Analytical Report based on the Land Matrix Database*. Hamburg: The Land Matrix Partnership.

Anseeuw, W. L., Alden, W., Cotula, L., & Taylor, M. (2012a). *Land Rights and the Rush for Land: Findings of the Global Commercial Pressures on Land Research Project*. Rome: International Land Coalition.

Arezki, R., Deininger, K., & Selod, H. (2011). *What Drives the Global Land Rush?* World Bank Policy Research Working Paper, No.WPS5864.

Ariyo, J. A., & Mortimore, M. (2011). *Land Deals and Commercial Agriculture in Nigeria: The New Nigerian Farms in Shonga District, Kwara State.* Paper Presented at the International Conference on Global Land Grabbing Organised by the Land Deals Politics Initiatives (LDPI) in Collaboration with the *Journal of Peasant Studies* and hosted by the Future Agricultural Consortium at the Institute of Development Studies, University of Sussex, UK.

Aruwa, A. (2011). *Kogi State Governor Launches Bio-fuels Refinery in "Itobe"*. Graphic Newspaper.

Borras, S. M. Jr, Franco, J., & van der Ploeg, J. D. (2013). Land concentration, land grabbing and people's struggles in Europe: introduction to the collection of studies. In J. Franco & S. Borras Jr., (Eds.), *Land Concentration, Land Grabbing and People's Struggles in Europe* (pp. 6–30). Amsterdam: Transnational Institute, European Coordination Via Campesina.

Borras, S. M. Jr, Hall, R., Scoones, I., White, B., & Wolford, W. (2011). Towards a better understanding of global land grabbing: An editorial introduction. *The Journal of Peasant Studies*, *38*(2), 209–216. doi:10.1080/03066150.2011.559005

Borras, S. M. Jr, McMichael, P., & Scoones, I. (2010). the politics of biofuel, land and agrarian change: Editors' introduction. *The Journal of Peasant Studies*, *27*(4), 575–592. doi:10.1080/03066150.2010.512448 PMID:20873025

Brüntrup, M. (2011). *Detrimental Land Grabbing or Growth Poles? Determinants and Potential Development Effects of Foreign Direct Land Investments. Technikfolgenabschätzung – Theorie und Praxis*. Bonn: German Development Institute. doi:10.14512/tatup.20.2.8

Cotula, L., Dyer, N., & Vermeulen, S. (2008). Fuelling Exclusion? The Biofuels Boom and Poor People's Access to Land. FAO/IIED.

Cotula, L., Toulmin, C., & Hesse, C. (2004). *Land Tenure and Administration in Africa; Lessons of Experience and Emerging Issues* (p. 50). London: International Institute for Environmental and Development.

Cotula, L., & Vermeulen, S. (2009b). Deal or no Deal: The Outlook of Agricultural Land Investment in Africa. *International Affairs*, *85*(6), 1233–1247. doi:10.1111/j.1468-2346.2009.00859.x

Cotula, L., Vermeulen, S., Leonard, R., & Keeley, J. (2009a). *Land Grab or Development Opportunity?* London: Agricultural Investment and International Land Deals in Africa. IIED/FAO/IFAD.

D'Odorico, P., & Rulli, M. C. (2013). The fourth food revolution. *Nature Geoscience*, *6*(6), 417–418. doi:10.1038/ngeo1842

Daniel, S., & Mittal, A. (2009). *The great land grab. Rush for world's farmland threatens food security for the poor*. Oakland, CA: The Oakland Institute.

Davis, K. F., D'Odorico, P., & Rulli, M. C. (2014). Land grabbing: A preliminary quantification of economic impacts on rural livelihoods. *Population and Environment*, *36*(2), 36. doi:10.100711111-014-0215-2 PMID:25400309

De Schutter, O. (2009). *Large-scale land acquisitions and leases: A set of core principles and measures to address the human rights challenges*. Retrieved from http://www.oecd.org/site/swacmali2010/44031283.pdf

De Schutter, O. (2011). How not to think of land-grabbing: Three critiques of large-scale investments in farmland. *The Journal of Peasant Studies*, *38*(2), 249–279. doi:10.1080/03066150.2011.559008

De Zoysa, R. (2013). *The Implications of Large Scale Land Acquisition on Smallholder's Food Security*. Development Planning Unity. Working Paper No. 156, p. 34.

Deininger, K., & Byerlee, D. (2011). *Rising global interest in farmland: Can it yield sustainable and equitable benefits?* Washington, DC: World Bank. doi:10.1596/978-0-8213-8591-3

Edelman, M. (2013). Messy hectares: Questions about the epistemology of land grabbing data. *The Journal of Peasant Studies*, *40*(3), 485–501. doi:10.1080/03066150.2013.801340

Elver, H. (2015). *Special Rapporteur on the Right to Food.* Interim Rep. of the Special Rapporteuron the Right to Food, U.N. Doc. A/70/287. Retrieved from https://www.un.org/en/ga/search/

European Coordination Via Campesina (ECVC). (2013). *Land Concentration and Land Grabbing, and People's Struggles in Europe*. Via Campesina.

Fisseha, M. (2011). *Commercial Pressure on Land in Africa: A regional overview of opportunities challenges and Impacts*. International Land Coalition.

Friis, C., & Reenberg, A. (2010). *Land grab in Africa: Emerging land system drivers in a teleconnected world*. GLP Report No.1. GLP-IPO.

Geary, K. (2012). *Our Land Our Lives, Time out on the Global Land Rush*. Oxford, UK: OXFAM GB for Oxfam International.

German, L., Schoneveld, G., & Mwangi, E. (2011). *Contemporary Processes of Large Scale Land Acquisition by Investors; Case Studies from Sub-Saharan Africa*. Occasional Paper 68, p. 56. CIFOR, Bagor, Indonesia.

Gobena, J. M. (2010). *Effects of Large-scale Land Acquisition in Rural Ethiopia; The Case of Bako-Tibe Woreda, Uppsala* (Master's dissertation). Swedish University of Agricultural Science.

GRAIN. (2008). *Seized! The 2008 land grab for food and financial security*. GRAIN Briefing. Available at: https://www.grain.org/article/93-seized-the-2008-landgrab-for-food-and-financial-security

GRAIN. (2015). *Dominion Farm's land grab in Nigeria*. Available at: https://www.grain.org/article/entries/5126-dominionfarm-s-land-grab-in-nigeria

Guerena, A., & Burgos, S. (2014). *Smallholder at risk, monoculture expansion, land, food and livelihoods in Latin America*. Oxford, UK: Oxfam International.

Halseth, G., & Ryser, L. (2006). Trends in Service Delivery: Examples from rural and small town Canada, 1998 to 2005. *Journal of Rural and Community Development*, *14*, 69–90.

Haralambous, S., Liversage, H., & Romano, M. (2009). *The Growing Demand for Land – Risks and Opportunities for Smallholder Farmers*. Rome: IFAD. Available at: www.ifad.org/events/gc/32/roundtables/2.pdf

Hilhorst, T., & Zomers, A. (2011). *How Can Large Scale Transnational Land Acquisition Contribute to Inclusive and Sustainable Growth?* University of Utrecht.

Huggins, C. (2012). *Consolidating Land, Consolidating Control: What Future for Smallholder Farming in Rwanda's Green Revolution.* Carleton University.

Kachika, T. (2010). *Land Grabbing in Africa: A review of the impacts and the possible policy response*. Available at: http://www.oxfam.org.uk/resources/learning/landrights/downloads/land_grabbing_in_africa_impacts_&_policy_responses.pdf

Kachingwe, N. (2012). *From Under Their Feet: A Think Piece on the Gender Dimensions of Land Grabs in Africa*. Johannesburg, South Africa: Action Aid International.

Kaloustian, F. C., Jaffrey, L., & Newman, I. (2009). *Land Grabbing for Food and Fuel Outsourcing: a rising Threat to the Right to Food*. International Human Right Clinic and University of San Francisco School of Law.

Kay, S. (2016). *Land grabbing and land concentration in Europe*. A research brief. Transnational Institute for HOTL. Available at: www.tni.org/files/publication-down loads/landgrabbingeurope_a5-2.pdf

Kosciejew, M. (2011). *The New Scramble for Africa: Land-grabbing Exploitation Presented as Economic Development*. Retrieved from http://www.caaglop.com/robbenisland-blog/economy/the-new-scramble-for-africa-land-grabbing-exploitation-presented-as-economic-development/

Kuwornu, J. K. M., Suleyman, D. M., & Amegashie, D. P. K. (2013). Analysis of Food Security Status of Farming Households in the Forest Belt of the Central Region of Ghana. *Russian Journal of Agricultural and Socio-Economic Sciences*, *1*(13), 17. doi:10.18551/rjoas.2013-01.04

Levien, M. (2011). Special Economic Zones and Accumulation by Dispossession in India. *Journal of Agrarian Change*, *11*(4), 454–483. doi:10.1111/j.1471-0366.2011.00329.x

Liu, P., Karoma, S., Arias, P., & Hallam, D. (2013). *Trends and Impacts on Foreign Investments in developing country agriculture. Evidence from case studies*. Rome, Italy: FAO.

Lonza, L., Hass, H., Maas, H., Reid, A., & Rose, K. D. (2011). *EU Renewable Energy Targets in 2020: Analysis of Scenarios for Transport*. Joint Scientific and Technical Report. Retrieved from http://ies.jrc.ec.europa.eu/uploads/jec/JECBiofuels%20Report_2011_PRINT.pdf

Lorenzo, C., Vermeulen, S., Leonard, R., & Keeley, J. (2009). *Land grab or development opportunity? Agricultural Investment and International land deals in Africa*. London: IIED/FAO/IFAD.

Lundius, R. V. (2009). *Smallholder Agriculture and Food Security in the 21st Century*. IFAD.

Makutsa, P. (2010). *Land Grab in Kenya: Implications for Smallholder Farmers*. Nairobi, Kenya: Eastern Africa Farmers Federation.

McElvenney, L. (2012). *Can Africa Feed the World? Land Acquisition as a Solution to Global Food Security* (Master Thesis). LUCSUS Lund University Centre for Sustainability Studies.

Merian Research. (2009). *The Ventures of Land Grabbing. The Involvement of European Financial Companies in Large Scale Land Acquisition abroad*. European Union.

Mittal, A. (2009). *The Great Land Grab: Rush for World's Farmland Threatens Food Security For the Poor*. Oakland: The Oakland Institute.

Oakland Institute. (2011). *Understanding land investment deals in Tanzania. Country Report* (p. 59). Oakland: The Oakland Institute.

Ojo, G. U., & Offiong, R. A. (2018). Land grabbing: Implications of 1978 Land Use Act on rural livelihood sustainability in Cross River State, Nigeria. *International Journal of Development and Economic Sustainability*, *6*(5), 60–69.

Oxfam International. (2012). *Our Land, Our Lives: Time Out on the Global Land Rush*. Oxfam Briefing Note 5. Retrieved from https://www.oxfam.org/sites/www.oxfam.org/files/bn-land-livesfreeze-041012-en_1.pdf

Polack, E., Cotula, L., & Côte, M. (2013). *Accountability in Africa's Land Rush: What Role for Legal Empowerment?* London: IIED/IDRC.

Practvuthy, M. (2011). *Land Acquisition by Non-Local Actors and Consequences for Local Development: Impact of Economic Land Concessions on the Livelihoods of Indigenous Communities in Northeast Provinces of Cambodia*. Royal University of Phnon Penh, Mekong Institute of Cambodia.

Rafiee, V., & Stenberg, E. (2018). *Land grabbing and its implications on rural livelihoods in Ghana and Ethiopia: A comparative study. Södertörns högskola*. Institutionen för Naturvetenskap.

Richards, M. (2013). Social and environmental impacts of agricultural large scale land acquisitions in Africa-with a focus on West and Central Africa. Rights and Resources Initiative.

Schaffnit, C. (2012). *Foreign investment in farmland: Frankfurt*. Deutsche Bank, AG.

Sindayigaya, W. (2012). *Foreign investment in agriculture-land grabbing*. PELUM Network.

Theting, H., & Brekke, B. (2010. *Land investments or land grab: A critical view from Tanzania and Mozambique*. Available at: http://www.spireorg.no/files/spire/documents/Land_investment_or_land_grab08112010.pdf

Thorpe, J. (2013). *Sugar rush, land rights and the supply chains of the biggest food and beverage companies*. Oxford, UK: OXFAM GB for Oxfam International.

Twene, S. K. (2016). *Land grabbing and rural livelihood sustainability: Experiences from the Buidam construction* (A published M.Sc Thesis). Available at: http://ir.knust.edu.gh/bitstream/123456789/10495/1/SAMUEL%20TWENE%20%28FINAL%20THESIS%29.pdf

Van Donge, J. K., & Pherani, L. (1999). Law and order as a development issue: Land conflicts and the creation of social order in Southern Malawi. *The Journal of Development Studies*, *36*(2), 373–399. doi:10.1080/00220389908422620

Veldés, A., Foster, W., Gustavo, A., Carlo, A., Covarrubias, K., Benjamin, D., ... Zezza, A. (2010). *A profile of rural poor. Background Paper for IFAD Rural Poverty Report 2010*. Rome, Italy: IFAD.

von Braun, J., & Meinzen-Dick, R. (2009). *Land grabbing by foreign investors in developing countries: Risks and opportunities. IFPRI Policy Brief 13*. Washington, DC: International Food Policy Research Institute.

Wahl, N., Jamnadass, R., Baur, H., Munster, C., & Liyama, M. (2009). *Economic Viability of Jatropha Plantations in Northern Tanzania-Assessing Farmers Prospects via Cost Benefit Analysis*. ICRAF Working Paper, World Agro forestry Centre.

Waterhouse, R., Lauriciano, G., & Norfolk, S. (2010). *Social analysis of selected projects – Issues Note and Case Studies Large-Scale Land Acquisition for Agricultural Production Mozambique*. World Bank.

World Bank. (2010). *Rising Global Interest in Farmland: Can it Yield Sustainable and Equitable Benefits?* Washington, DC: World Bank.

Yamano, T., & Deininger, K. (2005). *Land Conflict in Kenya: Causes; Impacts and Resolutions*. FASID Discussion Paper 2005-12-002, Foundation for Advanced Studies on International Development National Graduate Institute for Policy Studies, The World Bank.

Yusuf, N., Olatunji, A. G., & Issah, M. (2014). Factors influencing foreign land acquisitions and sustainable development in Nigeria. In E. Osabuohien (Ed.), *In-country determinants and implications of foreign land acquisition* (pp. 214–236). IGI Global.

Section 4
Management of Agricultural Value Chain Development

Chapter 8
Perspective of Policy Interventions in Rice and Cassava Value Chain Among Women in Nigeria

Eleojo Grace Adejo
Kogi State University, Anyigba, Nigeria

Olusola Jamiu Saliu
Kogi State University, Anyigba, Nigeria

Patrick Emmanuel Adejo
Kogi State University, Anyigba, Nigeria

ABSTRACT

The study investigated trends in rice grain and cassava tuber value addition through processing. Among the staple foods in Nigeria, rice and cassava have gained special prominence and priority attention by the government in terms of their production and value addition. The result indicated that the rice and cassava value chain is affected by different policy regimes. It was also found that women in the north central of Nigeria participated actively in rice and cassava value addition with some challenges. It is recommended that women processors of these commodities should have access to productive resources that can help add value to these commodities, training women on improved value added technologies and innovations by both public and private organizations, and most importantly, making these innovations and technologies affordable, adoptable, and adaptable will go a long way to boost their value added on these commodities through processing.

INTRODUCTION

Nigeria is the world's largest cassava producer and Africa's largest rice importer until the recent embargo

DOI: 10.4018/978-1-7998-2599-9.ch008

on importation of rice by the Federal Government Nigeria which has triggered local rice production and consumption. The development of an improved value chain on these crops will result in investment on additional processing facilities so that marketable surpluses can be pushed to women processors and farmers to reduce post-harvest losses thereby increasing farm income. Value addition can help farmers to claim part of the unexplored profit going unclaimed in the manufacture of food, fiber and industrial or other product from agricultural produce (Kehinde & Aboaba, 2016). Therefore, value addition in the production and processing of rice or cassava implies all the activities, processes or strategies and distribution of rice which in one way or the other contribute to benefit/ utility maximization (Ugwu, Mgbakor & Chitor, 2014). Under the Agricultural Transformation Agenda (ATA), the Government of Nigeria has expressed its determination to end the era of food imports, particularly rice, and develop cassava and rice value chains to produce and add value to these selected products and create domestic and export markets for farmers (FAO, 2018). Recently, Agricultural Promotion Policy (APP) was put in place to building on success from the ATA and has the following four key areas; food security, import substitution, job creation, and economic diversification (Odunze, 2019). However, diversification and transformation of the economy of Nigeria can only be actualized by initiating a business environment right from the farm to the other stages of the value chain especially the area of processing raw commodities into preservable and consumable forms.

It is on this premise that the study focus on the analysis of value addition through the processing of rice and cassava by women farmers in the country. On this note, the following specific objectives abound:

1. Review the trends in rice and cassava production in three policy regimes in the country;
2. identify the indigenous and improved value addition technologies adopted by women farmers on the selected crops within these policy regimes;
3. identify constraints militating against value addition activities in Nigeria.

SIGNIFICANCE OF THE STUDY

The importance of women participation in the post-harvest management of crops cannot be over emphasized. Especially in recent times, when the concept of value addition in the agricultural sector of Nigerian economy has been given priority by both government and non-governmental organizations towards improving the income of the rural communities particularly among the women folk.

Nigeria has a highly diversified agro-ecological condition which makes it possible for the production of a wide range of agricultural products. Despite Nigeria's potential competitive advantage of favorable agro-ecological and natural conditions in several agricultural commodities (including roots and tubers, cereals and legumes, tree crops and livestock), past and present agricultural policies and programmes have not been able to adequately and significantly address the constraints faced by small-scale farmers (Onwualu, 2012). Small-scale farmers in Nigeria are still confronted with, among other problems, poor access to modern inputs and credits, poor agricultural infrastructure, inadequate access to markets, land and environmental degradation (Mgbenka & Mbah, 2016).

Agriculture is a sector with room for significant productivity improvements, especially through agro-industry and agro-processing, given that the African market has accounted for 50% of growth in processed food exports from African countries since 2000 (Spore, 2018).

Agricultural activities in Nigeria and most African nations are characterized by gender division of labour (Adejo et al., 2015). Studies have shown that 79% of rural women take agriculture as their primary occupation and about 70-80% of them are involved in processing of agricultural commodities (Onuekwusi, Odoemelam, & Kanu, 2017; Odurukwe, Matthews-Njoku & Ejiogu-Okereke, 2006). Suggestion: Despite their immense contribution to the agricultural sector, they are always victims of neglect with limited or no access to productive resources such as land, credit and extension services.

The problem of how to preserve, and add value to the harvested crops by farmers have been worrisome situation, and of great concern to past and present governments in Nigeria. According to Okoruwa *et al.* (2008), Nigeria is losing about 2.4billion tonnes of food yearly to poor post-harvest handling of crops. Thus, losses associated with these crops limit the potential income of the farmers, threaten food security and exacerbate conditions of poverty among rural households, whose income stream depends on the ability to store excess farm produce for a later date. The African Post-harvest Losses Information System (APHLIS) reported that physical grain losses (prior to processing) range from 10 to 20 percent (Rick et al., 2014). Farmers, therefore, grow what they can easily sell or store and new production technologies remain unused. It is a problem that needs to be addressed at both on-farm and industry levels. In order to help address the problem of small-scale agriculture towards development into a modern production sector, strengthening the post-harvest sector or system is essential. Value addition" to agricultural products refers to processes such as cleaning, drying, grating, grinding, sorting, grading, packing, processing, packaging, and branding. According to Boland (2009), value-added agriculture is the process of increasing the economic value and consumer appeal of an agricultural commodity.

Enhancing the value chain can improve the livelihood of smallholder farmers, ensure competitiveness in the global market, and ultimately contribute to economic growth. Nigerian government over some years took cognizance of this fact and has established programmes, agencies and projects with the mandates to carry out research on post-harvest management and technologies in order to add value to crops. Notable among these agencies include Nigerian Stored Products Research Institutes (NSPRI) established in 1954 to conduct research in all aspects of post-harvest handling of crops and their products, pesticide development, residue analysis and mycotoxin survey on food items in Nigeria as stated in Decree 5 of 1977. Raw Materials Research and Development Council (RMRDC), Abuja has a mandate to carry out research on post-harvest management of agricultural commodities and a way of reaching the farmers through extension (Adejo et al., 2016).

The paradigm shifts from the focus on "quantity" of agricultural commodity production to "quality" which has imperative effects on large-scale industrial agricultural production in developed countries is yet to take its place Nigeria, meanwhile agriculture remains the integral sector of the nation's economy. It is important for the identification of indigenous practices and for the formulation of sustainable value-added strategies relevant to local conditions. Unfortunately, indigenous practices have been dismissed by some researchers in the absence of sufficient work to demonstrate their potential. The focus of this study for value addition is anchored on the need to increase rural incomes, employment and investment opportunities among women through the processing of Rice and Cassava. Nigeria is the continent's leading consumer of rice, one of the largest producers of rice in Africa and simultaneously one of the largest rice importers in the world. As well as an important food security crop, it is an essential cash crop for it is mainly small-scale producers who commonly sell 80 percent of total production and consume only 20 percent (Brandspurng, 2018). Rice generates more income for Nigerian farmers than any other cash crop in the country.

A range of policies and initiatives to strengthen cassava and rice value chains, from production to marketing are being put in place. However, because of the country's massive size and diversity, different regions may face different constraints because of a decentralized approach to designing industrial policies and initiatives that may not be in sync with the agricultural policies. Similarly, Worldstage (2018), online magazine reported that in a bid to eliminate supply bottlenecks in rice and cassava value chains the Federal Government have announced a partnership with the Nigeria Incentive-Based Risk Sharing System for Agricultural Lending (NIRSAL) and United Nations Development Programme (UNDP). It is also reported that the project which would be implemented through a public-private partnership arrangement had small holder farmers, youths and women as its target beneficiaries mostly in the North-central and South-west of the country.

METHODOLOGY OF THE STUDY

The study area is Nigeria. This study mainly used secondary data. Data were collected systematically from the various sources across the different agro- ecological zones of the country and subjected to content analysis based on the review of the subject matter. Data for the study were sourced from FAOSTAT (2018) and USDA (2018). Data collected were analysed using descriptive statistics such as Percentage, Bar Charts and line graph Trends analysis was done to examine the of rice and cassava production yield from 1960 – 2017. The linear model was hence used to predict the production for year 2030 i.e 13 years into the future. The intercept as well as the slope was estimated and the two parameters used to forecast production for 2030 for both produce.

RESULTS AND DISCUSSION

Analysis of Value Addition Through Processing of Rice Grains and Cassava Tubers in Three Policy Regimes

Trend in Rice and Cassava Production in Nigeria From 1960 - 2018

The Central Bank of Nigeria (CBN) has stated that the production of rice locally has increased to 70per cent following the introduction of the Anchor Borrowers Programme by the federal government (CBN, 2017). This is not unconnected with the ban on importation of rice into the country by the current government. Indeed a lot of states in Nigeria produce rice. Nigeria is the largest producer of rice in Africa as a whole. There is hardly a state in Nigeria where you don't find rice farmer cultivating and producing rice.

That is to say that rice is produced in all the six agro ecological zones of the country namely; North West, North East, North Central, South West, South East and South South agro ecological zones. It is important to understand that rice is produced in clusters in Nigeria. There are more than 170 rice clusters in Nigeria and 2,820 sub-clusters, all producing more than 15 million metric tonnes of rice in the country. However, there are states in the country where rice is dominantly produced. Therefore, states in Nigeria with an appreciable level of rice production are; Kebbi, Benue, Ebonyi, Ekiti, The federal capital territory, Jigawa, Kaduna, Kano, Kastina, Bauchi. Others are Nasarawa, Taraba, Kogi, Zamfara, Ogun, Niger, Kwara and Sokoto (Olawale, 2018). Oladimeji (2017) reported that local rice production in Nigeria varied from a maximum of 2,678,900 thousand metric tons (2010-2015) to a minimum of 268,840 thousand metric tonnes (1960-1969) with a standard deviation of 955, 045 thousand metric tonnes. Both rice imported and rice demanded increased from average of 39,350 thousand and 300,569 thousand metric tonnes in 1960-1969 to 2,678,900 thousand and 3,399,000 metric tonnes in 2010-2015 respectively. This report is in line with an update of milled rice in Nigeria as reported by United State Department of Agriculture (USDA, 2018).

Rice is cultivated in virtually all the agro-ecological zones in Nigeria, therefore successful cultivation of rice starts with choice of right rice variety suitable for the site. Because fields differ in their soil quality, the risk of flooding, or the risk of drought, a suitable variety must be selected for each field. Using suitable varieties minimizes the risk of crop loss or failure and ensures good yields. A suitable variety should give good yields, taste good, attract high market price, and many things more. Although the expected Local Rice Production (LRP) fluctuates both upward and downward from 1960 to 1980, Estimated Demand (ESD) rose sporadically and linearly since the 1980s to date. This connotes that rice demanded increases at increasing rate in the period 1960 to 2015 which was in consonance with the studies of Ayanwale and Amusan, (2012), FAO, (2013), Oladimeji and Ajoa, (2014) that affirmed increase in rice demand in Nigeria between 7.3% and 10.3%. According to Ewepu (2018), with efforts to boost food security by the Federal Government through value addition on rice, the nation's rice sub-sector will attract N250 billion investments based on high paddy rice production and increased milling capacity. It was also reported that there are presently 21 large integrated rice mills with a total processing capacity of 1.22 million metric tonnes annually, which spread across the country including Kano, Enugu, Ebonyi, Kebbi, Anambra, Edo, Nasarawa, Benue, Kwara, Jigawa, Niger and Kogi States, which the integrated rice mills have also employed about 2 million unskilled workers.

Moreover, the country is the largest producer of cassava in the world, with about 50 million metric tons annually from a cultivated area of about 3.7 million ha. Nigeria accounts for cassava production of up to 20 per cent of the world, about 34 per cent of Africa's and about 46 per cent of West Africa's. The national average yield of cassava is estimated at about 13.63 MT per ha, as against potential yield of up to 40 metric tons per ha. Close to two-thirds (66 per cent) of total production is in the southern part of the country, while about 30 per cent is in the north-central, and 4 per cent in other parts of the north (FAO, 2018). The crop is predominantly grown by smallholders on small plots for family consumption and local sale. Large scale commercial plantations are rare. According to Cadena (2017), women oversee 70% of the activities of cassava production and processing. Rice and Cassava Production Trend can be categorized under 3 subheadings based on policy implementation/ variations and inconsistencies in different period of agricultural development which have effects on the outputs of some agricultural commodities. Thus, they include:

Rice Production at Independence / First Decade After Independence Era (1960 – 1969)

The independence/first decade after independence era as reflected in Figure 1, covers the period 1960-1969. This period was characterized by the carryover of colonial policies and ideologies in terms of more attention giving to some selected cash crops (namely cocoa, coffee, ruber, oil palm, and groundnut) at the expense of food crops (rice, cassava, maize, sorghum, yam, etc) production. Rice production at this era was basically on small scale bases. Rice production was slow compare to its demand by the increasing population. Oladimeji (2017) attested to this fact that variation in rice imported had a rapid increase compared with local rice production and estimated demand that could be assumed to have a slow rise.

Figure 1. Rice production at independence / first decade after Independence Era (1960 – 1969)

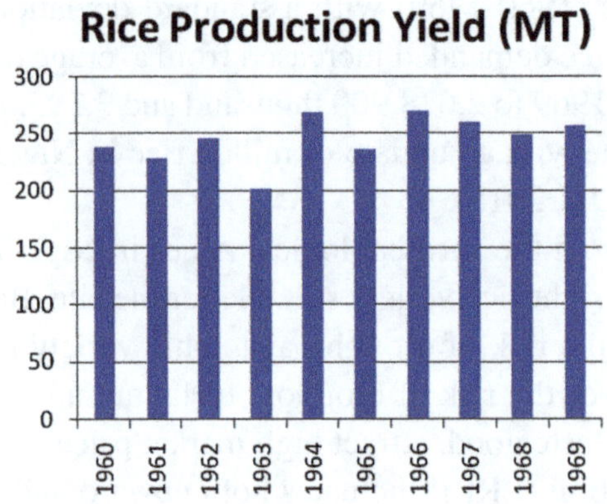

Cassava Production at Independence / First Decade After Independence Era (1960 – 1969)

At this period, the carryover of colonial stigmatization of cassava still manifested. British colonial policies forced indigenous farmers to plant cassava as a famine reserve measure and subsidized maize grown by settler farmers Cassava is a marginalized crop in food policy debates because it is burdened with the stigma of being an inferior, low-protein food that is uncompetitive with the glamour crops such as imported rice and wheat. Many food policy analysts consider cassava an inferior food because it is assumed that its per capita consumption will decline with increasing per capita incomes.

However, cassava production trend in this period witnessed some level of increments, but at an arithmetic rate compared to the geometric population rise. Cassava production rose from 6,980 MT in 1960 to 9040 MT in 1969 as indicated in Figure 2.

Figure 2. Cassava Production at independence / first decade after Independence Era (1960 – 1969)

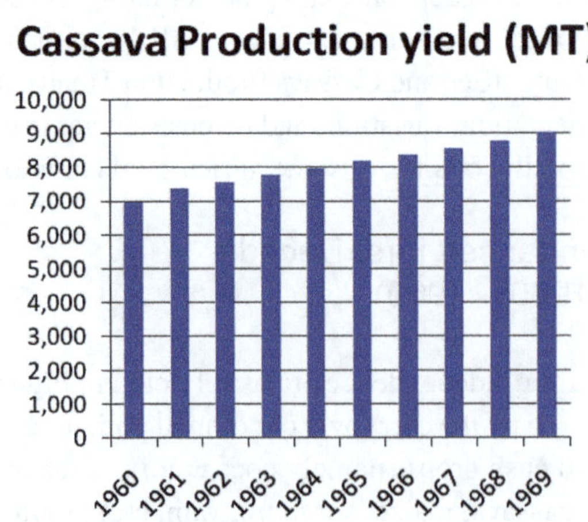

Rice Production in the Oil-Boom/Policy Reconstruction Era (1970 – 1985)

This era was a period when most agricultural policies were put in place by Nigerian governments. However, from 1970 to 1985, most governments in Nigeria have decided as a matter of policy to promote and reinforce research and development in rice and cassava improvement and production. Notably among them, were the establishment of research Institutes and programmes such as: National Cereals Reseaarch Institute (NCRI), Nigerian Stored Products Research Institute (NSPRI), Green Revolution (GR), Operation Feed the Nation (OFN), Agricultural Development Projects (ADPs), National Accelerated Food Production Programm (NAFPP), River Basin Development Authority (RBDA) among others (Jibowo & Ajayi, 2011; Eke-Okoro & Njoku, 2012). This era was notably deviations from colonial commodity policies.

It was apparently the strategies designed to boost agricultural production, but the oil-boom that had become the "oil-doom" for agriculture did not make it possible for the complete realization of the implementations of these policies and their subsequent manifestation in the development of agricultural sector. It was reported by FAO (2017) that cash crops, which earned significant revenue before the oil boom of the 1970s, have experienced low investment. Prioritizing the oil sector at the expense of agriculture resulted in leaving Nigeria highly vulnerable to fluctuating oil prices on the world market. Concerning trade, although Nigeriais one of the largest rice producers in Africa, it is also one of the largest rice importers.

Nigerian farmers may have responded to the increased demand for rice. Despite this, Nigerian rice production remains insufficient to fulfil demand as seen in Figure 3. This is partly because Nigerian production yields have barely changed since the 1970s especially compared to yields in South East Asia. From 1970, the Badeggi Research Station started making effort in varietal development by creating its own variation for selection and combined this effort with continued introduction of materials from other parts of the world. Similarly, the era between 1970s and early 1980s marked the evolution of semi -dwarf plant types with much higher grain yield and suitability to irrigated and shallow rain fed ecologies in Asia spear headed by International Rice Research Institute (IRRI) in Philippines (Biyi, 2005). FAROs 14 to 17 and 25 to 30 were materials developed internally through hybridization of exotic parent materials selection from segregating population of such crosses. FAROs 18 to 24 were introduction from IRRI. In 1986, the rice research programme of International Institute of Tropical Agriculture (IITA) started to contribute to the varietal selection and development in Nigerian. FAROs 35 to 37 were the pioneer varieties developed and released in Nigeria by IITA. Other varieties released in the 80s and early 90s were as a result of international rice germplasm exchange program than called international Rice Testing Program (IRTP). Also, was the activities of West Africa Rice Development Association WARDA now called Africa Rice Center (AfricaRice) with the national research institutions (NAREs), leading to the formation of Task Forces in different areas of rice research from breeding to natural resources management, and economic studies groups.

Cassava Production in the Oil-Boom/Policy Reconstruction Era (1970 – 1985)

The National Root and Tuber Crops Research Institute (NRCRI) umudike and IITA, Ibadan where the Institutions with the mandate on cassava production and value addition. According to Eke-Okoro and Njoku (2012), at this period, the initiation of research collaboration between national, regional and international research Institutions tremendously contributed to cassava development in Nigeria. Following the devastating effect of cassava bacterial blight in 1972 and the current devastating effect of new strain of mosaic virus ravaging cassava in the continent (Africa), which was detected by IITA, Ibadan, the Federal Government of Nigeria decided to reinforce research collaboration between IITA and NRCRI through policy. Hence, the development and spread of early-maturing, pest tolerant and high yielding cassava varieties have been promoted partly by demand for cassava products such as gari, fufu etc.

Rice Production and Value Addition in the Policy Stabilization Era (1986 – 2017)

This era was characterized by various programmes and strategies aimed at stabilizing already made catalogue of policies that were marred by inconsistencies and failure of implementations based on the target objectives of the policies. Despite this improved trend in rice production, it fails to meet the increasing demand for rice consumption in the country. Some of the reasons for this gap according to Biyi (2005) are connected with the improper production methods, scarcity and high cost of inputs, rudimentary post - harvest and processing methods, inefficient milling techniques and poor marketing standards particularly in terms of polishing and packaging. Also poor or low mechanization on rice farms means heavy reliance on manual labor to carry out all farm operations

Figure 3. Rice production in the Oil-Boom/Policy Reconstruction Era (1970 – 1985)

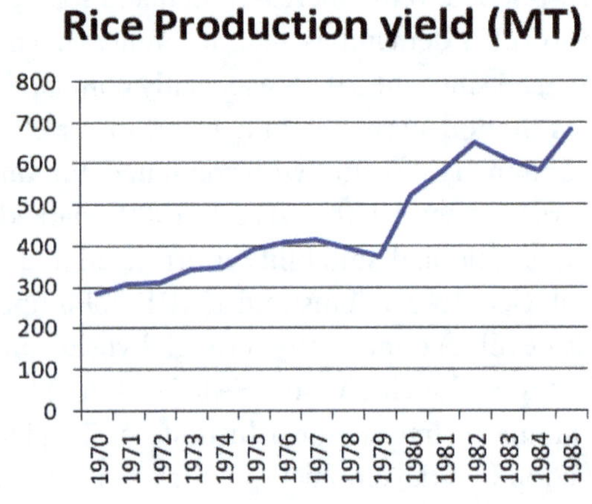

Figure 4. Cassava production in the Oil-Boom/Policy Reconstruction Era (1970 – 1985)

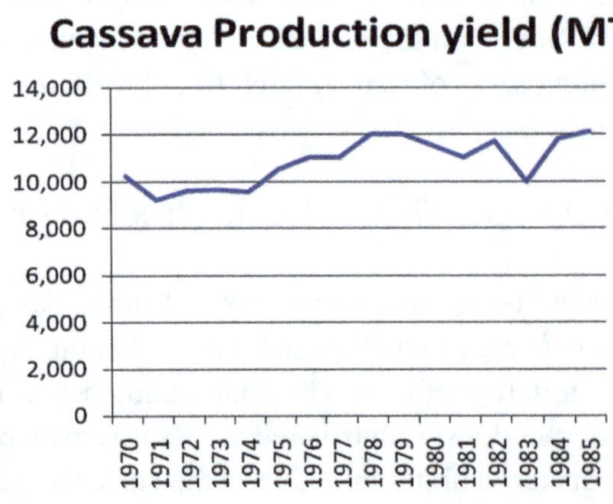

The earlier stabilizing policies initiated by successive governments in this era include: The Directorate of Food, Roads and Rural Infrastructure (DFRRI), National Fadama Development Projects (NFDP), Nigerian Agricultural Insurance (NAI), National Special Programme for Food Security (NSPFS). Others that were put in place as from 2004 till date include: National Agriculture and Food Security Strategy (NAFSS, 2010–2020), which is now embedded in the Agriculture Transformation Agenda (2013–2015). The agenda was launched in 2011 to diversify the economy and enhance foreign exchange earnings, with the objective of achieving a hunger-free Nigeria through an agricultural sector that drives equitable income growth and distribution, accelerates the achievement of food and nutrition security, generates decent employment and transforms Nigeria into a leading player in global food markets (FAO, 2017). National Agricultural Investment Plan (NAIP, 2011–2014), which seeks to enhance agro industrialization and employment. In 2016, the Government of Nigeria developed the Green Alternative: The Agriculture Promotion Policy (APP, 2016–2020). During the review period (2007–2017) Nigeria's agriculture sector underwent major reforms. The introduction of the Agricultural Transformation Agenda (ATA) reformed the input delivery system, strengthened famers' resilience to shocks and enhanced agricultural credit in order to boost agricultural production. Under the ATA, the Government of Nigeria expressed its determination to end the era of food imports, particularly rice, and develop cassava and rice value chains to produce and add value to these selected products and create domestic and export markets for farmers. A range of policies and initiatives to strengthen rice value chains, from production to marketing are being put in place. In terms of financial assistance to rice farmers to boost production locally, in 2006, Nigeria implemented the Agricultural Credit Support Scheme (ACSS) to enhance access to credit for small- and medium-scale farmers. In 2016, the government launched the Anchor Borrowers Programme (ABP), which is being managed by the Central Bank of Nigeria and provides farmers with financial assistance through bank loans. With these developments aimed at increasing rice farmers access to credits, The Central Bank of Nigeria (CBN) has stated that the production of rice locally has increased to 70per cent following the introduction of the Anchor Borrowers Programme by the federal government (Nzelu, 2017). This is one of the reasons for stability in local rice production and value addition on same couple with the ban in rice importation as seen between 2016 and 2018 in Table 1 with expectation for an improvement in the subsequent years.

Cassava Production and Value Addition in the Policy Stabilization Era (1986 – 2017)

In this period, the Collaborative Study of Cassava in Africa (COSCA) was prominently featured. The COSCA studies were carried out from 1989 to 1997 under the aegis of the IITA (International Institute of Tropical Agriculture) in Ibadan, Nigeria. The study focuses on cassava production systems, processing and food preparation methods, market prospects, and consumption patterns. From 1993 to 1997, COSCA researchers analyzed the field data and prepared a series of written reports on cassava production, processing, and consumption in the six COSCA study countries, culminating in early to mid-2002, 10 years after the original COSCA field studies, COSCA team conducted a follow-up survey of the COSCA farmers in Nigeria, financed by the IFPRI (International International Food Policy Research Institute). This research focused on the agronomic improvement and value chain of cassava. Another development in this era in Nigeria was that virtually all the processing steps in the production of *gari* were mechanized, thereby removing the drudgery of the most female rural cassava processors. The Federal Institute of Industrial Research (FIIRO), Lagos, successfully improved the technology, splitting the process into definite steps and optimizing each step with respect to time, product quality and unit cost. Notable areas of improvement include the use of a mechanical grater, mechanical or hydraulic press or centrifuge to replace the stone and milling to a uniform particle size. With the application of these modern operations, the quality of the product is adequately guaranteed

Several measures have been taken from colonial era to the present time to improve and extend the crop to Nigeria farmers and other end-users. NRCRI has contributed substantially to the development of cassava through the release of 33 improved varieties towards increased productivity in Nigeria (Eke-Okoro & Njoku, 2012). The presidential initiative on cassava in 2004 brought a turn-around in cassava area production and yield as many farmers started large-scale cassava production and value addition for instance, the establishment of the Cassava Bread Development Fund (CBDF) by the Nigerian government had further raised hopes for improving the cassava sector. This initiative led to the development and release of five new cassava varieties to check-mate the recent virulent mosaic virus strain that is ravaging cassava in Africa. This accounts for the change witnessed between 2004 – 2017 as reflected in Table 2.

APP has recognized the fact that Nigeria is one of the biggest producer of a lot of commodities like cassava, in the region and at the same time, the biggest importers of products such as ethanol, starch and other products from cassava and this is a direct effect of the lack of good processing and storage systems throughout the country (Odunze, 2019). The APP recognizes this, and programs and reforms are aimed at not only improving access to storage and processing facilities, but also improving the quality of cassava products. Adequate Storage is needed if more value chain players are to become entrepreneurial in the sector.

Rice and Cassava Production Yield Projection for 2030

In order to forecast and project the yield of rice and cassava in 2030 trend analysis was achieved using linear trend least square method. The result shows that Rice production yield prediction for year 2030 is 6,530.2 (MT), with the slope coefficient of 63.34, which indicates that each year there will be 63.34 increment. Similarly, Cassava production yield prediction for year 2030 is 112,233.64(MT), with slope coefficient of 958 which indicated that each year there is 958 increment. The forecasted graph is presented in Figure 5.

Figure 5. Forecasted rice and cassava production yield for 2030

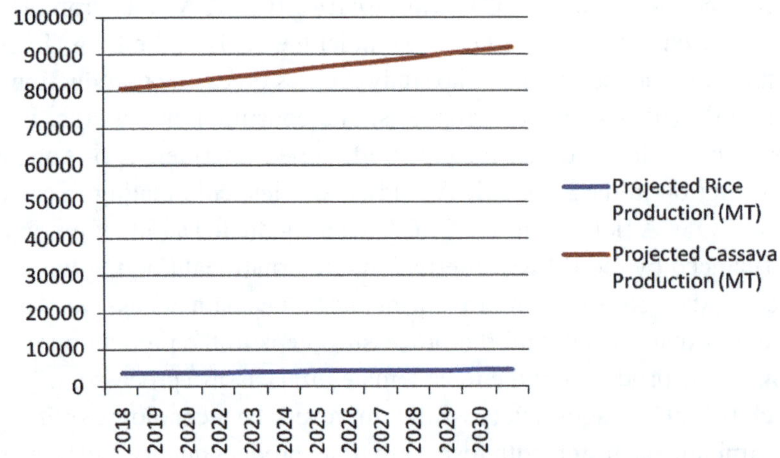

Table 1. Rice production yield (MT) during the Policy Stabilization Era (1986 – 2017)

Year	Rice Yield (1000 MT)	Growth Rate (%)
1986	630	-
1987	1184	87.94
1988	1249	5.49
1989	1982	58.69
1990	1500	-24.32
1991	1911	27.40
1992	1956	2.35
1993	1839	-5.98
1994	1456	-20.83
1995	1752	20.33
1996	1873	6.91
1997	1961	4.70
1998	1965	0.20
1999	1966	0.05
2000	1979	0.66
2001	1651	-16.57
2002	1757	6.42
2003	1870	6.43
2004	2000	6.95
2005	2140	7.00
2006	2546	18.97
2007	2008	-21.13
2008	2632	31.08
2009	2234	-15.12
2010	2818	26.14
2011	2906	3.12
2012	3423	17.79
2013	3038	-11.25
2014	3782	24.49
2015	3941	4.20
2016	3780	-4.09
2017	3780	0.00
2018	3780	0.00

Source: United States Department of Agriculture (USDA, 2018)

Table 2. Cassava production yield (MT) during the Policy Stabilization Era (1986 – 2017)

Year	Cassava Yield (1000 MT)	Growth Rate (%)
1986	12,388	-
1987	13,876	12.01
1988	15,439	11.01
1989	17,404	12.72
1990	19,043	9.41
1991	26,004	36.55
1992	29,184	12.22
1993	30,128	3.23
1994	31,005	2.91
1995	31,404	1.28
1996	31,418	0.04
1997	32,050	2.01
1998	32,695	2.01
1999	32,697	0.00
2000	32,010	-2.10
2001	32,068	0.18
2002	34,120	6.39
2003	36,304	6.40
2004	38,845	6.99
2005	41,565	7.00
2006	45,721	9.99
2007	43,410	-5.05
2008	44,582	2.69
2009	36,822.248	-1.17
2010	42,533.18	15.50
2011	46,190.248	8.59
2012	50,950.292	10.30
2013	47,406.77	-6.95
2014	56,328.48	18.81
2015	57,643.271	2.33
2016	57,134.478	-0.88
2017	80, 500	40.89
2018	80, 500	0.00

Source: United States Department of Agriculture (USDA, 2018)

Indigenous and Improved Value Addition Technologies on Rice and Cassava in Nigeria

In Nigeria, modern agricultural technology has contributed significantly to agricultural development and the gap between developed and developing countries in the area of agricultural production can be attributed largely to differences in the level of technological development, adaptation and transfer process. In developed nations, there is an advanced level of technical know-how and widespread application of technological innovations resulting in high productive capability in agriculture as well as in industry (Odebode, 2008).

Value addition is important for the agricultural sector for Nigeria to be able to actualize the economic agenda of different governments towards increasing agricultural Gross Domestic Product and diversification of economic activities away from the oil sector (Kehinde & Aboaba, 2016). Value-added agriculture refers to increasing the economic value of a commodity through particular production processes, e.g., organic produce, or through regionally-branded products that increase consumer appeal and willingness to pay a premium over similar but undifferentiated products (World Bank, 2011).

According to United State Agency for International Development (USAID) (2010), Nigeria's rice sub sector is dominated by weak and insufficient producer – market linkage due to poor infrastructure and limited efficiency of distribution network which has resulted to low productivity and participation of farmers in the rice field. Compared with Thailand and Bangladesh, Nigerian rice millers are less efficient in making use of milling technologies and practices, as evidenced by their lower paddy-to milled conversion ratios and higher milled-to-paddy price ratios (Michael, Hiroyuki, Kwaben, & Oluyemisi, 2013).

Adding value to locally produced rice to enable it compete favorably with imported rice and improve income of local farmers has been the greatest challenge of governments. This value addition can only come through improved rice milling technology to suit consumers' taste. Rice meeting this standard has to produce whole white rice kernels that are sufficiently milled, free of impurities and contain a minimum number of broken kernels. In order to reduce the rate of rice importation, Rice is produced in Nigeria using a variety of rice production systems and technological levels coexisting together. It is a fact that most operations at the milling machine are most supported by womenfolk.

In Nigeria, rough paddy goes through parboiling before it is milled because this enhances the taste and texture for preferred local rice dishes. Rice processing involves many actors with varying degrees of skills and access to technologies. However, most post-harvest handling and processing in Nigeria is still a cottage industry made up of small scale operators (Federic et al. 2003). Rice mills are found all over the country. There are at least 3-5 mills in every rice producing community. Where none exists, a mill will be available within a radius of 10 km. Lafia, the capital of Nasarawa state has the highest concentration of rice mills per unit area in Nigeria.

There are well over 400 mills and 5 destoners at the milling complex. The quality of the local rice is a major concern for the future of the Nigerian rice sector. While part of the issue relates to the biophysical properties of the varieties locally produced, the major problem is the appearance and the cleanliness of the rice delivered to the market. While the milling technology has a great incidence on the technical performance, it is recognized that these attributes are greatly affected by the attention given to pre-milling and post-milling operations (Federic et al., 2003). These operations include winnowing paddy, drying, destoning, parboiling and eventually packaging. Parboiling paddy is the most important processing operation besides milling. It consists in soaking paddy in hot or cold water in a drum, followed by a rapid exposure of the soaked paddy to steam and a gradual drying for at least one day. The purpose of the operation is to respond to consumer preferences while it also has a positive effect on the grain milling properties (high recovery ratio) and on its nutritious properties. In Nigeria, all paddy processed is parboiled. Rice farmers, millers and specialized operators providing the service to producers or traders can equally take care of parboiling operation. It is recognized that the quality of the parboiling operation has a great influence on the technical performance of milling and therefore on the quality of rice. Accordingly, miller-traders generally preferred to parboil the purchased paddy themselves, while millers-only generally do not parboil themselves. Millers-only do not carry out any of the other pre- or post-milling operation, while almost half of the miller-traders winnow and dry the paddy they purchase, to ensure a better conservation and/or to increase the quality of the product

Traditionally, after harvest rice paddy is winnowed and processed by parboiling, drying and milling. Ajala and Gana, (2015) asserted the traditional methods of processing rice paddy involve soaking of the paddy in water for 2 to 3 days to soften the kernel, followed by steaming of the soaked paddy for 5–10 minutes and dried in the sun, followed by pounding the dried paddy in a mortar and pestle device to remove the husk or use of simple machines for dehulling/milling; then the grain is cleaned using a winnowing basket. This agrees with that of Rathna (2016), who revealed that hand-pounding of paddy in a mortar with a pestle, is still practiced in some remote areas. Pounding the paddy induces upward and downward forces on grain against grain that removes the husk and some bran layers and final cleaning is done by winnowing and gravity separation by hand. According to NCRI, (2007) parboiling is a heat treatment given to paddy rice before drying and milling. Rice milling is usually on a separate location to that of parboiling. Women farmer simply bring in their rice ready for milling whereas parboiling is usually separate activity which is usually done by women who operate parboiling units located in their households as a cottage enterprise (Nwalieji & Uzaegbunam, 2012).

Drying and winnowing tend to be a combined operation often taking place on drying floors and often done by women. Drying is done by spreading the paddy across the pavement, preferable on a protective mat, and occasionally stirring it so the sun will dry it evenly (Tinsley, 2012). Result of Research carried out by Oyediran (2016) showed that rice were dried manually (100%) by the rice processors. and that Small-scale rice processors used tarpaulins spread on bare ground. or polythene for paddy sun-drying.

According to Ajala and Gana (2015), In modern methods, the rough rice or paddy is first cleaned to remove contaminants, and the husks are then removed by the so-called shellers; these are most commonly horizontally spaced rotating abrasive stones, but increasing use is being made of rubber roll or rubber belt made shellers. The rice and hulls are separated by aspiration and any paddy remaining with the rice is removed in a paddy separator. It is discovered that the main problem of Nigerian rice is the presence of stone in the rice grains. These steps are diagrammatized in the Figure 6.

Another improvement in rice processing in Nigeria reported by Propcom (2012) is the drying process. The traditional sun drying has been replaced by mechanical dryer or improved sun drying method. This type of dryer is found scattered all over the country. It can process about 3000 kg and remove 50% moisture of rice in 6 hrs. Apart from this mechanical dryer that uses diesel or electricity; other dryers have been developed such as solar dryer for drying rice paddy. A cross section of this modern rice processing machines is depicted in Figure 7, 8 and 9 from Confluence Rice Milling Factory established by the Kogi State Government. Though this Company has its own out-growers however, they cannot supply rice paddy commensurate to even 10% of its production capacity. The Company has always gotten its paddy supply from other neighboring states like Niger, Nassarawa, Benue and Kwara until its out-growing strategies is able to satisfy the production capacity of the company. According to a report by Dayo et al. (2018), a persistent problem in smallholder agricultural production is the inability of farmers to process own raw outputs. This has always led to sales at poor prices, and leaving most of the gains to those who buy process and sell to others within the value chain. Farmers gain or lose an average of N5, 795.96 on every 100 kg of rice grain processed (not processed). This is a significant value when viewed across millions of metric tonnes of harvested rice per season. Indeed, the difference between the farm-gate price of unprocessed rice (N11,091.18/100kg) and retail price of processed price (N22,748.35/100kg) is disturbingly high (N11,657.17/100kg), and underscore the enormity of value losses by farmers for not processing rice before selling. Oteng and Saint'Anna (2002) opined that in the free market system, where local products compete with high-quality imports, farmers' incomes will be determined by the quality of their produce. In this regard, efforts should be made to improve harvesting methods and post-harvest processing by discarding obsolete mills and introducing modern rubber-roller mills for high out-turn and a high percentage of head rice.

Figure 6. Modern management and processing of rice.
Adapted from Ashraf (2013).

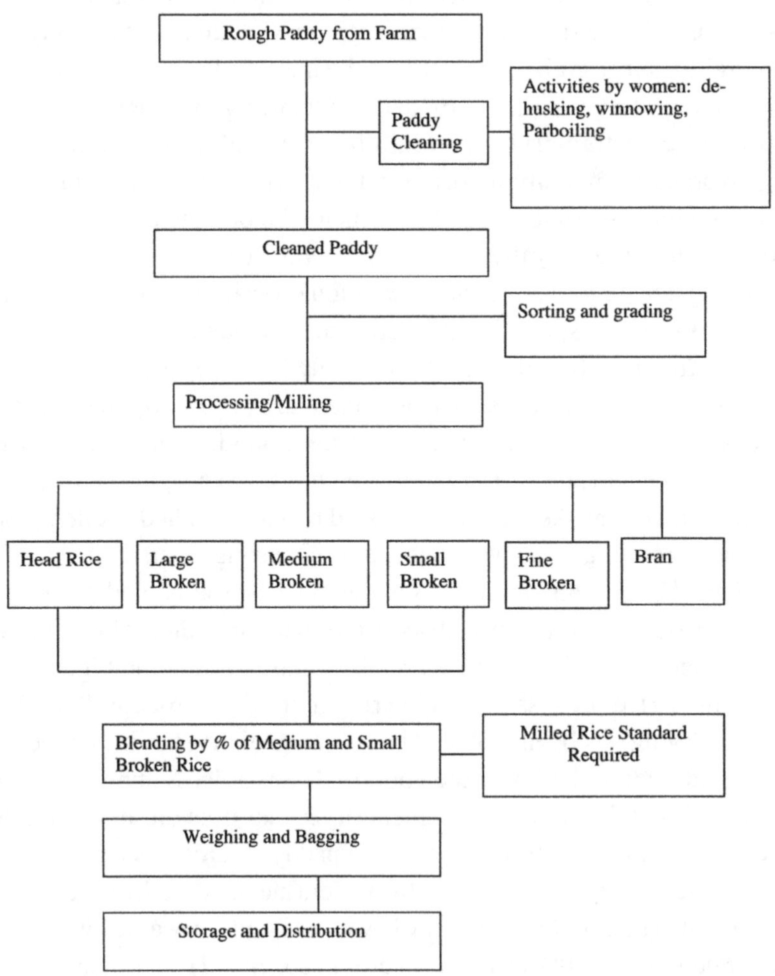

Figure 7. Improved milling machine at the Confluence Rice Processing Factory, Kogi State.

Figure 8. A woman (Staff) using the packaging section of the milling machine at the Confluence Rice Milling Factory, Kogi State

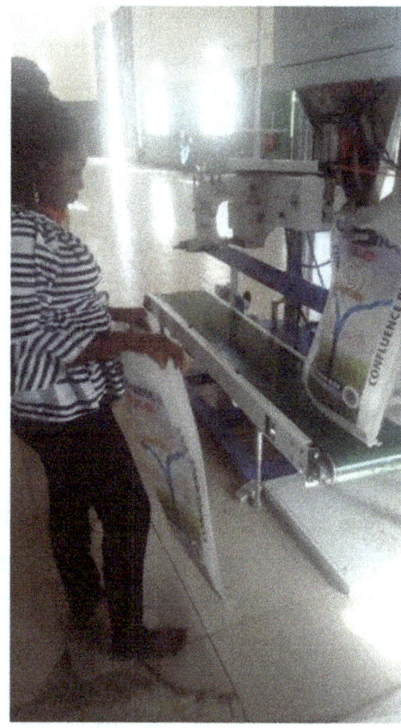

Figure 9. Section for parboling rice at the Confluence Rice Milling Factory

Cassava is an important crop in Africa. Nigeria is the largest producer of cassava tubers in the world with production of about 45 million mt in 2009 (PIND, 2011) It survives in poor soils, has a high yield of carbohydrates and good resistance to pest infestations, diseases and drought. Cassava roots are processed and eaten by 500 million people a day in Africa where it is a staple for 40% of the population (Kehinde & Subuola, 2015).Women play a central role in cassava production, harvesting, processing and marketing, contributing about 58 percent of the total agricultural labor in the Southwest, 67 percent in the Southeast and 58 percent in the central zones (FAO, 2004; Onyemauwa, 2012). According to Sampson (2014) who stated that as cassava matured for harvesting and processing, the level of engagement of several other agents reduces as the activity of women increases

Cassava value added technologies are the processes of changing or transforming the product from its original state to a more valuable state. Cassava roots are transformed into various forms in order to increase the shelf life because they cannot be stored for too long as they rot within 3-4 days of harvest. It helps to reduce bulk and improve product quality. It also facilitates transportation and marketing, reduce cyanide content and improve the product palatability (Odebode, 2008). Many cassava processing and value addition technologies have been developed by research centers and disseminated to farmers over the years. These technologies are aimed at reducing drudgery, increasing food forms and adding values to the crop (Nwakor et al, 2011). Traditional cassava processing methods in use in Africa probably originated from tropical America, particularly northeastern Brazil and may have been adapted from indigenous techniques for processing yams. The processing methods include peeling, boiling, steaming, slicing, grating, soaking or seeping, fermenting, pounding, roasting, pressing, drying, and milling (FAO, 2013).

The roots which are used for human consumption are processed into many food forms. The women are largely responsible for the work of processing it to make gari, fufu, tapioca and other products The uses of cassava are expanding, as further processing can produce chips, pellets, flour, alcohol and starch (Adebayo, 2009). Traditional cassava processing does not require sophisticated equipment. Processing cassava into gari requires equipment such as grater, presser and fryer. The traditional cassava grater is made of a flattened kerosine tin or iron sheet perforated with nails and fastened onto a wooden board with handles. Grating is done by rubbing the peeled roots against the rough perforated surface of the iron sheet which tears off the peeled cassava root flesh into mash. In recent years, various attempts have been made to improve graters. Graters which are belt-driven from a static 5 HP Lister type engine which are being extensively used in Nigeria. Its capacity to grate cassava is about one ton of fresh peeled roots per hour. (FAO, 2018). Peeling is a unit operation mostly done by women and children manually using knives. The peeled roots are grated by women, using a simple traditional grater. Mechanical and automated peelers and power driven graters are new technologies which have been designed and fabricated to reduce drudgery in the peeling and grating process of cassava. Taiwo *et al.* (2001) reported in a study assessing the capacity and technology needs of gari processors that mechanical grating of peeled cassava tubers has been fully adopted. According to Agbarevo and Okeke (2015), the adoption of the value added technologies by women farmers is designed not only to improve the value chain but also increase production and income of farmers from the crop, create jobs, etc. Nigerian government has created enabling investment opportunities for foreign countries to help develop the value chain through technologies that farmers can use to add value to their harvest crop especially cassava processing. Figure 10, 11, 12, and 13 show cassava processing machined made by Chinese and are already in use some places across the globe since 1991 (Nanyang Goodway Machine, 2015).

Figure 10. Cassava Peeler

Figure 11. Cassava Presser

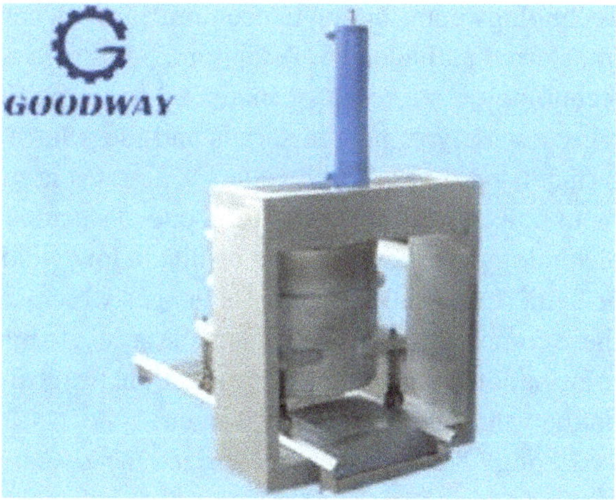

Figure 12. Cassava Gari Fryer

Figure 13. Cassava Chipper

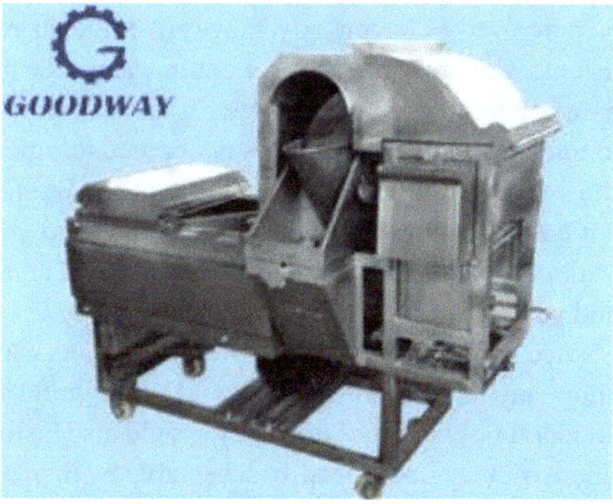

Compare to other countries of the world, Nigeria still has gap in terms of rating the value addition and quality assurance on rice processing. As it is indicated in Table 2, rating on the scale of "0-5" with regards to cleaning/polising of rice, destoning, improved product quality, improved milling in terms of use of modern machines, sorting/grading to standardize, storage, packaging and branding, Nigeria is rated with 53.33% and came 10th position among the major rice producing countries of the world with China taking the lead followed by Thailand and USA.

Constraints to the Adoption of Rice and Cassava Value Addition Technologies

Technology contribution to economic growth can only be realized when and if the technology is widely transferred and adopted. Adoption comes as a result of decision being made by the adopters to use a technology. As a result of increased in scientific research and improved methods of communication, a great variety of new materials and ideas have been generated and brought to the doors of Nigerian farmers and other rural dwellers. The rates at which these people learn of this innovations and adopt them however differ greatly from one place and circumstance to another (Ekong, 2010). Generally, low technology practice in Africa results in low yields, which are exacerbated by several challenges (Oteng & Saint'Anna, 2002). These challenges which still persist and inherent in the post-harvest handling of these commodities in Nigeria include access to innovative technologies, access to technical and economic information on improved value addition, the cost of using value added technologies, inadequate training on the value addition for the women.

Ekong (2010) further categorized factor that affect adoption of any innovation (Technology) as; (i) relative advantage: the degree to which an innovation is superior to the one it is meant to supercede. For instance, manual processing of oil palm fruits is tedious and inefficient. A farmer who is presented with a machine that can process palm fruit in a matter of minutes and produce more oil per unit measure of fruits, would see this as of greater advantage than hand processing. Any technology that is not of relative advantage to their current practice will not be adopted by the rural farmer. (ii) Cost: an innovation may be perceived as having advantages over the currently used practice but may not be adopted because of its cost. Cost may also be thought in terms of what the adopter is supposed to give up and what he is to gain in adopting the technology (iii) complexity: the degree to which an innovation is relatively difficult to understand or use. Farmers adopt technologies that are relatively simple to understand and use than those that are complex. (iv)Visibility: farmer tends to adopt technologies that their results or operations are easily seen. (v) Divisibility: the extent to which an innovation can be tried in parts or on limited scale. (vi) Compatibility: the extent to which an innovation is consistent with the existing values, norms and past experiences of the adopter.

According to Chidoze (2014), rural farmers are mostly poor resource farmers. Hence, to enable them adopt any innovation, funds should be provided and the provision of energy and time saving equipment and facilities to reduce drudgery should also be made available. Otherwise, any innovation that is labour intensive may not be readily acceptable by them. Also availability of organized market readily available as an incentive for those who adopt these innovations is important. It was reported in a study carried out by Apata (2015), that female cassava growers understand that there are market value added products like cassava flour and cassava chips . Furthermore, in order to exploit the markets there is need to promote appropriate cassava processing technologies such as grating, chipping and crushing by educating farmers on these technologies and facilitate acquisition of processing equipments. Also, the access of farmers to urban market with their value added cassava products can immensely boost their production and profitability.

CONCLUSION AND RECOMMENDATIONS

It was indicated that indigenous methods of cassava and rice processing give low output which is also of low quality. The demands for these commodities have been on the increase against their production and yield. Rapid urbanization in Nigeria increased dynamics in both rural and urban areas and the changing social environment have resulted in an unprecedented demand for variety of foods through value addition to harvested crops. In spite of the numerous challenges faced by farmers in carrying out post-harvest activities, there are opportunities to encourage them in the promotion of value addition to these commodities through equitable and sustainable policies affecting the processing and value addition of these commodities. Hence, the following recommendations shall go a long way to improve the processing and value addition on these commodities for local consumption and export couple with effective and practical policy implementation.

1. In spite of the Federal Government of Nigeria and other private sector (national and international) interventions in ensuring that there is improved rice and cassava production in quantity and quality, there should be political will power by successive government to sustain these efforts in order to achieve optimization of rice and cassava value addition strategies and potential of the nations endowment in the future.
2. Up-scaling indigenous methods of processing of rice and cassava through sustainable policy reform to meet both local consumption and export of the same are pertinent to be considered by policy makers in their interventions in the value addition to these commodities. Proper implementation and sustainability of APP will help in this regard.
3. There should be policy in place aimed at creating an enabling environment and training on innovative technologies on these two commodities in order to invest in more innovative value added products of these crops; with the view to making these innovations and technologies affordable, adoptable and adaptable
4. The target to improve value addition to rice and cassava should address issues peculiar to farmers' access to productive resources, like technical and economic information on improved processing technologies, credit facilities, and land. Most of the obstacles faced by women can be jointly addressed by private-public sectors' interventions.

REFERENCES

Adam, A. G., Bidoli, T. D., Ammani, A. A., & Oduehie, T. C (2018). Gender Participation in Rice Processing Value Chain in kebbiand Sokoto States, Nigeria. *Global Journal of Science Frontier Research: Agriculture and Veterinary, 18*(2).

Adebayo, K. (2009). Dynamics of technology adoption in rural-based cassava processing enterprises in South-West Nigeria. *International Journal of Agricultural Economics and Rural Development, 2*(1), 15–24.

Adejo, P. E., Okwu, J. O., & Edoka, M. H. (2015). Assessment of Gender Participation in Post-harvest Management Practices of Maize in Kogi State, Nigeria. *Proceedings of the 49th Annual National Conference of the Agricultural Society of Nigeria "Agriculture: The Nigerian Economy Beyond Oil"*.

Adejo, P. E., Okwu, J. O., & Saliu, O. J. (2016). Assessment of Postharvest Information Needs ofMaize Farmers in Kogi State, Nigeria. *Journal of Agricultural & Food Information, 17*(4), 260–273. doi:10.1080/10496505.2016.1221766

Agbarevo, M. N., & Obinne, C. P. O. (2008). Evaluation of the effect of Agricultural extensiondelivery on cassava production in Nigeria. A case study of Cross-River state of Nigeria. *Nigerian Agricultural Journal, 39*(1&2), 16–21.

Agbarevo, M. N. B., & Okeke, S. O. (2015). The Effect of Adoption of Cassava Value added Technologies on Farmers' Production in Abia State, Nigeria. *European Journal of Physical and Agricultural Science, 3*(1), 48–53.

Ajala, A. S. & Gana, A. (2015). Analysis of Challenges Facing Rice Processing in Nigeria. *Journal of Food Processing.*

Anyiro, C. O., & Onyemachi, A. D. (2014). *Adoption of Cassava Value Added Innovation and Its Implication on Rural Livelihood: A Case of Rural Women in Abia State, Nigeria.* Retrieved from https://www.researchgate.net/publication/296183628

Apata, T. G. (2015). *Analysis of Cassava Value Chain in Nigeria, From a Pro-poor and Gender Perspective of Farming Households in Southwest, Nigeria.* Retrieved from www.agrodep.org RV_Apata_T_final

Ayodele, D., Udah, A., Elechi, N., Oriuwa, C., Tijani, G., & Sanni, L. (2011). Cassava Value Chain Analysis in the Niger Delta. A Report by Foundation for Partnership Initiatives in the Niger Delta (PIND).

Biyi, D. (2005). *Government Policies and Competitiveness of Nigerian Rice Economy.* Paper presented at the `Workshop on Rice Policy & Food Security in Sub-SaharanAfrica' organized by WARDA, Cotonou, Benin.

Boland, M. (2009). *What is Value added Agriculture?* Retrieved from http://www.agmrc.org/business_developm ent/getting_prepared/valueadded_agriculture/articles/index.cfm

Brandspurng. (2018, Feb 13). Agriculture in Nigeria: Food Production and Consumption Rates. *Brandspurng.* Retrieved from https//brandspurng.com/category/agro-news/

Cadena, P. M. (2017). *Stimulating cassava growth in Nigeria and Tanzania.* Retrieved from https://agroecology.ifas.ufl.edu/2017/09/06/august-30-2017-simulatin-cassava-growth-in-nigeria-and-tanzania/

CBN. (2017). *CBN: Local Rice Production Records 70% Increase.* Retrieved from https://s3west1.amazonaws.com/leadersandco/wpcontent/uploads/2016/07/25051645/CentralBank-of-Nigeria-CBN.jpg

Dayo, P., Olumuyiwa, O. J., Yarma, N., & Fatunbi, O. (2018). Innovation Opportunities in the Rice Value Chain in Nigeria. *FARA Results, 2*(3). Retrieved from https://research4agrinnovation.org/wpcontent/uploads/2018/03/Volume-2-No- 3_2018.pdf

Echebiri, R. N., & Edaba, M. E. I. (2008). Production and Utilization of Cassava in Nigeria: Prospects for Food Securit and Infant Nutrition. *Production Agriculture and Technology, 4*(1), 38–52.

Effiong, J. B., Ijioma, J. C., & Okolo, L. C. (2015). Participation of women farmers in rice production in bende local government area, abia state. *International Journal of Agricultural Extension and Rural Development Studies, 2*(2), 1–9.

Eke-okoro, O. N., & Njoku, D. N. (2012). A Review of Cassava Development in Nigeria from 1940-2010. *Journal of Agricultural and Biological Science, 7*(1), 59–65.

Ekong, E. E. (2010) Diffusion and Adoption of Agricultural Innovations in Nigeria: Rural Sociology – Third Edition. Uyo: DOVE Educational Publishers.

Erhie, E., Iwelurio, M., Agbeyi, E., Oladipo, O., Oyaniran, T., Akinbiyi, A., & Adegunle, E. (2018). *Boosting Rice Production through increased Mechanization*. Retrieved from http://www.pwc.com.ng

Ewepu, G. (2018, April 11). Nigeria Rice Sub-sector to attract N250bn Investment. *Vanguard Newspaper*. Retrieved from https://www.vanguardngr.com/category/agric/

FAO. (2013). *Save and Grow: Casssava "Harvest, Post-harvest and Value addition"*. Retrieved from http//www.fao.org/ag/save-and-grow/cassava/index_en.html

FAO. (2017). *Country fact sheet on Food and Agricultural Policy Trends*. Article prepared by Food and Agriculture Policy Decision Analysis (FAPDA). Retrieved from www.fao.org/in-action/fapda/tool

FAO. (2018). *Nigeria at a glance*. Retrieved from www.fao.org/nigeria/programmes-and-projects/en/

FAOSTAT. (2018). *Cassava Production (tons) for Nigeria from 1960 – 2018*. Retrieved from www.factfish.com/country/Nigeria

Frederic, L., Olaf, E., Akande, S. O., Titilola, S. O., Akpaokodje, G., & Ogundele, O. O. (2003). The Nigerian Rice Economy in a Competitive World. Constraints, Opportunities Strategic Choice. Imported Rice Retailing and Purchasing in Nigeria. A Report of a Survey funded by USAID and jointly Implemented by WARDA and NISER.

Gujarati, D. N. (2005). *Basic Econometrics*. Tat McGraw-Hill Publishing Company Ltd.

Ja'afar-Furo, M. R., Bello, K., & Sulaiman, A. (2011). Assessment of the prospects of value addition among small-scale rural enterprises in Nigeria: Evidence from North-eastern Adamawa State. *Journal of Development and Agricultural Economics*, *3*(3), 144–149.

Jatto, N. A., & Okebiorun, E. O. (2017). Value Addition in Cassava Processing: Evidence from Women in Ilesa West Local Government Area of Osun State. *Agriculture and Food Sciences Research*, *4*(1), 30–36. doi:10.20448/journal.512.2017.41.30.36

Jibowo, A. A., & Ajayi, A. O. (2011). Agricultural Extension in Nigeria: History of Agricultural Extension in Nigeria. Agricultural Extension Society of Nigeria (AESON).

Kehinde, A. L., & Aboaba, K. O. (2016). Analysis of value addition in the processing of cassava tubers to "garri" among cottage level processors in southwestern Nigeria. Invited poster presented at the 5th International Conference of the African Association of Agricultural Economists, Addis Ababa, Ethiopia.

Longtau, S. R. (2003). African Agriculture: Nigeria Case Study Report on Rice Production. Prepared by Eco-System Development Organsiation (EDO), Jos, Nigeria for Overseas Development Institute (ODI).

Maji, A. T., Ukwungwu, M. N., Nahemiah, D., & Bakare, S. O. (2017). Rice: History, Research & Development in Nigeria. In D. Nahemiah, I. Nkama, A.T. Maji, M.N. Ukwungu (Eds.), Rice in Nigeria: Traditional Recipes and Research Needs (pp. 1-10). RONAB GRAPHIX PRINTS.

Mgbenka, R. N., & Mbah, E. N. (2016). A Review of Smallholder Farming in Nigeria: Need for Transformation. *International Journal of Agricultural Extension and Rural Development Studies*, *3*(2), 43–54.

Michael, J., Hiroyuki, T., Kwabena, G. B., & Oluyemisi, K. S. (2013). *Policy Options for Accelerated Growth and Competitiveness of the Domestic Rice Economy in Nigeria: Nigeria Strategy Support Program II*. Retrieved from www.ifpri.org

Nanyang, G. M. (2015). *Industrial Cassava Processing machine from China*. Retrieved from http://www.cassavamachine.cn/ProInfo_4.html

NCRI. (2007). *Definition of Ofada rice qualities through varietal identification and testing*. NCRI.

Nwakor, F. N., Ifenkwe, G. E., & Asumughha, G. N. (2011). Root and Tuber Crops Research for Food Security and Empowerment. In C.O. Amadi, K.C. Ekwe, G. O. Chukwu, A. O. Olojede., & C.N. Egesi (Eds.), Potentials in Adoption of Improved Cassava Varieties for increased Food Security in Nigeria (pp. 563-571). National Root Crops Research Institute Umudike.

Nwalieji, H. U., & Uzuegbunam, C. O. (2012, December). Effect of climate change on rice production in Anambra state, Nigeria. *Journal of Agricultural Extension, 16*(12), 81–91.

Nzelu, C. (2017, Dec 13). CBN: Local Rice Production Records 70% Increase. *This Day Newspaper*. Retrieved from https://www.thisdaylive.com

Odebode, S. O. (2008). Appropriate Technology for Cassava Processing in Nigeria: User's Point of View. *Journal of International Women's Studies, 9*(3), 269–286.

Odunze, D. I. (2019). A review of the Nigerian Agricultural Promotion Policy (APP, 2016-2020): Implications for entrepreneurship in the agribusiness sector. *International Journal of Agricultural Policy and Research, 7*(3), 70–79.

Odurukwe, S. N., Matthews-Njoku, E. C., & Ejiogu-Okereke, N. (2006). Impacts of the women-in-agriculture (WIA) extension programme on women's lives; implications for subsistence agricultural production of women in Imo State, Nigeria. *Livestock Research for Rural Development, 18*(2). Retrieved from http://www.lrrd.org/lrrd18/2/odur18018.htm

Okoro, D., & Ujah, O. C. (2009). *Agricultural Policy and Budget Analysis in Nigeria (1999 –2007): Perspectives and Implications for SLSSFAN Project States*. OXFAM.

Okoruwa, V. O., Ojo, O. A., Akintola, C. M., Ologhobo, A. D., & Ewete, F. K. (2008). Post- harvest Grain Management Storage Techniques and Pesticides Use by Farmers in South- west Nigeria. *Journal of Economics and Rural Development, 18*(1), 53–72.

Oladimeji, Y. U. (2017). Food Production trend in Nigeria and Malthus theory Of Population: Empirical evidence from RiceProduction. *Nigerian Journal of Agriculture, Food and Environment, 13*(1), 126–132.

Oladimeji, Y. U., & Ajao, A. M. (2014). Structure and determinants of profit efficiency among small scale rice farmers in Kwara State, Nigeria. *International Journal of Applied Research and Technology, 3*(11), 27–33.

Olawale, S. (2018). *List of Rice Producing States in Nigeria*. Retrieved from https://naijaquest.com/rice-producing-states-in-nigeria/

Onuekwusi, G. C., Odoemelam, L. E., & Kanu, R. I. (2017). Capacity Building on Cocoyam Value Additions Training for Rural Women in Abia State, Nigeria. *Journal of Agriculture Research and Technology*, *6*(4), 2–6.

Onwualu, A. P. (2012). *Agricultural sector and national development: focus on value chain approach*. Presented at the 5th Edition of the Annual Lecture of Onitsha Chamber of Commerce at Sharon house, GRA, Onitsha.

Onyemauwa, C. S. (2012). Analysis of Women Participation in Cassava Production and Processing in Imo State, South-east Nigeria. *Journal of Economics and Sustainable*, *3*(5), 81–90.

Oteng, J. W., & Sant'Anna, R. (2002). *Rice production in Africa: current situation and issues*. Retrieved from http://www.fao.org/docrep/003/x2243t/x2243t00.htm#TopOfPage

Oyediran, W. O. (2016) Assessment of Paddy Rice (*Ofada*) Processing in ensuring Rural Women livelihood sustainability in Ogun State, Nigeria. *International Journal for Research in Agricultural and Food Science,* 82-92.

PIND. (2011). *Cassava Value Chain Analysis in the Niger Delta*. A Report by Foundation for Partnership Initiative in the Niger Delta.

Premium Times. (2015, Oct 19). *Nigerian women contribute 70 per cent of agricultural workforce – AfDB*. Retrieved from https://www.premiumtimesng.com/news/top-news/191777-nigerian-women-contribute-70-per-cent-of-agricultural-workforce-afdb.html

Pro-poor Opportunities in Commodity and Service Markets Project (Propcom). (2012). Retrieved from http://www.propcom.org/

Rathna. (2016). *Rice milling technology*. New Rathna Rice (pvt) ltd. Retrieved from http://newrathnarice.com

Rick, H., Bernard, M., & Rembold, F. (2014). *APHLIS – Postharvest cereal losses in Sub-Saharan Africa, their estimation, assessment and reduction*. A Technical Report for European Commission. Retrieved from https://www.google.com.ng/url?sa=t&rct=j&q=&esrc=s&source=web&cd=11&cad=rja&uact=8&ved=2ahUKEwjK5IG38_bmAhUhsaQKHf41AhEQFjAKegQIAxAC&url=https%3A%2F%2Fpublications.jrc.ec.europa.eu%2Frepository%2Fbitstream%2FJRC92152%2Flbna26897enn.pdf&usg=AOvVaw2TZCnu3x18iFqEqZ97J1hQ

Sampson, O. M. (2014). *Women's Contribution to Local Economic Development: A study of Women in Cassava Production and Processing in Central Tongu District* (Thesis). Kwame Nkuumah University of Science and Technology.

Singh, R. B., Woodhead, T., & Papademtriou, M. K. (2003). *Strategies to Sustain and Enhance Asia Pacific Rice Production*. Retrieved from www.fao.org/docrep/006/y475Ie0t.htmHTopOfpage

Somado, E. A., Guei, R. G., & Nguyen, N. (2010). *Nguyen Overview: Rice in Africa*. Retrieved from www.africarice,org

Spore, C. T. A. (2018). *A promising new continent-wide trade agreement for Africa: Trade and Marketing*. Retrieved from https://spore.cta.int/en/marketing/all/article/a-promising-new-continent-wide-trade-agreement-for-africa-sid0343958eb-d79b-4a03-9d46-44dfe69f9188b

Tadlock, C. (2002). *Value-Added Agricultural Enterprisesin Rural Development Strategies*. The Library of Congress.

Taiwo, K. A., & Fasoyiro, S. B. (2015). Women and cassava processing in Nigeria. *International Journal of Development Research*, *5*(2), 3513–3517.

Taiwo, K. A., Osunbitan, A. J., Sunmmonu, T. O., Ajayi, M. O., & Ajibola, O. O. (2001). Technology choice and technical capacity in gari production. *Food Reviews International*, *17*(1), 89–107. doi:10.1081/FRI-100000516

Tijani, S. A., & Umoh, I. E. (2016). Attitude of Women Cassava Farmers towards Participation in Growth Enhancement Support Scheme in Oyo state, Nigeria. *Journal of Agricultural Extension*, *20*(2), 89. doi:10.4314/jae.v20i2.7

Ugwu, J. N., Mgbakor, M. N., & Chitor, C. (2014). Assessment of Value addition in Rice Production and Processing in Adani inUzouwani Local Government Area of Enugu state, Nigeria. *International Journal of Agricultural Science and Veterinary Medicine*, *2*(4), 54–68.

USDA. (2018). *Nigerian Milled Rice Production by Year (1960 - 2018)*. Retrieved from https://www.indexmundi.com/agriculture/?country=ng&commodity=milledrice&graph=production-growth-rate

West African Rice Development Authority (WARDA). (2003). *The Nigerian rice economy in a competitive world: constraints, opportunities and strategic choices: strategy for rice sector revitalization in Nigeria*. A paper on rice sector study implemented WARDA. Presented at a Technical Workshop 20-21 August, Ibadan, Nigeria.

World Bank. (2011). *Missing Food: The Case of Post-harvest Grain Losses in Sub-Saharan Africa*. World Bank Report No: 60371-AFR. Retrieved from www.worldbank.org/rural

Worldstage. (2018). *Nigeria: Federal Government to improve supply value chains of rice, cassava by partnering with the Nigerian Incentive-Based Risk Sharing System For Agricultural Lending (NIRSAL) in collaboration with the united nations development project(undp)*. Retrieved from https://www.worldstagegroup.com/category/business/agriculture

Chapter 9
Training Needs Assessment of Palm Oil Processors in Ijebu North Local Government Area, Ogun State

Titilope Omolara Olarewaju
https://orcid.org/0000-0002-3144-3483
Forestry Research Institute of Nigeria, Nigeria

Felix Oaikhena Idumah
Forestry Research Institute of Nigeria, Nigeria

Agatha Itohan Oseghale
Federal University of Technology, Minna, Nigeria

Lucy Adeteju Orumwense
Forestry Research Institute of Nigeria, Nigeria

Olugbenga Simeon Oke
Forestry Research Institute of Nigeria, Nigeria

Elizabeth Funmilayo Okedeji
National Open University of Nigeria, Nigeria

ABSTRACT

The study assessed training needs of palm oil processors in Ogun State, Nigeria. A well-structured questionnaire was used to elicit information from 90 palm oil processors. The data was analyzed using both descriptive and inferential statistics. Majority of the respondents were young, married, and experienced in palm oil processing. Women are mostly involved in palm oil processing using manual method of processing with oil palm fruits sourced more from family farms. Respondents require training for manual and mechanized processing methods. Socioeconomic factors have significant influence on different stages of palm oil processing. Poor extension service, high cost of labour, and processing machine were the most perceived constraints to palm oil processing in the study area. The study therefore concludes that there is need for training in oil palm processing. Extension service providers should intensify efforts in this regard so as to boost the palm oil supply both within and outside the country.

DOI: 10.4018/978-1-7998-2599-9.ch009

BACKGROUND

The oil palm sub-sector of Nigeria agriculture presented itself as a potential productive sector that could be used to diversify the economy after years of neglect. The non-participation of Nigeria in oil palm plantation development until the late 2000 and government policy on plantation development are most certainly reasons for Nigeria losing her leading position to Malaysia whose total production are export oriented (Green, 2003). Nigeria oil palm production fell, and by 1999 only 10% of the total value of the country's annual exports remains (Bello et al., 2015). In the past, the Nigerian government had tried to implement large-scale oil palm plantations, most of which resulted in complete failures. The Cross River State palm projects of the 1960′s and the 1990′s European Union funded "Oil palm belt rural development programme" was abandoned in 1999 and reactivated in 2003 and the state governor's intention to privatize it was announced in 2010 (WRM, 2010).

Nigeria has been "the second largest recipient of World Bank palm oil sector projects, with six projects over the 1975 to 2009 period (WRM, 2010). The palm oil from the eastern region of the country was described as being of the highest quality and the people took pride in the work of their hands. The people were so good at it that the Malaysians like the three wise oriental kings followed the scent of the palm oil to Imo state to learn the fine art of palm oil production. However, with the crude mineral oil boom, laziness and indolence took the place of hard work and dignity of labour. Agriculture suffered a setback as it was relegated to the back burner. Between 2003 and 2005 Nigeria lost her leading place in palm oil export to Malaysia and Indonesia and regained it temporarily between 2005 and 2008. In 2004, according to a report by Friends of the Earth-Netherlands, Indonesia cultivated oil palm plantations which covered 5.3 million hectares of land. These plantations generated 11.4 million metric tons of palm oil with an export value of US$ 4.43 billion and brought in $42.4 million to the Indonesian treasury (Butler, 2006). Beyond this problem, there has been a steady decline in the Nigeria's domestic supply of palm oil.

Oil palm is appreciated by most people in the Southern part of Nigeria because of its level of utilization. Oil palm gives the highest yield of oil per unit area, compared to any other oil producing plant when processed, and it produces two distinct oils: palm oil and palm kernel oil which are of great importance in the industrial market (FAO, 2002).The two oils were once very vital to Nigeria's export trade, as Nigeria was a leading producer of oil palm products in the world (Ibitoye et al., 2011). Loss of foreign earning as well as local scarcity is now confronting the nation due to varying quality and drudgery involved in processing of palm fruits. Processing method generally accounts for low quantity and poor quality of palm oil. Although palm oil processing methods include manual and mechanized methods, manual method of processing is more prevalent among small scale processors and these small-scale processors are responsible for the bulk of palm oil processed in Nigeria (Olagunju, 2008). Majority (80%) of palm oil processors comes from dispersed smallholders who harvest semi-wild palm fruits and use manual processing techniques, a processing technique that is labour intensive and highly inefficient, with a low palm oil extraction rate and high free fatty acid content that can be up to 30% in some instances (Orewa et al., 2009; Ugwu, 2009).

Currently, Nigeria oil palm sector is under reactivation after it collapsed during the discovery of crude oil. For the palm oil processing industry to sustain competitive edge, continued research is very crucial with regard to appropriate processing technology to pave the way in shaping the future of the palm oil industry. It is therefore logical to examine training needs of palm oil processors. This study attempted to fill this gap and also answer other questions such as methods used for palm oil processing and factors that have significant influence on different stages of palm oil processing in the study area. Generally, the study determined the stage at which training is required along the processing chain as well as the influence of socio-economic factors at such stages of palm oil processing in the study area.

CONCEPTUAL REVIEW

Training is a term, which covers a wide range of activities. Its length can vary from short term training activities such as one-day demonstration, to longer-term professional courses that may last several months. Human Resource Development (HRD) is one of the many strategies in achieving the vision for development in any country. Training can be one of the best ways to develop human resources. It aims to develop people's potential and enable them to use this potential towards the achievement of their vision of self-reliance and self-sufficiency.

Training is the process of acquiring specific skills to perform a job better. It involves the processes of teaching, informing and educating people to become qualified and proficient in performing their duties (Obibuaku, 2008). Trainings are important tools for assisting government officials, development personnel, extension experts and agriculturalists in the realization of their program objectives and plans. Often people are faced with the need to change something or to implement a new way of doing something. It allows us to orient those who was involved in and/or affected by the change. We may also need to provide people with new knowledge and /or with new skills that are necessary to implement a change. Therefore, the development training, selected as a focus of this study, refers to such trainings of farmers and extension staff undertaken in conventional and/ or participatory models. Thus, in the system where the role of extension and communication- intervention was looked on as transferring and disseminating readymade knowledge from research to farmers, or from 'early adopters' to other farmers, which is often referred to as the 'transfer of technology' model of extension (Hagman et al., 2000), it is obvious that the methods and techniques used follow the same manner, which holds true for training approaches too.

Training need is the difference between the required level of individual competence and his present level of competence (Youdeowei & Kwarteng, 2010). A training need is a shortage of skills or abilities, which could be reduced or eliminated by means of education and development.

Training requirements hinder employees in the fulfillment of their job responsibilities or prevent an organization from achieving its objectives. They may be caused by a lack of skills, knowledge or understanding, or arise from a change in the workplace.

Several theories of training needs identification have been proposed and practiced by organizations. The three popular theories are skill-gap analysis; organizational and occupational analysis and critical incident theory (Jasim et al., 2016). Wentling (2016) describes the skill-gap analysis as a process which involves understanding the current skill levels of those who need training in order to focus on the desired and important skills. The intercept theory stipulates that the needs agreed by both extension agents and block extension supervisors based on the job description of the extension agents should be reduced or solved through training before embarking on other needs. The process involves rating of staff (trainee or beneficiary) by super ordinate staff (supervisors). Colleagues and another stakeholder can be involved in rating the training needs.

MATERIAL AND METHODS

The population for this study consists of farm households involved in palm oil processing in Ijebu North Local Government Area. Ijebu North is a Local Government Area in Ogun State, Nigeria. Its headquarters are in the town of Ijebu Igbo at 6°57′N 4°00′E (Wikipedia, 2019). It has an area of 967 km^2 and a population of 284,336 at the 2006 census. Also, several indigenes of the town engage in timber business so there are many sawmills in the town.

Multi stage sampling was used to select respondents from communities notable for oil palm production in the study area. Ijebu North local government was purposively selected at the first stage while five out of the twelve wards in the local government were randomly chosen at the second stage. These are Oru-awa-ilaporu, Osun, Ome, Ago-Iwoye, and Mamu/Ehin-Etiri. At the third and fourth stages, two communities were selected from each local government and nine processors were selected from each community to give a total of ninety respondents used for this study. The communities chosen include Ilaporu, Awa, Ajegunle, Mamu, Ago Iwoye, Tekole, Apoje, Osun, Ololo and Ajebo. Data used for this study were gathered through questionnaires and interview. Structured questionnaire personally administered was the main instrument of data collection. The instrument was divided into four sections. Section A contained information on demographic characteristic of palm oil processors in the study area, Section B had factors that affect palm oil processing taken on a four-point Likert scale which include strongly agreed, agreed, disagreed and strongly disagreed, while Section C assessed the method and training requirement for oil palm fruit processing in the study.

Data collected were analyzed using descriptive and inferential statistics. Descriptive statistics used include frequency, percentage and mean while Pearson correlation (inferential statistics) was used to test for significant relationship between socio economic characteristics and training needs at different processing stages.

RESULTS AND DISCUSSION

Socio Economic Characteristics of Palm Oil Processors

The results of socio-economic characteristics are presented in Table 1. The table showed that majority (56.7%) of the palm oil processors are between the ages of 31-40. This implies that majority of the oil palm processors are in their youthful years. It therefore implies that youth are involved in palm oil processing which will be an advantage for training and technological transfer.

Male constitute a minority (18.9%) of the respondents. This implies that oil palm processing is a female dominated practice. Furthermore, about half (41.1%) of the oil processors have secondary education indicating an appreciable level of education that can be depended upon for knowledge transfer. Although, high literacy level can aid adoption of technical knowledge, secondary school education is still useful for transfer of skill and useful information. The respondents have appreciable years of processing as majority (56.7%) of them have 11-20 years of processing experience with only a minority (5.6%) having ten years or below.

Majority (55.6%) of the processors have family lands with oil palm while only 8.9% of the respondents cultivate their oil palm on purchased lands. It can thus be deduced that most of the processors are natives or married to natives who has land inheritance in the study area. Their possibility of remaining in this

Table 1. Socio economic characteristics of palm oil processors

Socio Economic Characteristics		Frequency	Percent
Age	21 – 30	5	5.6
	31 – 40	51	56.7
	41 – 50	22	24.4
	51 and above	12	13.3
Gender	Male	17	18.9
	Female	73	81.1
Marital Status	Single	11	12.2
	Married	71	78.9
	Divorced	8	8.9
Educational attainment	No formal education	17	18.9
	Primary education	35	38.9
	Secondary school	37	41.1
	Tertiary	1	1.1
Processing experience	1-10	5	5.6
	11 – 20	51	56.7
	21 – 30	22	24.4
	31 – 40	12	13.3
Land Ownership	Purchased	8	8.9
	Hired	10	11.1
	Communal	22	24.4
	Family	50	55.6
Method of oil palm fruit processing	Manual	69	76.7
	Mechanized	21	23.3
	Total	90	100.0

Source: Field Survey, 2019

processing venture is therefore high. If their training need is known and successfully worked upon, palm oil industry will receive a boost in the country. Furthermore, the processors mostly (77.5%) make use of manual method of palm oil processing. This confirms the assertion that manual method of processing is more prevalent among small scale processors in Nigeria (Olagunju, 2008).

Training Needs of Palm Oil Processors by Stages in Manual Method of Processing

Table 2 shows the stages at which training is required by majority in manual processing. These are pounding (50.7%), mashing (66.7%) and separation of fibre from nuts (84.1%). Other stages include squeezing (71%), boiling (71%), skimming (71%) and filtering (75.4%).

This result shows the processors agree they require training at all stages of manual processing. The result cannot be disregarded as these are experienced in this activity.

Table 2. Training needs of manual palm oil processors

Processing Stages		Frequency	Percent
Pounding of fruit	No	34	49.3
	Yes	35	50.7
Mashing in hot or cold water	No	23	33.3
	Yes	46	66.7
Removal of fibre and nuts in small baskets	No	11	15.9
	Yes	58	84.1
Hand squeezing	No	20	29.0
	Yes	49	71.0
Filtering out residual fibre from the oil	No	17	24.6
	Yes	52	75.4
Manual boiling	No	20	29.0
	Yes	49	71.0
Skimming of palm oil	No	20	29.0
	Yes	49	71.0

Source: Field Survey, 2019

Training Needs of Palm Oil Processors by Stages in Mechanized Method of Processing

Table 3 shows that training needs of mechanized palm oil processors. Majority requires training in threshing (65%), sterilization (60%), digestion (60%) and pressing (55%). Further training needs of the processors include pressing (55%), heating (65%) and filtering (70%).

However, majority of the processors revealed that they do not require training in fermenting (45%) and extraction (45%). Although the percentage of manual processors that require training at different stages is greater than that of the mechanized processors, there is still exist for training even among the mechanized processors. This is true as the need for training was not in the majority for just two processing stage i.e. fermenting (45%) and extraction (45%).

Test of Association Between Socio Economic Characteristics of Palm Oil Processors and Stages in Manual Processing

Table 4 reveals the result of Pearson Correlation two tailed significance analysis. The analysis tests significant relationship between socio economic characteristics of palm oil processors and stages in manual processing. The result of the analysis shows that there is a significant negative relationship between training needs on mashing (r=0.246, p=0.05); removal of fibre (r=0.345, p=0.01); boiling (r=0.352, p=0.01); and age of processor. The aged therefore requires less training in mashing, removal of fibre and boiling in manual processing of palm oil. Increasing age reduces training need in these stages of processing. This may not be unrelated to experience which may also increase with age. Also, gender (r=0.264, p=0.05), educational status (r=0.351, p=0.01), and marital status of the respondents (r=0.244; p=0.01)

Table 3. Training needs of mechanized palm oil processors

Processing Stages		Frequency	Percent
Fermenting	No	11	55.0
	Yes	9	45.0
Threshing	No	7	35.0
	Yes	13	65.0
Sterilization	No	8	40.0
	Yes	12	60.0
Digestion	No	8	40.0
	Yes	12	60.0
Extraction	No	11	55.0
	Yes	9	45.0
Pressing	No	8	40.0
	Yes	12	60.0
Mechanized Boiling	No	9	45.0
	Yes	11	55.0
Heating	No	7	35.0
	Yes	13	65.0
Filtering	No	6	30.0
	Yes	14	70.0

Source: Field Survey, 2019

had positive and significant influence on boiling, boiling and pounding in manual palm oil processing respectively. The result implies that there is difference in training requirement within gender, educational status and marital status of palm oil processors using manual method.

Test of Association Between Socio Economic Characteristics of Palm Oil Processors and Stages in Mechanized Processing

Table 5 reveals Pearson Correlation analysis for test of association between socio-economic characteristics of mechanized palm oil processors and stages in mechanized oil palm processing. The result of the analysis shows that there is a negative significant relationship between filtering and age of the processors. An increase in age therefore influence increase in filtering training requirement. Gender also has positive significant relationship on fermenting (r=0.452, p=0.05), sterilization (r=0.612, p=0.01), extraction (r=0.452; p=0.05) and filtering (r=0.491, p=0.05). Gender thus has effect on these stages of mechanized oil palm processing.

Perceived Factors That Affect the Palm Oil Processing

A four-point Likert Scale (Strongly agreed, agreed, disagreed and strongly disagreed) and thirteen (13) factors that affect the palm oil processing were identified in this study. The perceived factors that affect

Table 4. Test of association between socio economic characteristics of palm oil processors and stages in manual processing

Variables		Pounding	Mashing	Removal of Fibre	Hand Squeezing	Filtering	Boiling	Skimming
Age	Pearson Correlation	.101	-.246*	-.345**	-.231	-.133	-.352**	.011
	Sig. (2-tailed)	.407	.041	.004	.056	.278	.003	.928
Gender	Pearson Correlation	-.104	.210	.195	.182	-.017	.264*	.101
	Sig. (2-tailed)	.394	.084	.108	.134	.887	.028	.411
Educational status	Pearson Correlation	.035	.070	.066	-.041	-.209	.351**	.090
	Sig. (2-tailed)	.772	.569	.587	.738	.085	.003	.464
Processing experience	Pearson Correlation	.146	-.203	-.061	-.128	-.048	-.044	-.086
	Sig. (2-tailed)	.230	.095	.617	.294	.694	.720	.482
Marital status	Pearson Correlation	.244*	-.150	.030	-.011	-.069	-.077	-.077
	Sig. (2-tailed)	.043	.218	.807	.931	.571	.527	.527

*Correlation is significant at 0.05 Level**Correlation is significant at 0.01 Level.

Table 5. Test of association between socio economic characteristics of palm oil processors and stages in mechanized processing

Variables		Fermenting	Threshing	Sterilization	Digestion	Extraction	Pressing	Mechanized Boiling	Heating	Filtering
Age	Pearson Correlation	-.256	-.111	-.178	-.051	-.381	-.304	-.119	-.371	-.447*
	Sig. (2-tailed)	.276	.642	.454	.832	.097	.192	.618	.107	.048
Gender	Pearson Correlation	.452*	.419	.612**	.102	.452*	.357	.302	.419	.491*
	Sig. (2-tailed)	.045	.066	.004	.669	.045	.122	.196	.066	.028
Marital status	Pearson Correlation	.406	.096	-.159	-.159	-.118	-.159	.379	.368	.369
	Sig. (2-tailed)	.076	.689	.502	.502	.621	.502	.099	.110	.109
Educational status	Pearson Correlation	.336	.178	.096	-.024	.100	.096	.018	.055	.141
	Sig. (2-tailed)	.148	.452	.688	.920	.674	.688	.941	.817	.554
Farming experience	Pearson Correlation	-.396	.206	-.327	.302	-.272	.302	-.099	-.052	-.295
	Sig. (2-tailed)	.084	.382	.160	.196	.246	.196	.678	.829	.206

*Correlation is significant at 0.05 Level, **Correlation is significant at 0.01 Level.

palm oil processing were coded as strongly agreed = 4, agreed = 3, disagreed = 2, strongly disagreed = 1. The Likert scale values were used to calculate the mean response on each perceived factor and this was used in discussing the results.

Table 6 shows that poor extension services was rated highest among the perceived factors that affect palm oil processing in the study area with a mean score of 3.03. It was not surprising that poor extension services were rated highest among other factors that affect palm oil processing in the study area. This is because processors are far away from agricultural innovations and some helpful information (Anderson, 2004). High cost of hired labour ($\overline{x}$ =2.85), high cost of milling machine/processing facilities ($\overline{x}$ = 2.78), insufficient fund for buying processing machine ($\overline{x}$ = 2.56) are also hindering palm oil processing. High cost of machine and insufficient fund to buy processing machine might be the reason majority of the palm oil processor in the study area are making use of manual method for palm oil processing in the study area.

Table 6. Perceived factors affecting palm oil processing

Factors	Mean
Poor extension services	3.03
High cost of hired labour	2.85
High cost of milling machine/ processing facilities	2.78
Insufficient fund for buying of processing machine	2.56
Inefficiency of processing methods	2.00
Lack of policy support to develop the production of quality palm oil	2.55
Scarcity of labour/ Shortage of labour	2.53
Presence of numerous actors and intermediaries involved in the value chain palm oil processing and marketing	2.48
Poor incentives to processors	2.46
Poor quality of palm oil produce	2.43
Poor access to good road network for easy transportation	2.38
Use of poor variety of oil palm	2.18
Lack of storage facilities	2.16

Source: Field Survey, 2019

SUMMARY, CONCLUSION AND RECOMMENDATIONS

The main objective of this study was to assess the training needs of oil palm processors in Ijebu North Local Government Area of Ogun state. The study was conducted on ninety (90) palm oil processors which were selected through a multistage sampling procedure. Data collected was analyzed using descriptive statistics, involving percentage, frequency, mean etc. and inferential statistics (Pearson correlation analysis). The simple percentage analysis was used to describe demographic variables while Pearson correlation was used to test association between socio demographic characteristics and different stages of palm oil processing. Constraints associated with palm oil processing in the area were also investigated. Based on evidence from the study, the following conclusions were drawn:

- Majority of the palm oil processors are young, married and experienced in palm oil processing. Women are mostly involved in palm oil processing and use manual method of processing. Both manual and mechanized palm oil processors require training at different processing stages.
- The study further reveals a significant negative relationship between training needs on mashing, removal of fibre, boiling and age of manual oil palm processor. Also gender and educational status had positive and significant influence on boiling while marital status of manual palm oil processor had positive and significant influence on pounding. Age has negative and significant relationship on filtering for mechanized palm oil processor while gender has positive and significant relationship on fermenting, sterilization, extraction and filtering.
- Poor extension service, high cost of labour and high cost of milling machine were the most perceived constraints to palm oil processing in the study area.

The following recommendations are proffered in light of the above findings and conclusion:

1. Government and non-governmental organization should promote palm oil processing among women by providing machines for mechanized processing.
2. Awareness should be increased for private plantation of oil palm so as to make more fruits available for processing.
3. Extension officers should intensify effort on training both manual and mechanized palm oil processing methods across different socio-economic characteristics.
4. Local engineering and construction companies should develop low cost labour saving technologies and locally fabricated machines for palm oil processing.

REFERENCES

Bello, R.S., Bello, M.B., Essien, B.A., & Saidu, M.J. (2015). Economic Potentials of Oil Palm Production in UDI, Enugu State. *Science Journal of Business and Management*. Special Issue: *Sustainable Entrepreneurial Developments in Agribusiness, 3*(5-1), 16-20. doi:10.11648/j.sjbm.s.2015030501.14

Butler, R. A. (2006). *Why is palm oil replacing tropical rainforests?* Retrieved from https://news.mongabay.com/2006/04/why-is-palm-oil-replacing-tropical-rainforests/

FAO. (2002). Small Scale Palm Oil Processing in Africa. FAO Agricultural Services Bulletin, 148.

Green, M. M. (2003). *Igbo village Affair* (2nd ed.). London: Frank Cass.

Hagman, J., Chuma, E., Murwira, K., & Connolly, M. (2000). Learning Together through Participatory Extension. *African Journal of Marketing Management, 3*(10), 253–260.

Ibitoye, O. O., Akinsorotan, A. O., Meludu, N. T., & Ibitoye, B. O. (2011). Factors Affecting Oil Palm Production in Ondo State of Nigeria. *Journal of Agriculture and Social Research*, 2.

Jasim, M. S., Norsida, M., Ahmad, H. L., Majeed, H. S., Salim, H., Nolila, M. N., & Bassim, H. K. (2016). A Review: Training Requirement of Agriculture Extension Officers in Iraq. *Asian Journal of Applied Sciences, 9*(2), 34–40. doi:10.3923/ajaps.2016.34.40

Obibuaku, L. O. (1983). Agricultural Extension as a Strategy for Agricultural Transformation. Nsukka: University of Nigeria Press.

Olagunju, F. I. (2008). Economics of Palm Oil Processing in South Western Nigeria. *International Journal of Agricultural Research*, *1*(2), 69–77.

Orewa, S. I., Adakaren, B., Ilechie, C. O., & Obulechei, S. (2009). An Analysis of the profitability of using the NIFOR small scale palm oil processing equipment (SSPE). *American-Eurasian Journal of Agronomy*, *2*(3), 192–300.

Ugwu, D. S. (2009). Problems and Prospects of Commercial Small and Medium Scale Cocoa and Oil Palm Production in Cross River State, Nigeria. *Journal of Applied Sciences Research*, *5*(7), 827–832.

Wentling, T. L. (1992). *Planning for Effective Training: A Guide to Curriculum Development*. Rome: Food and Agriculture Organization of the United Nations.

Wikipedia. (2019). *Ijebu North Local Government Area in Ogun State*. Retrieved from https://en.wikipedia.org/wiki/Ijebu_North

WRM. (2010, Dec). *World Rainforest Movement. Oil palm in Nigeria: shifting from smallholders and women to mass production.* Retrieved from https://oilpalminafrica.wordpress.com/2010/08/06/oil-palm-in-nigeria/

Youdeowei, A., & Kwarteng, J. (2006). *Tool Kit for the Production of Agricultural Extension Materials: Guidebook*. CTA Publishing.

Section 5

Gender and Its Implications on Sustainable Food Systems and Food Security

Chapter 10
Women Land Rights and Food Security Status of Farming Households in Oyo State, Nigeria

Abimbola O. Adepoju
Department of Agricultural Economics, University of Ibadan, Nigeria

Rahman A. Adewole
Department of Agricultural Economics, University of Ibadan, Nigeria

ABSTRACT

The dominance of men in decision-making processes and leadership positions within the communities has made land allocation, land use, and control skewed in favour of men. This study examined the effects of women's land rights on households' food security status using a sample of 300 representative farmers. Descriptive statistics, household food expenditure, logistic regression, and ordered logit models were the analytical tools used. Results revealed that about 35% of the rural women farmers had land use rights while the remaining 65% had land ownership rights. Women with ownership rights were more food secure, with the majority of the women having residual rights, while only a few had sell rights. Secure women land rights are germane to achieving and sustaining household and national food security. Strategies and instruments for protecting women rights should be developed and implemented, while efforts geared towards designing strategies, assessing multiple dimensions of women empowerment for improved food security status, and welfare of the households should be intensified.

INTRODUCTION

Women play a critical role in agricultural production in developing countries where they usually make up most of the agricultural workforce (World Bank et al., 2009). While their participation in agriculture in sub-Saharan Africa accounts for close to 70-80% of labour and 90% food processing and storage,

DOI: 10.4018/978-1-7998-2599-9.ch010

they own less than 20% of the land (Murisa, 2008). Land rights is defined as the indisputable ability of individuals and group of individuals to obtain, possess and utilize land at their discretion as long as their activities on the land do not impede on other individuals' rights (Adi, 2009). Under the customary land tenure system, which is still very much prevalent, the distribution of rights is based on socio-political system (the political history of the village and region from which the alliances and hierarchical relationships between lineages are derived) and on family relationships (access to land and resources depending on one's social status within the family). It is also worth noting that in most of these customary landholding systems, community level decisions about land are taken by chiefs or headmen on behalf of and in trust for the clan or family (Umezulike, 2004).

In Nigeria, the Land Use Act, enacted in 1978, was meant to standardise land administration systems across the country. It vested all urban land within a state in the state governor, and all non-urban land in the local governments in which they are found. The state governor and local government authorities are empowered by the Act to grant "statutory rights of occupancy". While both urban and rural land is secured in Nigeria through certificates of occupancy (instruments of title issued as evidence that the state has conferred on the holder of the certificate the statutory right to occupy the land for a defined period of time and deeds of assignment (agreement between the person with the rights to a piece of land and the person to whom the rights are being transferred), rules for transfer and succession depend on whether the person died with or without a will recognised by the courts and mainly by inheritance rights, which are primarily guided by native and customary laws (with variations across ethnic groups) and religious laws (sharia law, based on the Koran). For example, sharia laws, applicable across the 19 northern states, stipulate that female children get half of what males get and that children who are non-Muslims lose their inheritance rights (Africa Check, 2015).

However, secure access to productive land is critical to the millions of poor people living in rural areas, who depend on agriculture, livestock or forests for their livelihood. This is because it reduces their vulnerability to hunger and poverty; influences their capacity to invest in their productive activities and in the sustainable management of their resources; enhances their prospects for better livelihoods, and helps them develop more equitable relations with the rest of their society, thus contributing to justice, peace and sustainable development (International Fund for Agricultural Development [IFAD], 2010). Throughout Africa, most poor women (most of whom depend on land for their livelihood) are either landless or have limited and insecure rights to land. This reality has important consequences for sustainable socio-economic development of the continent as well as their food security status since they play a major role in agriculture (Odeny, 2013). In other words, when deprived of access to, ownership and use of land, women are left without the means to create stable and sustainable livelihoods and are food insecure.

However, food security, defined as a situation that exists when all people, at all times, have physical, social and economic access to sufficient, safe and nutritious food that meets their dietary needs and food preferences for an active and healthy life, has been linked to having secured land rights. This is particularly pertinent for women in ensuring equality of basic rights, reducing poverty and ensuring household food security (Jonckheere *et al.*, 2013). For instance, it had been shown that if women had the same access and ownership to productive resources as men, they would increase yields on their farms by 20-30 percent. This could raise total agricultural output in developing countries by 2.5 – 4.0%, which is enough to pull 100-150 million people out of hunger (Food and Agriculture Organization [FAO], 2010). Yet, women's land rights have been largely ignored mainly because of the highly patriarchal nature of most societies in the developing world.

Further, across the developing world women are less likely to own or operate land; when they do, the land they can access is often of poorer quality and in smaller plots. For example, in Kenya, men's landholdings are on average three times larger, and in Bangladesh, Ecuador and Pakistan they are twice the size of women's (FAO, 2011). These gender inequalities not only affect women's status; they have significant implications for food and nutrition security at the level of the household and community. Landesa (2012) reports that where women lack rights or opportunities to own land, there is an average of 60 per cent more malnourished children. They also report that when women have direct control over assets such as land and income, this increases their decision-making power and status, resulting in positive nutritional impacts for them and their families. Though state laws, including land titling, may protect women's rights to own land, customary laws often take precedence at the local level. A number of countries recognise both formal and customary land tenure systems and laws, but there are inherent contradictions in trying to accommodate both systems. For example, in Malawi and many other African countries, formal recognition of women's right to own land co-exists with an often contradictory parallel set of customary laws. When divorced women return to their natal villages they may only use land through male members of the family or are allocated a piece of land by the Chief or their clan members. In other cases, widows are chased away from their natal villages (ActionAid et al., 2012). In Nigeria, it is striking that there is no recognized formal category for the particular character of women's land access. Marriage is used as a determining variable in women's land rights because it is the major means by which women and men access land in Nigeria. However, whereas women's land rights are dependent on their relations with men, men's land rights are not dependent on their relations with women. Moreover, women are threatened with dispossession if divorced or widowed.

The justification of this study is rooted in the need to adequately respond to the problem of food security and peculiar difficulties women face in meeting their responsibilities as food providers particularly that of weak land rights as no literature has yet been sighted on women land rights in Oyo state. Considering that women have the tendency to grow food as opposed to cash crops and to spend income on family food, their security of tenure must be viewed as a key link in the chain from household food production to national food security (Mahoi, 2015). This study will thus take the initiative that will later stimulate further specialized researches on this problem. Also, closing the gender gap in secure access to land is fundamental not only for women's empowerment, but also for broader family food security, children's health and economic gains (FAO, 2011).

While there are many studies on land access, the relationship between women land rights and food security has not been well explored in literature. Thus, the justification of this study lies in its heuristic nature, implications for policy theory, research, and contribution to the body of data on gender, food security and secured women land rights. Based on this foregoing, this study therefore attempts to empirically assess what type(s) of land rights women in the farming households have; what factors influence the type of land rights of women in the farming households and the effects of women land rights on the food security status of farming households in Oyo state, Nigeria.

METHODOLOGY

The study was carried out in Oyo state Nigeria. Oyo State is one of the states in the south western part of Nigeria, It is located within Latitude 8^0 N, and Longitude 4^0 E with a total land mass of 28,454 square kilometres reaching a height of about 1,219 meters and a population of 5,591,589 according to (NPC,

2006) and it is bounded in the North by Kwara State, in the East by Osun State, in the South by Ogun State and in the Western part by Ogun State and partly by the Republic of Benin. The state has an equatorial climate with dry and wet seasons and relatively high humidity. The dry season lasts from November to March while the wet season starts from April and ends in October. Average daily temperature ranges between 25^{0}C (77.0 ^{0}F) and 35 ^{0}C (95.0 ^{0}F), almost throughout the year. The vegetation pattern of Oyo state is rain forest in the South and guinea savannah in the North. Thick forest in the south gives way to grassland, interspersed with trees in the North also, the climate in the state favours the cultivation of crops like maize, yam, cassava, millet, rice, plantain, cocoa tree, palm tree and cashew etc.

This study made use of primary data obtained through questionnaire administration to respondents. The information used for this study are the socio-economic characteristics such as age (years), gender, education, household size and marital status, food security variables such as expenditure on food as well as type of land rights that women have in the study area. A multi-stage random sampling technique was employed for the purpose of this study. The first stage was a random sampling of one Local Government Area (LGA) each from the three Senatorial Districts (SD) in Oyo State. Thus, Afijio LGA was randomly selected from Oyo Central SD, Iseyin LGA from Oyo North SD and Ibarapa East LGA from Oyo South SD. The second stage was the random selection of five communities/villages from each of the LGAs. This gave a total of 15 villages/communities. The last stage was a random sampling of 20 women farmers from each of the selected communities. In all, a total of 300 respondents were selected. However, only data from 293 farming households were analyzed as others were discarded for inconsistencies and or incomplete information .

Analytical tools used include Descriptive statistics, Logistic and Ordered Logit regression Models. Descriptive statistics was used to examine the socio-economic characteristics and type of land rights women have in the farming households while the Logistic Regression Model was used to examine the factors influencing the type of land rights of women in the farming households. Two separate logistic regression analysis were carried out for the types of land rights (Use and Ownership rights). The Logistic model is associated with a cumulative normal probability function and it is a type of regression model used to analyze binomial response variables (dichotomous variable).

The basic logistic model is specified as follows:

$$P\ (Yi{=}m) = 1/1 + e^{-z} \tag{1}$$

$$P/1\text{-}P = e^{z} \tag{2}$$

where

P is the probability of occurrence of the dependent variable Yi equal to a certain value, m;
z is the predictor variable and can be said to be a linear combination of the conversion factors;
e is the base of natural logarithm and
P is the estimated probability of occurrence of one point of the dependent variable.

From equation 2,

$$1\text{-}P = 1\text{-}\ 1/1 + e^{-z} \tag{3}$$

1-P is the probability of failure.

Given that Y =P/1-P (4)

Then, $Y = e^z = \exp(z)$ (5)

Y = P/1-P, represents the odd of the evaluative factors occurring for each explicative factors.

Assuming Z is a linear function of a set of predictor variables, then,

$$Z = \beta_0 + \beta_1 X_{1i} + \beta_2 X_{2i} + \beta_3 X_{3i} + \ldots B_n X_{ni} \quad (6)$$

If equation (6), then;

$$Y = e^{\beta 0 + \beta 1 X 1i + \beta 2 X 2i + \beta 3 X 3i + \ldots + \beta n X ni} \quad (7)$$

$$Y = \exp(\beta_0 + \beta_i \Sigma X_{1i} + \beta_2 \Sigma X_{2i} + \beta_3 \Sigma X_{3i} + \ldots B_n \Sigma X_{ni}) \quad (8)$$

where Y= Land rights (Ownership/Use rights =1, 0 otherwise). X_{is} include those socio-economic factors that influence the type of land rights. These include:

X_1 = Age (years), X_2 = No formal education, X_3 = Primary education, X_4 = Secondary education, X_5 =Christianity, X_6=Islam, X_7 = Married, X_8=Separated, X_9=Divorced, X_{10}=Widowed, X_{11}=Access to credit facilities, X_{12}=Income (₦),X_{13}=Membership of cooperative societies, X_{14}=Inheritance, X_{15} =Marriage, X_{16} =Gift, X_{17}=Farming experience, e_i = Error term or Disturbance term.

The Household Food Expenditure approach was used to classify households into their food security status. The share of the total household expenditure spent on food is an indicator of household food insecurity because the poorer and more vulnerable a household, the larger the share of household income spent on food. This implies that households that are very poor and consuming the lowest-cost foods will be unable to substitute cheaper foods and will be forced to spend more on basic staples, reduce the quality of their diets or even reduce the quantity consumed of the least expensive foods while also reducing non-food expenditures that may be equally needed (Lele et al., 2016).

This indicator is commonly calculated with data from Household Consumption and Expenditure Surveys (HCES) that include the monetary value of household consumption disaggregated into food and non-food items and is a measure of current economic vulnerability.

$$The\ share\ of\ household\ expenditure\ on\ food = \frac{Expenditure\ on\ food}{Total\ Expenditure} \times 100$$

Households spending over 75% of their income are considered very vulnerable and consequently highly food insecure whereas people spending between 65 and 75% are considered to be moderately food insecure, those spending 50 to less than 65% are moderately food secure and those that spend less than 50% of their income on food are considered to be food secure (Smith and Subandaro, 2007). Although there is no agreed international threshold, this indicator is one of the several indicators included in the ADePT-FSM (Food Security Module), developed by the Food and Agriculture Organization (FAO) and the World Bank that allows users to easily derive food security indicators from household survey data. This indicator is also included in the FAO suite of food security indicators (FAO, 2016)

The Ordered Logit Regression Model was used to examine the effects of Women Land Rights on the food security status of the farming households.

The Ordered Logistic Regression model is stated explicitly as follows:

$$y_i * = \beta i \; Xi + \varepsilon i \; -¥ < y_i * < -¥ \quad (9)$$

where

y_i*: Food security status
βi: Vector of parameters that would be estimated
xi: Observed vector of non-random explanatory variable which shows the characteristic of ith person
εi: Residual error which is logistically distributed.

Since y_i * is a latent variable, standard regression techniques are not applicable to estimate the sample size.

If y_i is considered as a discrete and observable variable which shows different levels of food security of the respondents, the relation between latent variable yi* and observable variable yi is obtained from the ordered logit model as follows:

$$y_i = 1 \text{ if } -¥ \leq y_i * < \mu_1, i = 1, \ldots, n, \quad (10)$$

$$y_i = 2 \text{ if } \mu_1 \leq y_i * < \mu_2 \; i = 1, \ldots, n, \quad (11)$$

$$y_i = 3 \text{ if } \mu_2 \leq y_i * < \mu_3 \; i = 1, \ldots, n, \quad (12)$$

$$y_i = J \text{ if } \mu_{J-1} \leq yi* < +¥ \; i = 1, \ldots, n,$$

in which "n" is the value for the sample size, "μ" and "s" are the thresholds that define observed discrete answers and would be estimated. The probability of y_i= J would be calculated by the following relation:

$$\Pr(y_i = J) = \Pr(y_i \geq \mu_{J-1}) = \Pr(\varepsilon i \geq \mu_{n-1} - \beta Xi) = F(\beta Xi - \mu_{J-1}) \quad (13)$$

The dependent variables Y_i include the levels of food security: Y_1=highly food insecure, Y_2=moderately food insecure, Y_3=moderately food secure and Y_4=food secure. Xi includes various socioeconomic and demographic variables that influence women's right to land. The explanatory variables include:X_1 = Age (years), X_2 = No formal education, X_3 = Primary education, X_4 =Secondary education,X_5=Christianity, X_6=Islam, X_7 = Married, X_8=Separated, X_9=Divorced, X_{10}=Widowed, X_{11}=Access to credit facilities, X_{12}=Membership of cooperative societies, X_{13}=Inheritance, X_{14}=Marriage, X_{15}=Tenancy, X_{16}=Leasehold, X_{17}=Gift, X_{18}=Purchase, X_{19}=Type of land right (ownership rights = 1, 0 otherwise),e_i = Error term or Disturbance term

RESULTS AND DISCUSSION

The result of the socioeconomic characteristics of the farming households (Table 1) shows that a larger percentage (33.4%) of the farmers were between the ages of 41 and 50 years with a mean age of about 44 ± 11.5 years. This implies that most rural women farmers in the study area are still productive and active. More than two-fifths of the respondents had no formal education, while about half had either primary or secondary education, but with majority having primary education. However, less than 5% had tertiary education. Majority of the farmers had household sizes of between 4 and 6 members with a mean of about 6 ± 3 members per household. More than four-fifths of the respondents were engaged in farming as their major occupation while a few were involved in the processing and marketing of farm produce. A larger percentage of the respondents are married and had access to credit facilities obtained mainly through informal sources such as thrift societies. According to Udoh (2005) in agricultural financing, informal credit sources are unquestionably the most popular. These informal sources are provided by traditional institutions that work together for the mutual benefits of their members (Ijere, 2000).

A wide extension agents-farmers gap was observed as about two-fifths of the respondents did not have access to extension agents at all and about one-fifths had access only bi-annually. This finding corroborates earlier findings in literature that only 5% of women farmers spanning 97 countries have access to extension services (FAO, 2013)

With respect to women's land rights in the farming households (Table 2), a little above half of the respondents farmed less than 1hectare of land while only less than 2% farmed more than 5 hectares. This situation is worrisome, given the leading role that women play in food production and provision for family consumption in developing countries. Women farmers' access to land is even more limited due to cultural, traditional and sociological factors. This is as a result of the customary nature of land distribution and ownership that operates in the study area. Such limited access is very tenuous and can be quickly lost (FAO, 2002; Mirtuse et al., 2006; Quisumbing et al., 1995). Also, about two-fifths of the respondents respectively acquired their lands through inheritance and marriage owing largely to customary laws governing land ownership in the study area. In other words, land ownership depicts affluence and power in the rural community and serves as home to ancestors in some places. Although women can purchase land, the institutional arrangement that exists in most rural societies does not always favour this, because women have to rely on their husbands before making a purchase and as such it is uncommon in the rural areas for a woman to purchase land in her own name with or without her husband's consent. Thus, a common scenario is for women to access land only through their husbands or senior male relatives (Ademola, 1994; Törhönen & Palmer, 2004; Elliss, 2000). Almost about two-fifths (35.1%) of women in farming households have land use rights as against 64.9% that have only ownership rights. Of those that had ownership rights, 48.5% had residual rights (the right of a woman to use land perpetually which can be redeemed later by the husband upon divorce, by her husband's relatives after her husband's death or by the community if such a woman commits an illicit act that is against the communal custom and tradition), 10.9% had transfer rights (the right of the woman to transfer the ownership rights to her husband, her children, her relatives or to whosoever she wishes as long as the beneficiary is an indigene of the community) while just 5.5% had sell rights to their lands (absolute right to own, use and control the land). Historically women's access to land is mainly based on status within the family and involved the right of use, not ownership (Juma & Ojwang, 1996).

Table 1. Socio-economic characteristics of the farming households

Variables		Frequency	Percentage (%)
Age (years)	<30	41	14.0
	31-40	74	25.3
	41-50	98	33.4
	51-60	50	17.1
Mean: 44.4 SD: 11.5	^{3}61	30	10.2
Educational status	No formal education	134	45.7
	Primary	95	32.4
	Secondary	50	17.1
	Tertiary	14	4.8
Household size	1-3	47	16.0
	4-6	138	47.1
	7-9	80	27.3
Mean: 5.9 SD: 2.8	^{3} 10	28	9.6
Main occupation	Farming	236	80.5
	Trading	12	4.1
	Civil servants	12	4.1
	Processing	31	10.6
	Marketing	2	0.7
Marital status	Single	11	3.8
	Married	214	73.0
	Separated	21	7.2
	Divorced	14	4.8
	Widowed	33	11.3
Access to credit	No	28	9.6
	Yes	265	90.4
Source of credit	Friends and relatives	42	14.3
	Thrifts/Esusu	191	65.2
	Cooperatives	26	8.9
	Commercial banks	6	2.0
Access to Ext. Agents	None	122	41.6
	Monthly	3	1.0
	Quarterly	54	18.4
	Bi-annually	60	20.5
	Annually	54	18.4

Source: Field Survey, 2017.

Table 2. Women land rights characteristics

Land Characteristic		Frequency	Percentage (%)
Size of land (Ha)	< 1	153	52.2
	1-2.99	126	43.0
	3-4.99	9	3.1
	^{3}5	5	1.7
Mode of acquisition	Inheritance	111	38.3
	Marriage	106	36.6
	Tenancy	17	5.9
	Leasehold	2	0.7
	Gift	33	11.4
	Purchase	21	7.2
Type of Right	Use right	103	35.1
	Residual right	142	48.5
	Transfer right	32	10.9
	Right to sell	16	5.5
	Total Ownership rights	190	64.9
Type of crops cultivated	Arable crops	111	37.8
	Tree/cash crops	22	7.6
	Vegetables	160	54.7

Source: Field Survey, 2017.

In Africa, custom excludes women from ownership which makes women rights to land to be secondary and dependent on their relationships with their husbands, male relatives and the headmen in their communities. Under the customary land tenure system which is very much prevalent in the study area, the distribution of rights is based on socio-political system and family relationships, thus, most of these customary landholding systems and community level decisions on land are taken by chiefs or headmen on behalf of and in trust for the clan or family. It is worthy of note that majority of land owned by these rural women farmers are in microplots and of poor fertility which do not favour large scale commercial farming, however land rights in the study area and Africa at large is said to be gender biased (Agwu et al., 2010).

Also, more than half of the rural women farmers in the study area cultivate vegetables as their primary crops, these include both fruits and leafy vegetables, the discrimination in crops cultivated however varies across the country as some crops are designated as female crops (IFPRI, 2005). For instance, in Southeast Nigeria, yam is the traditional male prestige crop while cassava and other crops like melon, cocoyam are female crops, while in the southwest part of the country vegetable crops are mostly female crops. It can also be observed that just 7.6% of the respondents cultivate tree and cash crops and these farmers inherited the tree/cash crops plantation mostly from their parents. Gender biased differences also existed in relation to the use of forest resources. This discrimination against women in crop cultivation in the study area is mainly due to the lack of secured land, required farm productive resources as well as non- availability of labour as a result of the labour-intensive nature of tree and arable crops (FAO, 2011).

Following the household food expenditure approach as shown in (Table 3), majority of the households were classified as food secure; 24.9% and 40.3% as food secure and moderately food secure respectively while 31.7% were classified as moderately food insecure and only 3.1% as highly food insecure. This result, although contrary to *a priori* expectation is however not surprising because about 65% of the respondents had land ownership rights (secured land rights), which according to literature improves their bargaining power, decision making on foods, improved household food and nutritional security as well as overall household welfare (FAO, 2012; Mitchell & Hanstad, 2004).

Table 3. Food security status of the rural women farming households

Food Security Status	Frequency	Percentage (%)
Food secure	73	24.9
Moderately food secure	118	40.3
Moderately food insecure	93	31.7
Highly food insecure	9	3.1
Total	**293**	**100.0**

Source: Field Survey, 2017.

A profile of food security status by type of land rights showed that women that have ownership rights were more food secure when compared to those that have land use rights. On the other hand, women that have land use rights were more food insecure relative to those that had land ownership rights. These variation in food security levels could be attributed to the fact that women that have land ownership rights are likely to have better control of their lands and are able to determine what to produce. As a result, they able to vary their diets and benefit from the forest resources either for their own family consumption or sales which will boost their source of income and hence improve their food security status. In summary, households where women have land rights are more food secure and have better decision-making power relative to those without secure land rights (Miggiano, 2010; Katz & Chamorro, 2002).

Table 4. Profile of the food security status of respondents by land rights

Food Security Status	Use Right		Ownership Rights	
	Frequency	Percentage (%)	Frequency	Percentage (%)
Food secure	4	3.9	69	36.3
Moderately food secure	41	39.8	77	40.5
Moderately food insecure	51	49.5	42	22.1
Highly food insecure	7	6.8	2	1.1
Total	**103**	**100.0**	**190**	**100.0**

Table 5 presents the results of the Logistic regression analysis of the factors influencing land use rights among women in farming households in the study area. The log likelihood value of -172.1660 and Chi-square value of 26.77 which was significant at 5% indicates that the model fits the data. Age had negative effects on the likelihood of having land use rights while marital status and access to credit facilities had positive effects. Marginal effects results of the analysis are discussed as follows:

Table 5. Factors influencing land use rights of women in farming households

Variables	dy/dx	Z-Value	P>/Z/
Age	-0.0065	-2.30**	0.021
No formal education	0.0531	0.29	0.772
Primary education	0.2001	1.09	0.274
Secondary education	0.1764	0.94	0.345
Christianity	-0.2255	-1.55	0.121
Islam	-0.1908	-1.29	0.197
Married	0.4292	2.41**	0.016
Separated	0.1726	1.10	0.272
Divorced	0.1521	0.71	0.478
Widowed	0.2684	1.41	0.158
Access to credit	0.1676	1.83*	0.068
Income	-6.52e-08	-0.20	0.842
Farming Experience	0.0038	0.92	0.355
Constant		0.56	0.574

Source: Field Survey, 2017. *** Significant at 1%, ** Significant at 5%,* Significant at 10%. Numbers of observations: 283; LR chi^2 (14): 26.77; Log likelihood: -172.1660; Prob> chi^2: 0.0206; Pseudo R^2: 0.5721.

The negative effect of age on the likelihood of having land use rights showed that a year increase in the age of rural women farmers reduced the likelihood of having land use rights by 0.0065 unit. This as earlier discussed could be owing to the customary land tenure system prevalent in the study in which women cannot generally use or own land except through familial relationship. Older women are particularly disadvantaged because of their voicelessness.

Marital status on the other hand had a positive effect on women land use rights implying that rural women who were married had a higher likelihood of having land use rights relative to their counterparts who are single. This is as a result of the land use rights conferred on a married woman (whose rights are secondary to her husbands), as a result of her union. Specifically, being married increased the likelihood of having use rights by about 0.4292 unit. In addition, access to credit facilities had a positive effect on women having land use rights. In other words, having access to credit facilities increased the likelihood of having land use rights by 0.1676 unit.

Table 6 presents the results of the logistic regression analysis of the factors influencing land ownership rights among women in farming households in the study area. The log likelihood value of -92.8545 and Chi-square value of 132.64 which was significant at 1% indicates that the model has a good fit. Marginal effects results are discussed as follows:

Table 6. Logistic regression results of the factors influencing land ownership rights

Variables	dy/dx	Z-Value	P>/Z/
Age	-0.0041	-1.79*	0.073
No formal education	-1.7989	-0.01	0.994
Primary education	-1.8956	-0.01	0.994
Secondary education	-1.8600	-0.01	0.994
Christianity	0.0975	0.76	0.449
Islam	0.1250	0.96	0.338
Married	0.2878	1.66*	0.097
Separated	0.1752	0.90	0.369
Divorced	0.2204	1.05	0.295
Widowed	0.2158	1.05	0.292
Access to credit	0.1414	1.56	0.118
Income	0.5730	2.18**	0.029
Membership of cooperative	0.0200	0.17	0.864
Inheritance	2.8529	0.01	0.995
Marriage	2.1979	0.00	0.996
Gift	2.2879	0.00	0.996
Farming experience Constant	-0.0052	-1.55 -.0.00	0.122 0.998

Source: Field Survey, 2017 *** Significant at 1%, ** Significant at 5%,* Significant at 10%. Numbers of observations: 249; LR chi^2 (14): 132.64; Log likelihood: -92.8545; Prob> chi^2: 0.0000; Pseudo R^2: 0.4167.

Age had negative effects on rural women's land ownership rights implying that a percentage increase in the age of rural women farmers will reduce their likelihood of having land ownership rights by 0.0041 unit. This may be owing to the customary land tenure system in which women are not included as beneficiaries of land inheritance and therefore do not have ownership rights. Conversely, being married increased the likelihood of having land ownership rights by 0.2878 unit as expected. This is because custom excludes women from ownership which makes women rights to land to be secondary and dependent on their relationships with their husbands. This finding is in line with the findings of Ademola (1994) that 67% of women acquired lands through marriage while 23% through inheritance. Marital status is thus a key factor when it comes to land ownership in Nigeria.

The positive relationship between level of income and women land ownership, indicates that a unit increase in the level of income of a rural woman farmer will increase her likelihood of having land ownership rights by 0.5730 unit, as an improved status in the community will confer on such women the ability to purchase land(s) and have full control over such land(s).

The results of the ordered logit regression analysis of the effects of women land rights on the food security status of the farming households is presented in Table 7. The log likelihood value of -287.0139 and Chi-square value of 119.65 which was significant at 1% indicates that the model is well fitted. The table reveals a negative effect of age on food security status of the households. Specifically, a unit increase in the age of rural women farmers reduced the probability of being food secure by 0.0052 unit, of being moderately food secure by 0.0008 unit and of being highly food insecure by 0.1%, while it increased

Table 7. Effects of women land rights on the food security status of the farming households

Variables	Food Secure		Mod. Food Secure		Mod Food Insecure		Highly Food Insecure	
	dy/dx	Z	dy/dx	Z	dy/dx	Z	dy/dx	Z
Age	-0.0052	-3.29***	-0.0008	-1.82*	0.0051	3.34***	0.001	2.40**
No Edu	-0.3099	-2.69***	-0.0532	-1.49*	0.3019	2.59***	0.0612	2.06**
Pry Edu	0.3946	3.05***	0.0601	1.55*	-0.3409	-2.90***	-0.0691	-2.21**
Sec Edu	0.3946	3.43***	0.0677	1.64*	-0.3844	-3.28***	-0.0779	-2.35**
Married	0.2685	2.81***	0.0461	1.53*	-0.2615	-2.70***	-0.0530	-2.11**
Separated	0.2678	2.37**	0.0459	1.44*	-0.2609	-2.30**	-0.0529	-1.91*
Divorced	-0.1409	-1.12	-0.0242	-0.91	0.1373	1.10	0.0278	1.05
Widowed	0.2892	2.60***	0.0496	1.50*	-0.2817	-2.52***	-0.0571	-2.02**
Christianity	-0.0289	-0.32	-0.0049	-0.31	0.0282	0.32	0.0057	0.32
Islam	-0.0179	-0.19	-0.0031	-0.19	0.0175	0.19	0.0035	0.19
Credit Acc	0.1420	2.20**	0.0244	1.62*	-0.1384	-2.25**	-0.0280	-1.87*
Inheritance	0.0189	0.11	0.0033	0.11	-0.0185	-0.11	-0.0037	-0.11
Marriage	0.0262	0.16	0.0044	0.16	-0.0255	-0.16	-0.0052	-0.16
Tenancy	0.0989	0.55	0.0169	0.54	-0.0964	-0.55	-0.0195	-0.54
Leasehold	0.2431	0.983	0.0417	0.87	-0.2369	-0.94	-0.0480	-0.90
Gift	-0.0442	-0.26	-0.0076	-0.26	0.0431	0.26	0.0087	0.26
Purchase	-0.0056	-0.03	-0.0009	-0.03	0.0055	0.03	0.0011	0.03
Right type	0.2711	5.65***	0.0465	2.11**	-0.2641	-6.30***	-0.0535	-2.90***

Source: Field Survey, 2017. *** Significant at 1%, ** Significant at 5%,* Significant at 10%. Numbers of observations = 293; LR chi^2 (20): 119.65; Log likelihood: -287.0139; Prob> chi^2: 0.0000;Pseudo R^2: 0.6725.

the likelihood of being moderately food insecure by 0.51%. This finding contradicts earlier findings in literature that the likelihood of food insecurity reduced with increase in age, because older people have better experience in subsistence agriculture and are able to accumulate wealth better than their younger counterparts (Olagunju et. al., 2012; Benjamin & Joseph, 2012; Bogale & Shimelis, 2009).

Further, results showed that women with no formal education have a lower likelihood of being food secure and moderately food secure by 0.3099 unit and 0.0532 unit respectively and a higher likelihood of being moderately food insecure and highly food insecure by 0.3019 and 0.0612 units respectively. On the other hand, those with at least primary education had a higher probability of being food secure with reducing effects on the probability of being food insecure. The positive signs of the marginal effects of being food secure and moderately food secure for primary and secondary education indicate that an increase in educational attainment increases the chances of the household being food secure and moderately food secure respectively while the negative sign of the marginal effects of moderately food insecure and highly food insecure reflects a reducing effect on the probability of being food insecure. The lack of education limits the ability to understand written instructions, rules and by-laws, and also access to markets, technology, training, finances, infrastructure and information that could help improve income, food security and by extension, the overall wellbeing of households (Siqwana-Ndulo, 2013; Mukudi, 2003; Battersby, 2011; Rose & Charlton, 2002; Haile et. al., 2005).

Marital status had a positive effect on food security status of the households. The positive signs for married, separated and widowed shows that such women farmers have higher chances of their households being food secure by 0.2685, 0.2678 and 0.2892 units respectively and moderately food secure by 0.0461, 0.0459 and 0.0496 units respectively. This could be owing to the benefits of being married or having been married at some point in time, in terms of land ownership, as such women are likely to have acquired land from their husbands for farming purposes (Jacobs, 2000). They also have the opportunity of deploying their children to their farms as family labour to increase agricultural productivity and by extension improve their food security status. The marginal effects of these variables are however negative for the moderately food insecure and highly food insecure categories, implying that married women, separated women as well as widows have lower likelihoods of being moderately food insecure or highly food insecure.

The marginal effects of access to credit showed that having access to credit increased the likelihood of households being food secure or moderately food secure by 0.1420 and 0.0244 units respectively while it reduced the likelihood of households being moderately food insecure and highly food insecure by 0.1384 and 0.0280 units respectively. This could be attributed to the fact that households with access to credit facilities have better access to farming inputs and other capital-intensive farm resources to boost productivity. In most developing countries, agricultural credit is considered an important factor for increased agricultural production and food security because, it enhances productivity and promotes standard of living by breaking the vicious cycle of poverty of small scale farmers (Adebayo & Adeola, 2008). Households that have the opportunity to receive microcredits would build their capacity to produce more and enhance their food security status through the use of improved seeds and adoption of improved technologies (Bogale & Shimelis, 2009).

Land ownership rights increased the likelihood of households being food secure and had negative effects on the likelihood of households being food insecure. The marginal effects results revealed that having land ownership rights by women increased the likelihood of households being food secure and moderately food secure by 0.2711 and 0.0465 units respectively while it reduced the likelihood of being moderately food insecure and highly food insecure by 0.2641 and 0.0535 units respectively. This is because land ownership leads to increased household agricultural productivity which implies increased food consumption and ultimately increased incomes which enables the purchase of more and better quality foods (Landesa, 2012). This finding is in line with that of Miggiano (2010) in five Asian countries that among the four levels of food security, the food secure groups had the largest percentage of owner cultivators at 70%.

CONCLUSION AND POLICY RECOMMENDATIONS

Land is a key agricultural input and also an important source of security against poverty in the developing world but gender has become a critical issue in women's land rights due to the fact that there is a direct relationship between accessing land resources, having secured land rights, achieving food security and overcoming poverty (Gashaw, 2015). This study therefore concluded that majority of the rural women farmers have land ownership rights but these lands are in microplots, not secured and limits the type of crops they cultivate as well as their outputs. The study also revealed that access to credit facilities and increased income played significant roles in land ownership rights as they empower the women farmers financially which will improve their productivity, food security status and ultimately the welfare of their households. Further, having at least primary education, land ownership rights and access to credit increased the likelihood of being food secure.

Therefore, the study recommends that Government at all levels should develop and implement strategies and instruments for recognising and protecting women rights. Policies should be enacted by the government to secure the right to land for women especially in rural areas as they are involved greatly in farming activities and are key to improved food security of their households. Also, efforts of government and other stakeholders could be focused on strategies that assess multiple dimensions of women's empowerment such as human capital development and provision of support services in accessing credit, that will increase their income, enhance the food security status and ultimately the wellbeing of their households.

REFERENCES

ActionAid International., CARE., Christian Aid., Concern Worldwide., Find Your Feet., Oxfam Practical Action., … University of East Anglia. (2012). What works for women: Proven approaches for empowering women smallholders and achieving food security. *Oxfam Policy and Practice: Agriculture. Food and Land*, *12*(1), 113–140.

Adebayo, O. O., & Adeola, R. G. (2008). Sources and uses of agricultural credit by small scale farmers in Surulere Local Government Area of Oyo State. *Anthropologist*, *10*(4), 313–314. doi:10.1080/09720073.2008.11891069

Ademola, A. (1994). Women farmers in Ondo State, Nigeria. *Journal of Anthropological Research*, *50*(3), 311–326.

Adi, D. N. (2009). *Critical Mass Representation in Uganda. Independent Study Project (ISP) Collection Paper 674*. Retrieved from http://digitalcollections.sit.edu/isp_collection/674

Africa Check. (2015). *Who owns the land in Nigeria? Africa Check factsheet*. Retrieved from https://africacheck.org/factsheets/factsheet-who-owns-the-land-in-nigeria/

Agwu, K., Amasiatu, O. G., & Onuoha, O. U. (2010). Land Rights Characteristics and Access to Land: Implications on Food Security in Nigeria. *Journal of Agriculture and Environment For Developing Countries*, *2*(2&3), 146–152.

Aluko, B. T., & Amudu, A. (2006, March). *Women and Land Rights Reforms in Nigeria*. [Paper presentation]. The 5th FIG Regional Conference, Accra, Ghana. Retrieved from https://www.fig.net/resources/proceedings/fig_proceedings/accra/papers/ts09/ts09_04_aluko_amidu.pdf

Battersby, J. (2011). *The state of urban food insecurity in Cape Town: Urban food security series No 11*. Retrieved from http://www.afsun.org/wp-content/uploads/2013/09/AFSUN_11.pdf.

Benjamin, C. A., & Joseph, U. C. (2012). *Food insecurity determinants among rural farm households in Nigeria*. International conference on ecology, agriculture and chemical engineering, Thailand.

Bogale, A., & Shimelis, A. (2009). Household level determinants of food security in rural areas of Dire Dawa, Eastern Ethiopia. *African Journal of Food, Agriculture, Nutrition and Development*, *9*(9) 1914-1926.

Chamorro, J., & Katz, E. (2002). *Gender, Land Rights and the Household Economy in Rural Nicaragua and Honduras, 11*. Paper prepared for the Regional Workshop on Land Issues in Latin America and the Caribbean (USAID).

Ellis, F. (2000). *Rural livelihood diversity in developing countries: Evidence and policy implications*. London: Overseas Development Institute; Retrieved from http://hdl.handle.net/10535/4486

Food and Agricultural Organization. (2010). *Gender and land rights database*. Rome: Food and Agriculture Organization of the United Nations. Retrieved from http://www.fao.org/gender/landrights/en/

Food and Agricultural Organization. (2011). *The State of food and agriculture: Women in agriculture closing the gender gap for development*. Rome: Food and Agriculture Organization of the United Nations; doi:10.1109/PVSC.2008.4922754

Food and Agricultural Organization. (2012). *Investing in agriculture for a better future*. Rome: Food and Agriculture Organization of the United Nations. Retrieved from https://doi.org/10.18356/3e6e633a-en

Food and Agricultural Organization. (2013). *FAO Policy on Gender Equality. Attaining Food Security Goals in Agriculture and Rural Development*. Rome: Food and Agriculture Organization of the United Nations. Retrieved from http://www.fao.org/3/a-i3205e.pdf

Food and Agricultural Organization. (2016). *Food Security Indicators*. Rome: Food and Agriculture Organization of the United Nations. Retrieved from http://www.fao.org/economic/ess/ess-fs/ess-fadata/en/#.WKOLFhIrLR0

Food and Agriculture Organization. (2002). *Gender and access to land. FAO Land Tenure Studies 4*. Rome: Food and Agriculture Organization of the United Nations. Retrieved from ftp://ftp.fao.org/docrep/fao/005/y4308e/y4308e00.pdf

Gashaw, T. A. (2015). Women's Land Use Right Policy and Household Food Security in Ethiopia. *International Journal of African and Asian Studies*, *2*, 2409–6938.

Haile, H. K., Alemu, Z. G., & Kuhlande, G. (2005). *Causes of household food insecurity in Koredegaga pleasant association, Oromiya zone, Ethiopia*. Bloemfontein: University of Free State.

Ijere, M. O. (2000). *Microfinance and Economic Development*. Ibadan: University of Ibadan Press.

International Food Policy Research Institute. (2005). *Land rights and common properties*. Washington, DC: International Food Policy Research Institute.

International Fund for Agricultural Development. (2010). *Annual Report 2010*. Retrieved from https://www.ifad.org/documents/38714170/39309757/annual+report+2010+e.pdf/da09a822-4ed9-4106-a646-6947fbfa5be7

Jacobs, S. (2000). Zimbabwe: Why Land Reform is a Gender Issue. *Sociological Research Online*, *5*(2), 301–314. doi:10.5153ro.485

Jonckheere, S., Musirimu, E., & Liversaage, H. (2013, April). *Legal empowerment to secure women's land rights in Burundi*. [Paper presentation]. The Annual World Bank Conference on Land and Poverty. The World Bank, Washington DC, United States.

Juma, C., & Ojwang, J. (1996). *In land we trust: Environment, private property and constitutional change*. Nairobi: Initiative Publishers.

Katz, E., & Chamorro, J. S. (2002). *Gender, Land Rights and the Household Economy in Rural Nicaragua and Honduras. Paper prepared for USAID/Broadening Access and Strengthening Input Market Systems (BASIS) Collaborative Research Support Program*. CRSP.

Kidane, H., Alemu, Z. G., & Kundhlande, G. (2005). Causes of household food insecurity in Koredegaga peasant association, Oromiya zone, Ethiopia. *Agrekon*, *44*(4), 543–560.

Landesa (2012). *Women's secure rights to land: Benefits, barriers and best practices. Center for Women's Land Rights Issue Brief.* Seattle, WA. Retrieved from https://cdn.landesa.org/wp-content/uploads/Landesa-Women-and-Land-Issue-Brief.pdf

Lele, U., Masters, W. A., Kinabo, I., Meenakshi, J., Ramaswami, B., Tagwireyi, J., . . . Goswami, S. (2016). Measuring Food and Nutrition Security: An independent technical assessment and user's guide for existing indicators. http://www.fao.org/fileadmin/user_upload/fsin/docs/1_FSIN-TWG_UsersGuide_12June2016.compressed.pdf.

Mahoi, I. (2015). *An Assessment on Gender Equity and Land Ownership, and it's Impact on Food Security in Sierra Leone*. Universita' Cattolica Del Sacro Cuore, Milano. Retrieved from https://tesionline.unicatt.it/bitstream/10280/12223/1/tesiphd_completa_mahoi.pdf

Miggiano, L. (2010). *Links between land tenure security and food security (Annex III). Contribution of the International Land Coalition to the updating of the High Level Task Force's (HLTF) Comprehensive Framework for Action (CFA)*. Rome: International Land Coalition.

Mirutse, D., Gebregiorgis, H., & Selam Ataklt, H. (2006). *Female-headed households and livelihood intervention in four selected Weredas in Tigray, Ethiopia. DCG report, vol.44*. Retrieved from https://www.utviklingsfondet.no/dcg/assets/documents/Report-44.pdf

Mitchell, R., & Hanstad, T. (2004). *Small home garden plots and sustainable livelihoods for the poor. Food and Agriculture Organization of the United Nations Livelihood Support Programme (LSP)*. Rome: Food and Agriculture Organization of the United Nations.

Mukudi, E. (2003). Nutrition status, education participation, and school achievement among Kenyan middle-school children. *Nutrition (Burbank, Los Angeles County, Calif.)*, *19*(7–8), 612–616.

Murisa, T. (2008). *Agrarian Reforms in Southern Africa: Contradictions of Neo-liberal Prescriptions*. Extrait du suite Foundation Gabriel Peri. Retrieved from http://www.gabrielperi.fr/

NPC. (2006). *Nigeria National Census Report*. Abuja: National Population Commission.

Odeny, M. (2013). *Improving Access to Land and Strengthening Women's Land Rights in Africa*. Paper presented at the World Bank conference on Land and Poverty, Washington, DC.

Olagunju, F. I., Oke, J. I., Babatunde, R. O., & Ajiboye, A. (2012). Determinants of food security in Ogbomoso Metropolis of Oyo State. *Nigeria. Production Agriculture and Technology*, *8*(1), 111–124.

Quisumbing, A. R., Brown, L. R., Feldstein, H. S., Haddad, L. J., & Pena, C. (1995). *Women: The Key to Food Security*. Washington, DC: International Food Policy Research Institute.

Rose, D., & Charlton, K. E. (2002). Prevalence of household food poverty in South Africa: Results from a large, nationally representative survey. *Public Health Nutrition*, *5*(3), 383–389. doi:10.1079/PHN2001320 PMID:12003648

Siqwana-Ndulo, N. (2013). The Informal Sector in South Africa: Women Street Traders in Durban. Consultancy Africa Intelligence. Retrieved from http://www.polity.org.za/article/the-informal-sector-in-south-africa-women-street-traders-in-durban-part-2-2013-05-14

Smith, L. C., & Subandoro, A. (2007). *Measuring Food Security Using Household Expenditure Surveys. Food Security in Practice Technical Guide Series*. Washington, D.C.: International Food Policy Research Institute; doi:10.2499/0896297675

Törhönen, M., & Palmer, D. (2004, April). *Land administration in post conflict Cambodia*. [Paper presentation]. The Symposium on Land Administration in Post Conflict Areas. Geneva, Switzerland.

Udoh, E. J. (2005). Demand and control of credit from informal sources by rice producing female farmers of Akwa Ibom State Nigeria. *Journal of Agriculture and Social Sciences*, *1*(2), 152–155.

Umezulike, I. A. (2004). *The Land Use Act – More than two Decades After and Problems of Adaptive Strategies of Implementation. College of Law, Ugbinedion University*. Enugu: Snaap Press ltd.

World Bank, Food and Agricultural Organization., & International Fund for Agricultural Development. (2009). Module 4: Gender Issues in Land Policy and Administration. Gender in Agriculture Source-book. Washington, DC: World Bank. Food and Agricultural Organization., & International Fund for Agricultural Development

Chapter 11
Does Gender Inclusion Really Matter in Sustainable Food Systems?

Sarah Edore Edewor
Federal University of Agriculture, Abeokuta, Nigeria

Agatha Osivweneta Ogbe
Federal University of Agriculture, Abeokuta, Nigeria

ABSTRACT

Over the past decades, the food systems in developing countries have transformed rapidly. However, the rise in social inequalities has negatively affected, the vulnerable groups as the benefits associated with these transformations are still skewed. This chapter examined the role of gender inclusiveness in promoting sustainable food systems. Employment trends revealed that agricultural employment was higher among males. Five asymmetries (assets, access to agricultural market, access to technology, resilience and risks, and decision making) were identified as limitations to sustainable food systems stemming from the gender differentiated roles. The gender action learning system methodology was adopted using strategies such as empowering men and women through community action learning during catalyst workshops, gender mainstreaming for innovation and institutional change at organizational level, and through advocacy network for policy improvement at the national level. The study concluded that gender inclusion played a crucial role in achieving sustainable food systems.

INTRODUCTION

Globally, men and women play critical but distinct roles in productive activities. Findings has revealed that women account for over 50% of those contributing to food production (FA0, 2011; Doss, 2014; Akter et al., 2017) either directly or indirectly (through the provision of about 60% of agricultural labour force), despite the fact that they engage in other activities related to taking care of the home and their children (International Labour Organisation [ILO], 2016).

DOI: 10.4018/978-1-7998-2599-9.ch011

Women have had less access to productive inputs such as land and capital as compared to what is obtainable among men due to several factors and they are also not opportune in decision making relating to agricultural resource inputs use (Afolabi, 2008). These factors could be economic, social, political or cultural in nature. Some of the limitations are due to the inheritance pattern, cultural norms and productive or child bearing roles amongst others. In Nigeria, the pattern of inheritance is patrilineal and in some parts of the country, widowed or divorced women are sometime threatened with dispossession (Small, 1997). In the northern part of Nigeria, a divorced Muslim woman is entitled to take all her personal property, inclusive of land and landed property (Uzodeke, 1993).

Land ownership varies across various ethnic groups in Nigeria. For instance among the Edos (Benin people), the land tenure being practiced is 'primogeniture' in nature, this implies that the first son inherits the land upon the death of the landowner. Similarly, in the south-western part of the country in some chiefdom among the Ekitis, Ijeshas and Ondos, a regent who is usually the first daughter of the immediate deceased chief or king is appointed during a period of interregnum. This position confers authority on all matters including land management, including family, stool or community lands (Adegboye, 1993; Aluko, 2001).

Women in rural areas have been known to play crucial roles in achieving all the food security pillars: availability, access and utilization from production on the family farms through distribution within the household and to food preparation. However, these roles are generally undervalued and constrained by the limitations faced by the women due to access to resources, services as well as labour market opportunities. Findings (Folbre, 2006; Kabeer, 2012) have shown that in most regions of the world, the bulk of unpaid labour in both the care economy (child care, cleaning, and caring for the sick and elderly, fetching water and wood, purchasing and preparing food,)and agricultural production is performed largely by women.

In 2018, evidence revealed that women still had limited access to education, insecure land rights and had less political representation (FAO, 2018). The Global Gender Gap Report (2016) by the World Economic Forum reported that gender disparity between men and women is on the increase and in order to achieve the United Nations Sustainable Development Goal (SDG) 5 of gender equality and the empowerment of all women and girls (United Nations, 2015), it then becomes necessary to end all forms of discrimination against women by empowering them. Gender equality is not just a fundamental human right; it will help improve the food production systems since women account for a large proportion of those involved in the various segments of the food production system (FAO, 2011).

Gender systems are complex and highly diverse. Mason and Smith (2003) reported that they are determined by community values and norms. The nature and extent of gender inequity as well as the necessary conditions to empower women vary across communities, countries and regions (Jejeebhoy & Sathar, 2001; Alkire et al., 2013; Akter et al., 2017). For instance in South East Asia, women are generally more empowered compared with women in other developing regions (Mason & Smith, 2003; IFAD, 2013).

If gender equality is established, this can actualize the SDG 2 of ending hunger, achieving food security, improving nutrition and promoting sustainable agriculture. It has been confirmed that gender parity at the household and community level could lead to improved agricultural development outcomes (Farnworth & Colverson, 2015; United Nations Development Programme, 2015). Mukasa and Salami (2016) conducted a comparative study on closing the gender productivity gap in Nigeria, Tanzania and Uganda and affirmed that the production gain was 2.8%, 8.1% and 10.3% respectively.

Food and Agriculture Organisation (FAO, 2011) reported that if women had access to the same level of productive resources that their male counterparts had, agricultural productivity will increase from

2.5% to 4% and the number of malnourished people will reduce from 17% to 12%. The empowerment of women will protect the environment, increase agricultural productivity, reduce food losses and waste and improve dietary diversity (FAO, 2017).

Subject to this, this study examined the effect of gender inclusion in achieving sustainable food systems. This chapter specifically described the trend of agricultural employment, the implication of gender inclusion in designing food system policies and the appropriate responses to bridge the gender gap so as to achieve gender equality in the agricultural sector.

This chapter is categorized into three sections. The first section is the introduction followed by the conceptual review while the third section contains the methodology and the findings.

CONCEPTUAL REVIEW

Gender

FAO (1999) defined gender as ‘the perceptual and material relations between men and women. It encompasses the economic, sociocultural and political attributes, constraints as well as opportunities associated with being male or female. It refers to the social roles and relationships, behaviour, attitudes, personality traits, values and relative power coupled with the influence that society ascribes to the males and females on a differential basis (USAID, 2010). It reveals the roles that man and woman play and the power relations between them which often times have profound effect on natural resource usage and management. It is not based on the biologically sexual characteristics of either women or men, or the biological differences between men and women, but it is constructed socially and shaped by social relations, culture, and natural environments (UNESCO, 2003).

Stemming from the FAO definition of gender, gender issues is largely centered on women and the relationship that exists between women and men, their roles, their access to and control over resources, interest and needs and division of labour. Bravo-Baumann (2000) stated that gender relations will, affect household planning, production, household food security, family wellbeing and other aspects of life.

Gender concept is based on the facts that are held about the characteristics, aptitudes and likely behaviours of both men and women (femininity and masculinity). Gender idea also reveals how women’s subordination (or men’s domination) is socially constructed (UNESCO, 2003). The roles of women and men are seen in different labour responsibilities open to them, decision making processes and knowledge as well as how this affects their needs. This is basically because women and men use and manage of resources differently. Thus, implying that gender is an acquired identity that is learned which changes over time and varies widely within and across cultures. It is noteworthy to note that gender roles affect the opportunities (social, economic, political and ecological) and constraints women and men face and this is because it is subject to the social environment in which food systems are entrenched.

Gender analysis is often required to examine the relationship that exists between men and women. Gender analysis explores this relationship based on their roles, their access to and control of productive resources as well as the constraints faced by them (OCHA, 2012). It explores the differences between men and women policies, programmes and projects targeted at men and women so as to identify and meet their respective needs. It also facilitates the strategic use of distinct knowledge and skills possessed by both women and men (UNESCO, 2003). Gender analysis usually examines the following:

1. The differences that exists between women and men status as well as their differential access to assets, resources, opportunities, and services;
2. Gender roles and norms influence on time division between paid employment, unpaid work (including subsistence production and care for family members), and volunteer activities;
3. Gender roles and norms influence on leadership roles and decision-making; constraints, opportunities, and entry points for narrowing gender gaps and empowering females;
4. Potential differential impacts of development policies and programs on males and females, including unintended or negative consequences.

Gender analysis seeks to clearly state how gender roles and relations can create opportunities or obstacles in achieving development objectives. It plays a crucial role in creating gendered food systems innovations. It enhances the processes and outcomes of food systems innovations thus implying that it adds value to research and increases the likelihood that food systems innovations will achieve their overall goals of poverty reduction, nutrition and food security promotion and sustainability (UNESCO, 2015).

Gender Inclusion

Women and men have been described to have different labour responsibilities. They are also faced with different decision making processes and knowledge and differences in how they use and manage resources. Based on these differences, gender inclusion posits that men and women should not be limited in the roles they play and each gender (feminine and masculine) should have the opportunity of having equal playing ground in decision making, control and use of productive resources (UNESCO, 2003).

The aim of gender inclusion is gender equality. Gender equality refers to equal valuing by society of the roles played by men and women as well as the similarities and the differences between men and women (UNESCO, 2003). It involves both women and men assuming full partnerships in their homes, their communities and their societies. It also connotes that women and men have equal conditions for realizing their full human rights and contribute to and benefit from social, economic, political and cultural development (UNESCO, 2003).

In developing countries, the roles women play in agriculture cannot be overemphasized. These roles include agricultural production, processing, distributing, and food marketing mainly as unpaid workers or family farmers (FAO, 2011). Despite these roles, numerous challenges (productive and care roles) still hinder full participation in the food systems. Inability to fully participate in the various activities associated with the food system may probably result into reduced output, low technology adoption and reduced income amongst others.

FAO (2017) reported that strengthening of the food system which helps in fighting hunger and malnutrition as well as improving the lives and livelihoods of rural populations can be achieved through gender equality and women's empowerment. The focus of the 2030 agenda for Sustainable Development is based on the fundamental importance of achieving gender equality and enhancing women's empowerment by reflecting and mainstreaming that gender is established across all the 17 Sustainable Development Goals (FAO, 2017).

It will be impossible to achieve the sustainable development goal of gender equality if about half of the world population is still been denied access to full human rights and associated opportunities. To address goal five, equal access must be readily available to both girls and women through education, access to productive inputs and political representation as is available to their male counterparts at all levels.

The potential role of gender as an accelerator or inhibitor of food systems innovations has been recognized by Global Center for Food Systems Innovation (GCFSI). The GCFSI is committed to pro-actively pursuing the use of appropriate innovations that has strong potential to change unequal gender relations and empower women in the food system so as to increase their food security, reduce poverty, and improve their nutritional outcomes (Me-Nsope, 2017). Improving rural women's access to assets, resources, technologies, services and opportunities could create significant benefits in terms of agricultural productivity and rural peoples' livelihoods. Effective participation of women in agriculture translates into improved agricultural productivity, reduction in hunger, malnutrition and poverty (FAO, 2017). Without gender equality and rural women's economic, social and political empowerment, food security, nutrition and sustainable agriculture will not be achieved (FAO, 2017).

Sustainable Food Systems

Food systems involves the interactions between and within the biophysical and human environment which determines a set of activities related to the production, processing and packaging of food, distribution and retailing of food, preparation and consumption of food, the outcomes of which contribute to food security, environmental security and social welfare (Ericksen, 2008). A food system can thus be defined as a system that encompasses all elements (environment, people, processes, inputs, institutions, infrastructure, markets and trade) and activities related to the production, processing, distribution and marketing, preparation and consumption of food and the outputs of these activities, including socioeconomic and environmental outcomes.

FAO (2018) defined food systems as a "system that covers the entire range of actors in conjunction with the value-added activities involved in the production, aggregation, processing, distribution, consumption and disposal of food products originating from agriculture, forestry or fisheries, and parts of the broader economic, societal and natural environments in which they are embedded". The food system comprises of sub-systems which interacts with other key systems. Thus, a change in the food system could be brought about by a change in another system. Sustainable food system (SFS) refers to the food system that provides nutrition and food security to everyone in such a manner that does not compromise the nutrition and food security of future generations (HLPE, 2014). The implication of this is that it is beneficial to the society (social sustainability), it is absolutely profitable (economic sustainability) and it positively impacts the environment (environmental sustainability). Figure 1 shows the interaction within a typical local food system.

Combs et al. (1996) stated that food systems are entrenched in environments which vary according to factors such as climate, agroecology, health, policy, social aspects and economics. These varying environments affect the food systems activities and outcomes. DGAC (2015) stated embracing fundamental values (food security support, healthy living culture) when developing food systems is cogent for sustainable and healthy diet.

Theory of Change and Gender

The Theory of change (ToC) is an approach that explains how an invention(s) is meant to bring about change in specific development. They are important because they assist in identifying solutions to challenges that retard progress as well as serve as guide to enhance decision making on relevant approaches to be embarked upon. With respect to gender issues, a theory of change can help practitioners think through

Figure 1. Local food system model
Source: Douglas County Food Policy Council and reformatted by the Public Health Law Centre at William Mitchell College of Law, St. Paul, Minnesota

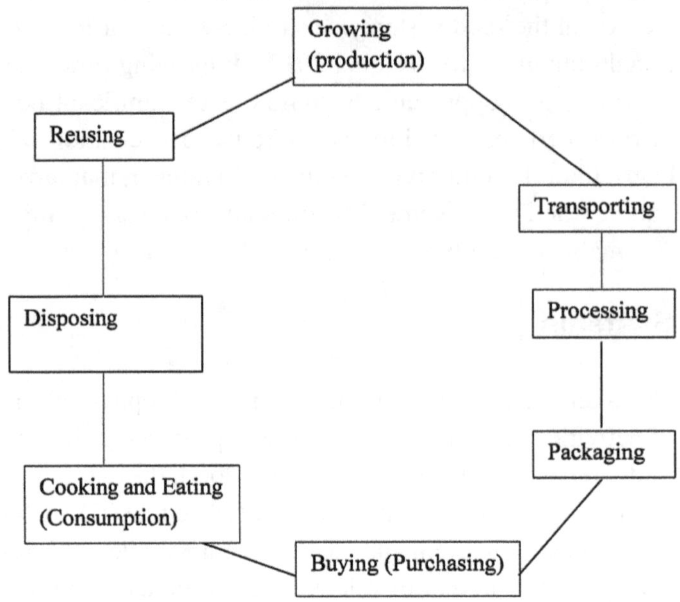

many underlying gender related issues and the root causes and how they influence gender participation in agriculture and sustainable food systems. The Food and Agriculture Organisation of the UN (FAO) in 2011 showed that if women are given the same access to productive resources and information as men, significant impacts would be achieved on agricultural productivity and hunger globally (Jost et al., 2015). There are two approaches to the development of the theory of change:

1. Those that focus on how projects or programmes are expected to bring change.
2. Those that explore how change happens more broadly and its implication for programme intervention – including advocacy and influencing.

The Gender theory of change developed in this chapter takes into consideration the following:

1. It is targeted towards changes that benefit the marginalised and most vulnerable individuals (women).
2. It addresses related issues of inequality and discrimination among the marginalised groups (women)
3. It identifies the solution that will target the need of vulnerable and marginalised group (women).

In this chapter, the gender theory of change developed is composed of four dimensions (inputs, outputs, outcomes and impact) which indicate a causal relationship in Figure 2.

Figure 2. Gender theory of change
Source: Authors' concept, 2019.

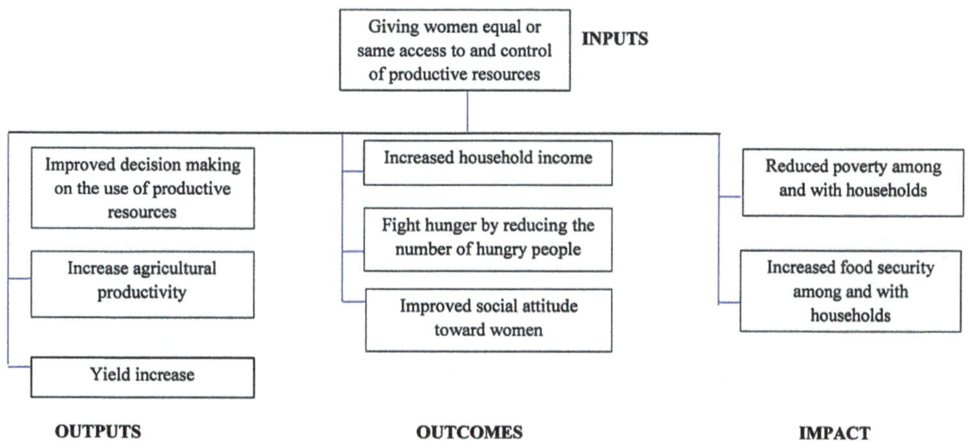

METHODOLOGY AND RESULTS

Methodology

The study was conducted in Nigeria. This study made use of secondary data. The study data include information on female and male employment in agriculture in Nigeria from 1991 to 2017. Other information was sourced from published and unpublished articles, mimeographs, working papers and seminal papers. The data on female and male employment in agriculture in Nigeria were obtained from the United Nations Development Programme (UNEP) website.

Trend of Women Participation in Agricultural Employment in Nigeria

In Nigeria, the level of women's productivity per hectare is much lower than the males. This is because women have been perceived to have limited access to land, other productive inputs and technology and the reverse is what is obtainable among men as Peterman et al. (2014) reported that they had access to more inputs such as fertilizers, improved seed varieties and extension services. FAO (2011) argued that the yield associated with women's productivity per hectares would increase by 20 percent to 30 percent.

Differences in productivity can also be attributable to the locations in the country as Oseni *et al.* (2015) reported that women produced food crops 28 percent less in Northern Nigeria as compared to men and there was no noticeable difference in Southern Nigeria when they controlled for the factors of production. Productivity difference was also noticed in Ethiopia between women and men (Aguilar et al., 2015) while in Uganda, child care responsibility was the main difference associated with agricultural productivity of women (Ali et al., 2016).

The computation from Figure 3 shows the trend of employment in agriculture by gender in Nigeria from 1990 to 2017. The trend shows that more males were gainfully employed in farming and agricultural related activities as compared to their female counterpart. The reason for the higher proportion of males as compared to females may be attributable to the fact that agriculture is a labour intensive sector. In most cases in the agricultural sector, females account for a higher proportion of those involved in

Figure 3. Employment in agriculture by gender, Nigeria

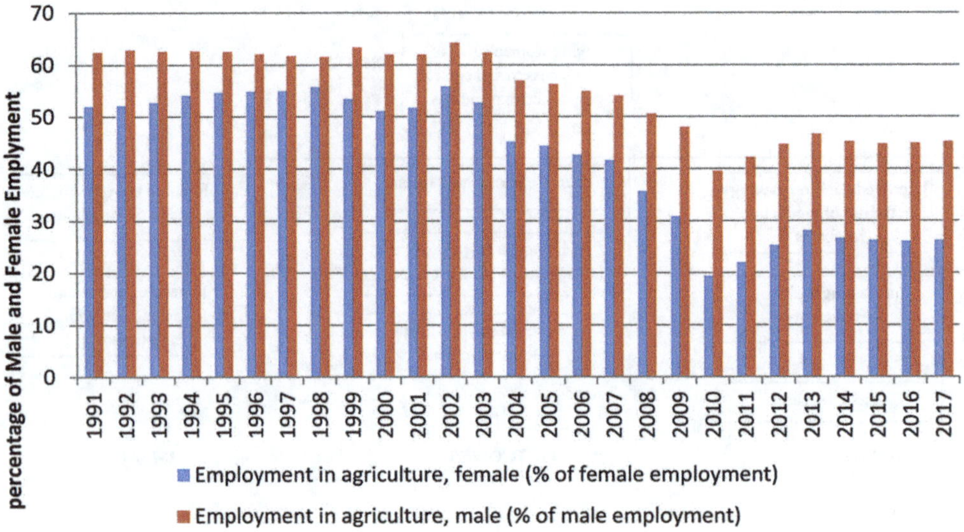

agro processing and agricultural marketing as compared to the males. From the trend, male and female employment in agriculture gradually declined from 2003. This may be due to occupational mobility of labour into other sectors such as services and construction. The year 2010 witnessed the lowest number of both males and females that were employed. This may be as a result of the food price crisis in 2007/2008 which later reoccurred in 2010/2011. A gradual increase began from 2011, and this period coincides with the Agricultural Transformation Agenda (ATA) which focused on building the value chains of selected agricultural commodities as well as reducing oil dependence (Fankun & Evbuomwan, 2017).

Globally, despite the fact that women account for about half of those involved in the agricultural sector, agricultural labour is believed to be dominated by males. In richer economies, there has been rekindled interest in gender roles in agriculture. For instance, in the United Kingdom (UK) between 2010 and 2013, farms owned and managed by women increased to about 10 percent while the total agricultural workforce comprised of 28 percent females. Within the educational sector, the same result is evident as more women register in agriculture related courses as compared with their male counterparts. This has been evident in Australian agricultural universities since 2003 as female enrollment has been higher than males because of the understanding that agriculture is now gender neutral (Pratley, 2017). Similarly in the UK, women registration in agriculture related courses in colleges and universities has increased by 25 percent compared to men in 2015 (UK DEFRA, 2016).

Gender Differentiated Roles and Agriculture

Women have been known to constitute a large part of the agricultural workforce due to their active participation on family/joint farms and informal trade. Agboola (2000) reiterated the fact that rural women constitute over 70 percent of agricultural production workforce and they are believed in most cases to site their farms close to their homestead due to the reproductive and domestic roles played by them in the households and the society (Adesina, 2003). Women have also been reported to engage more in the production and processing of arable crops (Afolabi, 2008).

Moser (1998) classified gender roles into four (4):

1. Reproductive role: This includes childbearing and rearing responsibilities as well as the domestic chores carried out by women. The reproductive roles are usually unpaid and are not only limited to biological reproduction but include the care of the workforce (the male partner, the, children who are employed and oneself) as well as the future workforce (infants and school-going children).
2. Productive role: This describes the payment (cash or kind) to men and women for work done. It comprises of potential exchange value, market production and home production with actual use-value. The work in this case could be paid (usually underpaid) and unpaid. Women may work as wage workers, peasant wives or independent workers in agricultural production
3. Community managing role: This typifies roles that women engage in apart from their reproductive roles at the community level which involves providing and managing of scarce resources (water, health care, education and energy sources) for collective consumption. This role is usually unpaid and carried out in 'free' time.
4. Community politics role: At the community level, this is primarily undertaken by men who organize formal politics usually within the framework of national politics. This role involves paid work either indirectly or directly attained through status or power.

Responses to Bridge the Gender Gap in the Agricultural Sector

A key target of the SDG 3 is targeted at boosting agricultural productivity. In the literature, several ways have been proposed to address this issue. Russel et al. (2015) suggested that in order to facilitate gender equitable control over resources and assets, focus should be more on collaborations between women and men rather than focusing solely on women. In a study conducted in the eastern part of the Democratic Republic of Congo on joint participation of women and men in an agricultural extension programme, the findings from this study revealed that the adoption of technology (row planting and the planting of improved legume varieties and mineral fertilizers) was very high compared among the group that participated jointly in the extension programmes as compared to what was obtainable among female farmers that solely participated in a female only extension programme (Lambrecht et al., 2014).

Similarly, other researchers suggest that there is a need to factor in gender-specific responsibilities, constraints and resources by agricultural development programmes to be beneficial to both male and female farmers. Since most women performed higher share of household chores and activities as well as other non-household duties; then the argument of Doss (2017) was that effort should be made to enhance women's agricultural productivity through interventions that can substantially reduce the drudgery they faced; such as water access improvement, child health and education improvement.

Gender inequality is one of the main causes of poverty. In order to address the issue of gender injustice, since there are five areas in the food systems in which gender inequality occurs, then it becomes very important to address these areas. This will be done through gender-smart greening of food systems which implies that efforts are made to remove the five gender asymmetries. The inter-linkages of the five areas of gender justice with the asymmetries related to the food system are presented in Figure 4 (Groverman & van der Wees, 2016).

1. **Asset**: In order to address this challenge, access to and control over economic resources, financial resources, human resources, social capital and material resources are needed for the equitable benefits from greening the food systems.
2. **Agricultural Market**: This is associated with the products, inputs, labour (in agricultural production and agribusiness), land, financial and water markets). The food systems do not function well because the men and women who participate in the market have different opportunities. As with value addition to a product, the benefits and risks associated with it are not distributed equitably among the men and women.
3. **Technology**: This includes 'soft' technology such as knowledge. These asymmetries focus on issues such as market information, risks and legal rights information, services and skill development provision amongst others. Gender asymmetries may intensify or be reinforced if access to assets and markets coupled with risk and vulnerability is exacerbated by access to technology
4. **Resilience and Risks**: Vulnerabilities to risks are experienced by both women and men. This may be in the form of natural or economic risks arising from socio-economic and cultural factors.
5. **Decision Making**: The feminine gender is often relegated to the background in household and business based decision making. This to a great extent affects their quality of life.

Figure 4. Five key areas of gender justice with the asymmetries related to the food system
Source: (Groverman & van der Wees, 2016)

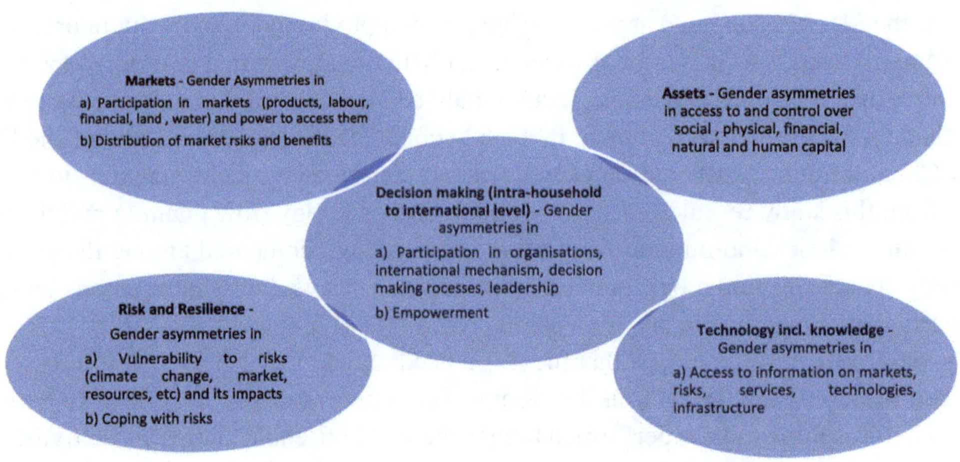

Groverman and van der Wees (2016) adapted the methodology developed by Linda Mayoux to green food system and tackle the five asymmetries. The methodology called Gender Action Learning System (GALS) is targeted at promoting the equitable distribution of reproductive and productive duties by promoting equal property ownership, improving market access for farmers, enhancing access to information and innovative technologies as well as the promotion of equal decision making. This methodology is a sustainable community-led planning methodology based on participatory tools.

The methodology involves three strategies:

1. Empowerment through community action learning at catalyst workshops: This is done at the community level
2. Gender mainstreaming for innovation and institutional change at organisational level: This is done at the institutional level
3. Advocacy network for policy improvement at the national level

Hannah and Jost (2011) suggested that it should involve participation which will help to empower men, women, boys and girls to find solutions to their own development challenges. This will help encourage learning, flexibility and discovery. Women need to be actively involved in leadership position as well as in the decision making process.

CONCLUSION

This chapter sought to examine whether the inclusion of gender was really important in achieving a sustainable food system. It sought to do this specifically by describing the trend of women participation in agricultural employment, identify the areas of gender asymmetries and key actions required to bridge gender gap so as to achieve sustainable food systems. The findings from this study revealed that males were more employed in the agricultural sector compared to their female counterparts and the level of agricultural productivity associated with female farmers was lower compared to that of males probably due to the reproductive and productive roles that women perform.

Five asymmetries (assets, access to agricultural market, access to technology, resilience and risks and decision making) were identified as the limitations to sustainable food systems. The Gender Action Learning System (GALS) methodology was adopted by the study to green food systems by empowering through community action learning at catalyst workshops, by mainstreaming gender for innovation and institutional change at organizational level and through an advocacy network for policy improvement at the national level.

The study concluded that gender inclusion played a critical role in sustainable food systems and recommends that any programme, intervention or policy targeted at sustainable food systems should be targeted at men or women directly but should involve collaborations between women and men. Also, there is a need to factor in gender-specific responsibilities, constraints and resources in agricultural development programmes so that it is beneficial to both male and female farmers. In the short-term, gender centered action should be addressed at identifying agreements on measurable indicators which will increase the acceptance of women's economic roles and aid in developing gender-responsive policies that will in the long-term ensure that women and men's priorities and needs are reflected at all levels.

REFERENCES

Adegboye, R. A. (1993). Agricultural land: its use, tenure and conservation in Nigeria. In E. A. Aiyelani, E. O. Lucas, M. O. Abatan, & O. A. Akinboade (Eds.), *Fundamentals of Agriculture* (pp. 9–13). Ibadan: Afrika-Link Books.

Adesina, H. A. (2003). *The gender variable in agricultural research.* Macmillan Publishers.

Afolabi, M. M. (2008). *Women as pillars of national economy in Nigeria: A study of economic activities of rural women in six local government areas of Ondo State*. IAFFE summer conference, International Association for Feminist Economics, Toronto, Italy.

Agboola, T. A. (2000). Water management in Fadama: a report on Fadama implementation. Academic Press.

Aguilar, A., Carranza, E., Goldstein, E., Kilic, T., & Oseni, G. (2015). Decomposition of gender differentials in agricultural productivity in Ethiopia. *Agricultural Economics*, *46*(3), 311–334. doi:10.1111/agec.12167

Akter, S., Rutsaert, P., Luis, J., Htwe, N. M., San, S. S., Raharjo, B., & Pustika, A. (2017). Women's empowerment and gender equity in agriculture: A different perspective from Southeast Asia. *Food Policy*, *69*, 270–279. doi:10.1016/j.foodpol.2017.05.003

Ali, D., Bowen, D., Deininger, K., & Duponchel, M. (2016). Investigating the gender gap in agricultural productivity: Evidence from Uganda. *World Development*, *87*, 152–170. doi:10.1016/j.worlddev.2016.06.006

Aluko, B. T. (2001). *Relationship between family and chieftancy land in Ibokun, LLB long essay*. Ibadan: Faculty of Law, University of Ibadan.

Balmori, H. H. (2003). *Gender and Budgets Overview Reports. BRIDGE (Development – Gender)*. Brighton: Institute of Development Studies, University of Sussex.

Bravo-Baumann, H. (2000). *Capitalisation of experiences on the contribution of livestock projects to gender issues. Working Document*. Bern: Swiss Agency for Development and Cooperation.

Dietary Guidelines Advisory Committee (DGAC). (2015). Dietary Guidelines Advisory Committee: Advisory Report to the Secretary of Health and Human Services and the Secretary of Agriculture.. *Scientific Reports*, 2015.

Doss, C. (2014). If women hold up half the sky, how much of the world's food do they produce? In A. R. Quisumbing, R. Meinzen-Dick, T. L. Raney, A. Croppenstedt, J. A. Behrman, & A. Peterman (Eds.), *Gender in Agriculture*. Springer. doi:10.1007/978-94-017-8616-4_4

Doss, C. R. (2017). Women and agricultural productivity: Reframing the issues. *Development Policy Review*, *36*(1), 35–50. doi:10.1111/dpr.12243 PMID:29263585

Ericksen, P. J. (2008). What is the vulnerability of a food system to global environmental change? *Ecology and Society*, *13*(2), 14. doi:10.5751/ES-02475-130214

Fankun, D. S., & Evbuomwan, G. O. (2017). An evaluation of agricultural financing, policies and initiatives for sustainable development in Nigeria, in the 21st Century: 1990-2014. *IOSR Journal of Economics and Finance*, *8*(3), 32–38. doi:10.9790/5933-0803033238

Farnworth, C. R., & Colverson, K. E. (2015). Building a Gender Transformation Extension and Advisory Facilitation System in sub-Saharan Africa. *Journal of Gender, Agriculture & Food Security*, *1*(1), 20–39.

Folbre, N. (2006). Measuring Care: Gender, Empowerment, and the Care Economy. *Journal of Human Development*, *7*(2), 183–199. doi:10.1080/14649880600768512

Food and Agriculture Organisation. (2017). *Leaving no one behind: Achieving gender equality for food security, nutrition and sustainable agriculture*. FAO Conference 40th Conference, Rome, Italy.

Food and Agriculture Organisation – FAO. (1997). *Gender: the key to sustainability and food security*. SD Dimensions, May 1997. Retrieved from www.fao.org/sd

Food and Agriculture Organisation - FAO. (2018). *Sustainable Food Systems: Concepts and Framework*. Retrieved from http://www.fao.org/3/ca2079en/CA2079EN.pdf

Food and Agriculture Organization of the United Nations - FAO. (2011). *The State of Food and Agriculture 2010–11: Women in Agriculture: Closing the Gender Gap for Development*. Rome: FAO. Retrieved from www.fao.org/docrep/013/i2050e/i2050e00.htm

Groverman, V., & van der Wees, C. (2016). How to Green Food Systems in a Gender Smart Way: A matter of Insight and Smart Interventions. Academic Press.

Hannah, H., & Jost, C. (2011). Public Health Participatory Epidemiology. *Introductory Training Module, manual for trainees*. Available at: https://cgspace.cgiar.org/bitstream/handle/10568/24715/OneHealth-Manual.pdf?sequence=1

High Level Panel of Experts on Food Security and Nutrition - HLPE. (2014). Food losses and waste in the context of sustainable food systems. A report by the High Level Panel of Experts on Food Security and Nutrition of the Committee on World Food Security, Rome.

IFAD (International Fund for Agricultural Development). (2013). *Gender and Rural Development Brief Southeast Asia*. Retrieved from www.ifad.org/gender/policy/ gender_e.pdf

International Labour Organization - ILO. (2016). *Women at work: Trends 2016*. International Labour Office. Retrieved from http://www.ilo.org/wcmsp5 /groups/public/—-dgreports/—-dcomm/—-publ/documents/publication/wcms_457317.pdf

Jejeebhoy, S. J., & Sathar, Z. A. (2001). Women's Autonomy in India and Pakistan: The Influence of Religion and Region. *Population and Development Review*, *27*(4), 687–712. doi:10.1111/j.1728-4457.2001.00687.x

Jost, C.C. Kristjanson, P., & Ferdous, N. (2015). *Lessons in Theory of Change: Gender and Inclusion*. CCSL Learning Brief No. 14.

Kabeer, N. (2012). *Women's Economic Empowerment and Inclusive Growth: Labour Markets and Enterprise Development*. SIG Working Paper 2012/1. London, UK: Department for International Development (DFID) and Ottawa, Canada: International Development Research Centre (IDRC) – Supporting Inclusive Growth (SIG) program.

Lambrecht, I., Vanlauwe, B., & Maertens, M. (2014). *What is the sense of gender targeting in agricultural extension programs? Evidence from eastern DR Congo*. Leuven, Belgium: University of Leuven, Department of Earth and Environmental Sciences, Division of Bioeconomics (Bioeconomics Working Paper 2014/4).

Mason, K. O., & Smith, H. L. (2003). *Women's Empowerment and Social Context: Results from Five Asian Countries. Gender and Development Group*. Washington, DC: World Bank.

Me-Nsope, N. (2017). Gender Practice in Food Systems Innovations: Approaches, Lessons, and Challenges. Michigan State University.

Miruka, M., Njuki, J., Starr, L., Kruger, E., & Hillenbrand, E. (2016). Addressing the multidimensional nature of gender dynamics in agriculture: Measuring gender relations and women's empowerment. In *Transforming Gender and Food Security in the Global South* (pp. 21-46). Routledge Studies in Food, Society and Environment. Retrieved from www.routledge.com/books/series/RSFSE/

Moser, C. (1998). *Unit 1: A conceptual framework for gender analysis and planning*. International Labour Organization. Retrieved from http://www.ilo.org/public/english/region/asro/mdtmanila/training/unit1/moserfw.htm

Mukasa, A., & Salami, A. (2016). A Gender Equality in Agriculture: What are really the Benefits for sub Saharan Africa? *Chief Economist Complex, AEB*, *7*(3).

OCHA. (2012). Gender Definitions and Mandates 2. *OCHA Gender Toolkit*. Available at: https://www.unocha.org/sites/dms/Documents/GenderToolkit1_2_GenderDefinitionsandMandates.pdf

Oseni, G., Corral, P., Goldstein, M., & Winters, P. (2015). Explaining gender differentials in agricultural production in Nigeria. *Agricultural Economics*, *46*(3), 285–310. doi:10.1111/agec.12166

Pratley, J. E. (2017). *Agriculture: from macho to gender balance*. Paper for 18th Australian Society of Agronomy Conference, Ballarat, Australia.

Rubin, D., Manfre, C., & Barrett, K. N. (2009). *Promoting Gender Equitable Opportunities in Agricultural Value Chains: A Handbook*. Prepared under the Greater Access to Trade Expansion project, Women in Development IQC (Contract No. GEW-I-00-02-00018-00, Task Order No. 02). Washington, DC: USAID. Retrieved from http://www.culturalpractice.com/site/wpcontent/downloads/4-200916.pdf

Russell, N., Karlsson, K., Ashby, J., & Mascarenhas, M. (2015). *Change in the Making: Progress Reports on CGIAR Gender Research: Issue No. 1. Toward gender-equitable control over productive assets and resources*. Cali, Colombia: CGIAR Gender and Agriculture Research Network, CGIAR Consortium Office and International Center for Tropical Agriculture.

Small, J. (1997). Women's Land Rights: a case study from the Northern Transvaal. In S. Meer (Ed.), *Women, Land and Authority: Perspectives from South African*. London: Oxfam.

United Kingdom Department for Environment, Food Rural Affairs - UK DEFRA. (2016). *Environment Secretary Salutes Britain's Women Farmers*. London: Department for Environment, Food and Rural Affairs. Retrieved from www.gov.uk/government/news/environment-secretary-salutes-britains-womenfarmers

United Nations. (2015). *Transforming our world: The 2030 Agenda for Sustainable Development*. United Nations. Retrieved from https://www.un.org/pga/wp-content/uploads/sites/3/2015/08/120815_outcome-document-of-Summitfor-adoption-of-the-post-2015-development-agenda.pdf

United Nations Development Programme. (2015). Gender, Climate Change and Food Security. *Training Module*, *3*, 1–36.

United Nations Educational, Scientific and Cultural Organisation – UNESCO. (2003). *Baseline Definitions of Key Concept and Terms*. UNESCO's Gender Mainstreaming Implementation Framework. Retrieved from http://www.unesco.org/new/fileadmin/MULTIMEDIA/HQ/BSP/GENDER/PDF/1.%20 Baseline%20Definitions%20of%20key%20gender-related%20concepts.pdf

United States Agency for International Development - USAID. (2007). *Gender Terminology*. Retrieved from https://pdf.usaid.gov/pdf_docs/Pnadl089.pdf

United States Agency for International Development - USAID. (2010). *Guide to Gender Integration and Analysis: Additional Help for Automated Directive System (ADS)*. Washington, DC: United States Agency for International Development.

United States Agency for International Development - USAID. (2012). *Gender Equality and Female Empowerment Policy*. Washington, DC: USAID.

United States Agency for International Development - USAID. (2013). *Integrating Gender Equality and Female Empowerment in USAID's Program Cycle. ADS Chapter 205*. Washington, DC: USAID. Retrieved from https://www.usaid.gov/ads/policy/200/205

United States Agency for International Development - USAID. (2017). *ADS Chapter 205 Integrating Gender Equality and Female Empowerment in USAID's Program Cycle*. Retrieved from https://www.usaid.gov/sites/default/files/documents/1870/205.pdf

Uzodike, E. N. U. (1993). Women's Rights in Law and Practice: Property Rights. In A. O. Obilade (Ed.), *Women in Law*. Lagos: Southern University of Lagos.

Whitehead, A., & Tsikata, D. (2003, January). Policy Discourses on Women's Land Rights in Sub-Saharan Africa: The Implications of the Re-turn to the Customary. *Journal of Agrarian Change*, *3*(1&2), 67–112. doi:10.1111/1471-0366.00051

Women Engage for a Common Future – WECF. (2018). *The Gender Impact Assessment and Monitoring Tool*. Retrieved from www.women2030.org

Section 6

Migration, Remittances, and Food Security

Chapter 12
Do Foreign Remittances Matter in Attaining Zero Hunger?
Evidence From Rural Nigeria

Oluwakemi Adeola Obayelu
Department of Agricultural Economics, University of Ibadan, Nigeria

Rebecca Funmi Akinmulewo
https://orcid.org/0000-0002-1462-5621
Department of Agricultural Economics, Faculty of Agriculture, University of Ibadan, Nigeria

ABSTRACT

Foreign remittance has remained a major source of income and a means to reduce hunger for many poor people in developing countries. The contribution of foreign remittances to food insecurity status of rural households in Nigeria was assessed using data from 2015/2016 Living Standard Measurement Study-Integrated Surveys on Agriculture (LSMS-ISA). Food insecurity status was achieved using the household food insecurity access scale. Data were analysed using descriptive, ordered, and nested logit models. Female-headed households residing in south-east zone with 51 to 70 years old heads and more than six members had greater access to remittances but were severely food insecure. Drivers of food insecurity were age, gender, marital status, education of the household head, membership of cooperatives, access to extension, farm size and per capita income, and living in the north central geo-political zone. Foreign remittances had a positive effect on the food insecurity status of rural households.

INTRODUCTION

Nigeria is the most populous country in Africa (over 160 million people), but 61 percent of her population are malnourished (World Bank, 2012; Momodu et al., 2011). Nigeria is characterized by threat of hunger, and among 113 countries, the nation ranked 92 by the Global Food Security Index in 2017 (The Economists, 2017). This is because with the advent of commercial oil exploration in the early 1970s, the fortunes of agriculture started to dwindle with a resultant downward decline in productivity and

DOI: 10.4018/978-1-7998-2599-9.ch012

consequently food insecurity (Vision, 2010). Food insecurity problem in Nigeria has been identified to be mainly access dominated. Food access has been defined as the ability of individuals or households to acquire sufficient quantities and quality of food to meet all households' members' nutritional requirements (Langsworthy, et al., 2003).

The rural households take the larger share of food insecurity in Nigeria. The greatest challenge faced by rural communities is concentrated mainly on food insecurity and nutritional concern. In investigating household food security strategies, migration and remittance are highlighted as possible pathways out of household food insecurity (Kuuire et al., 2013). Generally, it is assumed that households that receive remittance are more likely to withstand risk of food insecurity (Nguyen & Winters, 2011; Mango, et al., 2014). Dia (1992) described it as a very efficient strategy to promote agricultural investments and reduce food insecurity and income risks by families in Senegal. This financial aid of remittances seems to flow directly to the people who really needed it and does not require a costly bureaucracy on the sending side. Luginaah, et al. (2009) highlighted the importance of food remittances and transportation networks as important aspects of food security in the Northern part of the Ghana. Combes, et al. (2014) also asserted the power of remittances in reducing the effects of food price shocks in low-income countries and sub-Saharan African countries. Households spend the vast majority of remittances received on their basic needs such as food, clothing and shelter. This consumption, combined with investment in health care and education, constitutes 80-90 per cent of remittance spending (IFAD, 2007). Increased remittances an equivalent of a private "welfare payment" sent from abroad are seen to help recipients during an economic downturn and also help to smoothen consumption of the recipients (Martin, 2005).

Foreign remittances to Nigeria did not increase from 2008 ($10.6 billion) to 2011. However, it grew only marginally in 2015, as weak oil prices and other factors strained the earnings of international migrants and their ability to send money home to their families (World Bank, 2016). Decreased purchasing power caused by high food prices has been compounded by reduced remittance streams (FAO, 2009). Also this soaring price of food items has led to food insecurity status of many households. Although foreign remittances into Nigeria exceed Foreign Direct Investments (FDI) and Overseas Development Assistance (Fonta et al., 2011), only a small proportion of the population is having access to receiving remittances (Olowa et al., 2013). Rural households in the Nigerian receive significantly less remittances than their urban counterparts (Africa-Focus, 2010). Less than 40 per cent of all remittances to Africa were destined to rural whereas the greater per cent of the population dwell there (IFAD, 2009). Thus, the contribution of remittances towards zero hunger attainment in rural Nigeria is unclear. A relatively little policy efforts have been made to utilize the welfare and developmental potentials of remittances but concrete policies that could encourage the flow and effective use of remittances are generally lacking (Maphosa, 2007). The reasons for this might be the absence of concrete up-to-date empirical evidences regarding the effects of remittances (Shaw, 2007). Several research findings have been conducted on effect of remittances in relation to income and welfare in Nigeria (Odozi et al., 2010; Olowa & Shittu, 2013; Etowa, 2015), as well as its effects on food poverty in Nigeria (Adeyemo and Olajide, 2013). Remittance has also been found to promote agricultural investments and reduce food insecurity (Luginaah, 2009; Mango et al., 2014; Nguyen & Winter, 2011). Studies on the effects of foreign remittances on food insecurity status of households other countries abound in literature (Abadi et al., 2017; Ebadi et al., 2018) However, a micro-study on the effect of remittances on food security and nutrition in Kwara state was undertaken by Babatunde (2018). Owing to a dearth of studies on effect of remittances on food insecurity of rural households in Nigeria, using a nationally-representative data, the effect of foreign remittances on food insecurity status of rural households in Nigeria was therefore investigated.

METHODOLOGY

The scope of this study was rural agricultural households Nigeria. Nigeria is made up of 36 states and a Federal Capital Territory (FCT), grouped into six (6) geopolitical zones: North central, North east, North west, South east, South south, and South west. The Living Standard Measurement Survey (LSMS) collected in 2015/2016 by the National Bureau of Statistics in affiliation with the Federal Ministry of Agriculture and Rural Development (FMARD) was used for the study. The data was collected in response to the needs of the country given that the dependence of a high percentage of households on agricultural activities in the country. The LSMS-ISA 2016 is the third wave of the General Household Survey-Panel and it was carried out in two visits (post-planting visit in September – November 2015 and post-harvest visit in February-April 2016). The Wave 3 sample size for households interviewed in both post-planting and post-harvest visit is 4581 which comprises of 1449 urban households and 3132 rural households. However, only 2809 rural households with consistent information were used for this study. Data on food insecurity status (dependent variable) and other independent variables were gotten from from the post-harvest visits. The reason behind using post-harvest data is that the period is considered as the lean period when the rural households may not have much food to feed on. During this period, rural households are especially vulnerable to chronic food shortages owing to adverse weather and the unavailability of enough food from home production (Obayelu & Idowu, 2019).

Descriptive statistics were used to profile remittances and food insecurity status by demographic and economic characteristics of the respondents. Household Food Insecurity Access Scale (HFIAS) to determine the food insecurity status of rural households it is sensitive to changes in households' food situation over time and can reveal hidden hunger. Coates et al., (2007) developed this scale, and it is being used extensively to measure the perception of food insecurity of households within a four-week recall period. The HFIAS score is a continuous measure of the degree of food insecurity (access). The respondent was first asked an occurrence question i.e. whether the condition in the question happened at all in the past four weeks (yes or no), also a frequency-of-occurrence question is asked to determine whether the condition happened rarely (once or twice), sometimes (three to ten times) or often (more than ten times) in the past four weeks. First, a HFIAS score variable is calculated for each household by summing the codes for each frequency-of-occurrence question. The maximum score for a household is 27, the minimum score is zero. The higher the score, the more food insecurity (access) the household experienced. The lower the score, the less food insecurity (access) a household experienced. HFIAS is easy to use in food security data collection relative to other food security measures such as dietary recalls or anthropometric indicators (Kabunga et al., 2014). Although the scale can be created as a continuous variable, it is mostly an ordered variable with four categories namely food secure, moderately food insecure, mildly food insecure, and severely food insecure (See Appendix 1).

Ordered logit model was employed in this study to identify the determinants of food insecurity status among the rural households. Food security is regarded as an ordered variable which is grouped into four categories namely, food secure, moderately food insecure, mildly food insecure, and severely food insecure. The model is specified below:

$$y_i^* = \beta' X_i + \varepsilon_i \tag{1}$$

where, y_i^*= latent variable or unobserved food security; β'= Vector of regression coefficients estimated; X_i= Observed vector of non-random explanatory variable that affect food insecurity status of rural households (See Appendix 2); and ε_i= the independent error term for households, which is normally distributed with constant variance. y_i is considered the discrete random variable whose values ranges from 1 to 4. The relationship between latent variable y_i* and observable y_i which showed the alternatives are defined below:

$$y_i=1; \text{ if } y_i^* \pounds \boldsymbol{\mu_1} \tag{2}$$

$$y_i= 2; \text{ if } \boldsymbol{\mu_1} < y_i^* \pounds \boldsymbol{\mu_2} \tag{3}$$

$$y_i=3; \text{ if } \boldsymbol{\mu_2} < y_i^* \pounds \boldsymbol{\mu_3} \tag{4}$$

$$y_i= 4; \text{ if } \boldsymbol{\mu_3} \pounds y_i^* \tag{5}$$

$\mu_{1,}$ μ_2, μ_3 and μ_4 are the threshold parameters in the logit model. The threshold variables are unknown and they indicate the discrete category that the latent variable falls into.

$$\log\left[\frac{P\left(Y_{ij} \leq 1\right)}{1-P\left(Y_{ij} \leq 1\right)}\right] = \propto_0 + \sum_{k=1}^{p-1} \propto_{jk} X_{ijk} + V_{ij}, C = 1 \ldots \Omega - 1 . \tag{6}$$

Where α and Ω-1 are the intercept terms that help model the marginal frequencies in the ordered categories, α_{jk} is the coefficient term, X_{ijk} are the explanatory variables, V_{ij} is the error term in the model, $P(Y_{ij}$£1) is the probability given that a household will fall into a food insecurity category, (1- $P(Y_{ij}$£1) is the probability that a household will not fall into the food insecurity category, k=1 is the first explanatory model and p-1 is the last explanatory variable in the logistic model (Hedeker, et al., 2000).

The nested logit model was used to isolate the contribution of remittances to food insecurity status of rural households that received foreign remittances in Nigeria. The food insecurity is the dependent variable. The nested logit model according to Shefer et al., (2004) is specified as:

$$P_i = \frac{\exp\left(V_i + \left(\mu - 1\right)\right) \ln \sum_{k \in j_m} \exp\left(V_k\right)}{\sum_{j \in J} \exp\left[V_j + \left(\mu - 1\right) \ln \sum_{k \in J_m} \exp\left(V_k\right)\right]} . \tag{7}$$

Where μ is a scale parameter, which represents the nesting coefficient. To be consistent with random utility theory, this parameter should lie between 0 and 1. When μ is equal to 1, the model collapses to the multinomial collapses to the multinomial logit model. The inferences tools for hypothesis testing include the Wald likelihood ratio and Lagrange multiplier tests and tools for discrete choice analysis, including built-in procedures for testing independence of irrelevant alternatives (IIA) assumption of the multinomial logit model

$$V_{mifis} = \alpha_0 + \alpha_{mifis} X_i + \delta Ri + \varepsilon_2$$
$$V_{mfis} = \delta_0 + \delta_{mofis} X_i + \delta R_i + \varepsilon_3 \quad (8)$$
$$V_{sfis} = \lambda_0 + \lambda_{sfis} X_i + \lambda R_i + \varepsilon_4$$

Where:

V_{fs}= Received foreign remittances and food secure households
V_{mifis}= Received foreign remittances butmildly food insecure households
V_{mofis} = Received foreign remittances but moderately food insecure households
V_{sfis} = Received foreign remittances but severely food insecure households
ε_i= error term

RESULTS AND DISCUSSION

A minority of the respondents (4.1%) had access to remittances while a majority of households that received foreign remittances were headed by unmarried (62.6%), female-headed (59.1%) aging (82.6%) heads with no formal education (52.2%) and less than four members (52.2%) (Table 1). This suggests that most of the respondents with access to remittances were no longer in their economic active age and rely more on remittances. Female-headed households which were not married had more access to remittance than their male-headed counterparts. Further, the highest percentage (56.5%) of households with remittances had elderly heads (above 60 years). This buttresses the findings of Randazzo and Piracha (2014) that the probability of receiving one of remittances increases with the age of the household head. Household heads with no formal education also had higher access to remittances than household heads with any form of formal education. The majority of the rural household heads (91.6%) were primarily engaged in farming, while the highest proportion of rural households with foreign remittances resided in South East (59.1%).

A higher proportion of households that owned land (98.3%) and cultivated less than one hectare (96.5%) received remittances. They had no access credit (80.0%), agricultural extension services (98.3%) and were non-members of cooperative societies (81.7%) (Table 2).

Food Insecurity Profile

The HFIAS indicator categorizes households into four levels of household food insecurity (access namely food secure, mildly food insecure, moderately food insecure and severely food insecure. Households are categorized as increasingly food insecure as they respond affirmatively to more severe conditions and/or experience those conditions more frequently. Two out of every five households were food secure while one out of every three households was severely food insecure (Table 3). About one out of every three households with aging heads (above 60 years) were severely food insecure (38.19%) when compared to other age categories. This is line with a priori expectation. This is because household heads within this age group are no longer in their economic active age productivity decreases with increase in age and that older household heads lived in great deprivation than the younger household heads (Tijani & Ndukwe, 2012). This is consistent with life-cycle hypothesis and is consistent with the findings of

Table 1. Remittances profile by household demographic characteristics and location

Demographic Characteristics		Remittances		Total
		Yes (115)	No (2694)	
Age of household head (in years)	20-30	1(0.87)	101(3.75)	102(3.63)
	31-40	8(6.96)	439(16.30)	447(15.91)
	41-50	11(9.57)	630(23.39)	641(22.82)
	51-60	30(26.09)	613(22.75)	643(22.89)
	>60	65(56.52)	911(33.82)	976(34.75)
Gender	Male	47(40.87)	2239(83.11)	2286(81.38)
	Female	68(59.13)	455(16.89)	523(18.62)
Marital status	Married	43(37.39)	1961(72.79)	2004(71.34)
	Unmarried	72(62.61)	733(27.21)	805(28.66)
Household size	1-3	60(52.17)	714(26.50)	774(27.55)
	4-6	16(13.91)	607(22.53)	623(22.18)
	>6	39(33.91)	1373(50.97)	1412(50.27)
Education level of the household head	No formal education	60(52.17)	1123(41.69)	1183(42.11)
	Primary education	16 (13.91)	655(24.31)	671(23.89)
	Secondary education	10(8.70)	422(15.66)	432(15.38)
	Tertiary education	29(25.22)	494(18.34)	523(18.62)
Primary Occupation	Farming	111(96.52)	2463(91.43)	2574(91.63)
	Non-farming	4(3.48)	231(8.57)	235(8.37)
Geopolitical Zones	North central	10(8.70)	484(17.97)	494(17.59)
	North-east	8(6.96)	488(18.11)	496(17.66)
	North-west	10(8.70)	649(24.09)	659(23.46)
	South-east	68(59.13)	565(20.97)	633(22.53)
	South south	4(3.48)	316(11.73)	320(11.39)
	South-west	15(13.04)	192(7.13)	207(7.37)

Gebre (2012) that households with elderly heads had higher severity index of food insecurity. This is line with the findings of Gebre (2012) that household with elderly heads had higher severity index of food insecurity. This is because productivity decreases with increase in age (Tijani & Ndukwe, 2012) and consequently food insecure. They therefore rely on remittances for their living. Further, 40.1% and 28.8% of the female-headed and male-headed households were severely food insecure, respectively. This suggests the vulnerability of female-headed households to food insecurity than their male counterparts, possibly owing to inaccessibility to production assets.

One out of every two rural households had more than six members and this proportion is similar accorss all the food insecurity groups (Table 3). Large households face difficulties in meeting the basic requirements such as education for the children proper nutrition and adequate housing, all of which tend to reinforce food insecurity among households that fail to cope with them (Etim, et al., 2011). Further, a larger proportion of the household heads had no formal education and they represented the

Table 2. Foreign remittances profile by assets of rural households

Assets		Remittances		
		Yes	No	Total
Land ownership	**Yes**	113(98.26)	2659(98.70)	2772(98.68)
	No	2(1.74)	35(1.30)	37(1.32)
Access to loans	Yes	23(20.00)	488(18.11)	511(18.19)
	No	92(80.00)	2206(81.89)	2298(81.81)
Access to extension	Yes	2(1.74)	175(6.50)	177(6.30)
	No	113(98.26)	2519(93.50)	2632(93.70)
Membership of cooperatives	Yes	21(18.26)	520(19.30)	541(19.26)
	No	94(81.74)	2174(80.70)	2268(80.74)
Farm size(ha)	< 1	111(96.52)	2434(90.35)	2545(90.60)
	1.001-3	4(3.48)	226(8.39)	230(8.19)
	3.001-6	0(0.00)	26(0.97)	26(0.93)
	6.001-11	0(0.00)	8(0.30)	8(0.28)

Table 3. Food insecurity status profile by households' demographic characteristics

Demographics		Food Insecurity Status				Total (N=2809)
		Food Secure (N=1184)	Mildly Food Insecure (N=362)	Moderately Food Insecure (N=391)	Severely Food Insecure (N=872)	
Age of household head (in years)	20-30	57(4.81)	13(3.59)	7(1.79)	25(2.87)	102(3.63)
	31-40	220(18.58)	45(12.43)	57(14.58)	125(14.33)	447(15.91)
	41-50	278(23.48)	81(22.38)	98(25.060	184(21.10)	641(22.82)
	51-60	265(22.38)	94(25.97)	79(20.20)	205(23.51)	643(22.89)
	>60	361(30.74)	129(35.64)	150(38.36)	333(38.19)	976(34.75)
Gender	Male	997(84.21)	305(84.25)	325(83.12)	659(75.57)	2286(81.38)
	Female	187(15.79)	57(15.75)	66(16.88)	213(24.43)	523(18.62)
Marital status	Married	855(72.21)	270(74.59)	283(72.38)	596(68.35)	2004(71.34)
	Unmarried	329(27.79)	92(25.41)	108(27.62)	276(31.65)	805(28.66)
Education level	No formal education	525(44.34)	160(44.20)	169(43.22)	329(37.73)	1183(42.11)
	Primary education	240(20.27)	94(25.97)	108(27.62)	229(26.26)	671(23.89)
	Secondary education	216(18.24)	38(10.50)	38(9.72)	140(16.06)	432(15.38)
	Tertiary education	203(17.15)	70(19.34)	76(19.44)	174(19.95)	523(18.62)
Occupation	Farming	1092(92.23)	330(91.16)	358(91.56)	794(91.06)	2574(91.63)
	Non-farming	92(7.77)	32(8.84)	33(8.44)	78(8.94)	235(8.37)
Household size	1-3	325(27.45)	87(24.03)	97(24.81)	265(30.39)	774(27.55)
	4-6	255(21.45)	79(21.82)	82(20.97)	207(23.74)	623(22.18)
	>6	604(51.01)	196(54.14)	212(54.22)	400(45.87)	1412(50.27)

Source: Author's computation from GHS data, 2015

highest proportion of food insecurity groups. household heads with no formal education were severely food insecure than households with primary, secondary or tertiary education. This result implies that households who have household heads with relatively better education are more likely to be food secure than those headed by uneducated (illiterate) household heads. Household heads with secondary education had lowest severity of food insecurity since some level of education is expected to positively affect the household's food resources. Similarly, a higher percentage of the rural households were primarily farmers, owned land and without access to credit and extension services (Table 4).

About 63. 6% of households in the North central (known as the food basket of the nation) were food secure while almost a half of households in the South east were severely food insecure. This is line with the findings of Jabo et al., 2014 that rural households in South-east had highest incidence of food insecurity in Nigeria. The North central has a favourable climate and soil for production of both local and exotic agricultural species. This favours biodiversity and consequently food security in the region. Further, households with remittances were more found to have higher severity of food insecurity (Table 4). However, a third of rural households in Northeast and 41. 6% of those in the South south were severely food insecure. This is possibly owing to adverse effects of Boko Haram insurgency on livelihood activities in the North east zone as well as soil and water degradation due to oil spillage in the South south zone. The highest percentage of moderately food insecure households was also found in the North west, located in the sudan savannah, which limits the extent of agrobiodiversity in the region and consequently increases food insecurity level.

About 46.9% of households that received foreign remittances were severely food insecure while 42.7% of those without foreign remittances were food secure. Slightly above a quarter (28.7%) of households with foreign remittances were food secure. This suggests that most rural households were food sufficient without foreign remittances. This corroborates the findings of Atuoye (2015) that rural households that receive remittances were found to be in higher category of food insecurity.

Determinants of Food Insecurity Status Among Rural Households in Nigeria

Coefficients and standard errors from the ordered probit model, as well as marginal effects for the four levels of food security are presented in Table 5. Discussion focuses on marginal effects[1] since coefficients from ordered probit do not have a direct interpretation. The Log likelihood was significant at percent indicating that the model fits the data. The first cut point -0.7245 was partition between food secure and mildly food insecure range, -0.1590 was the partition of range between mildly and moderately food insecure range while 0.4910 was the partitions of range moderately and severely food insecure range.

A year increase in the age of the household heads decreased the probability of the households being food secure by 0.14% but increases the probability of being moderately food insecure and severely food insecure by 0.02% and 0.12%, respectively. The positive relationship implies the probability of a household being food insecure increased with age of its head. Following the life-cycle hypothesis, productivity first increases and reaches its peak up to middle-age and declines thereafter. Aged household heads are thus less productive and they live on rents and remittances. This is in conformity with the findings of Ojogho 2010 that the probability of food insecurity increases with age of household head. Being a female-headed household also reduces the probability being food secure but increases the probability of being severely food insecure. Women constitute a majority of food insecure population partly owing to gender inequality in access to basic human rights. They often lack access to own property, productive inputs and services required for rural livelihoods (SIDA, 2015).

Table 4. Food insecurity status profile by capital assets and location

Variables		Food Secure	Mildly Food Insecure	Moderately Food Insecure	Severely Food Insecure	Total
Land ownership	Yes	1167(98.56)	360(99.45)	387(98.98)	858(98.39)	2772(98.68)
	No	17(1.44)	2(0.55)	4(1.02)	14(1.61)	37(1.32)
Access to loans	Yes	201(16.98)	69(19.06)	65(16.62)	176(20.18)	511(18.19)
	No	983(83.02)	293(80.94)	326(83.38)	696(79.82)	2298(81.81)
Access to extension	Yes	98(8.28)	14(3.87)	26(6.65)	39(4.47)	177(6.30)
	No	1086(91.72)	348(96.13)	365(93.35)	833(95.53)	2632(93.70)
Membership of cooperatives	Yes	233(19.68)	70(19.34)	64(16.37)	174(19.95)	541(19.26)
	No	951(80.32)	292(80.66)	327(80.63)	698(80.05)	2268(80.74)
Farm size (ha)	0-1	1058(89.36)	324(89.50)	357(91.30)	806(92.43)	2545(90.60)
	1.001-3	107(9.04)	34(9.39)	30(7.67)	59(6.77)	230(8.19)
	3.001-6	16(1.35)	1(0.28)	3(0.77)	6(0.69)	26(0.93)
	6.001-11	3(0.25)	3(0.83)	1(0.26)	1(0.11)	8(0.28)
Zones	North central	314(26.52)	77(21.27)	49(12.53)	54(6.19)	494(17.59)
	North-east	199(16.81)	62(17.13)	66(16.88)	169(19.38)	496(17.66)
	North-west	296(25.00)	83(22.93)	104(26.60)	176(20.18)	659(23.46)
	South-east	152(12.84)	83(22.93)	97(24.81)	301(34.52)	633(22.53)
	South-south	118(9.97)	27(7.46)	42(10.74)	133(15.25)	320(11.39)
	South-west	105(8.87)	30(8.29)	33(8.44)	39(4.47)	207(7.37)
Foreign Remittances	Received remittances	33 (28.70)	15(13.04)	13(11.30)	54(46.96)	115 (4.09)
	No remittances	1151 (42.72)	347(12.88)	378(14.03)	818(30.36)	2694 (95.91)

Household size was negatively related to the probability of being food insecure but it is positively related to the probabilities of being moderately food insecure and being severely food insecure. This implies that the larger the household size, the less the probability of being food insecure. An additional household member increases the likelihood of being food secure by 1.01% but decreases the log-likelihoods of being moderately food insecure and being and severely food insecure by 0.16% and 0.87%, respectively in adult equivalent. This result contradicts the findings of Haile *et al.,* 2005 that increased in household size increases food insecurity. However, Sule *et al.,* (2002) found that household size has a great role to play in the provision of family labour for agricultural sector and its output.

Educational level of respondents is expected to reduce household food insecurity status (Ahmed et al., 2015; Maziya et al., 2017; Ehebhamen et al., 2017). It enhances better livelihood choices with attendant better income, which consequently shift both the budget line and the indifference curve upwards. Households will therefore have a higher propensity to consume and move towards food security line. Households whose heads had secondary (high school) education would likely neither be moderately food insecure nor be severely food insecure (Table 5). Conversely, households with access to foreign remittances had a higher probability of being food insecure which buttresses the findings of Castaldo and Reilly 2007 that Albanian households who received remittances from abroad spent less on non-durable goods and has greater consumption on durable goods compared to households who did not receive remittances.

In line with apriori expectation, a unit increase in per capita income reduced the probability of being food insecure by -4.00E-06. Thus, increasing per capita income of a person in a household would increase the log-likelihood of being food secure. This could be expected because increased income, other things being equal leads to increase in quantity and quality (access) to food which is also a sure way of combating food insecurity. The marginal effect of household per capita revealed that a marginal increase in per capita income of a person in the food secure category will increased the likelihood of being food secure by 9.69E-07 unit but reduced the log-likelihood of being moderately food insecure and being severely food insecure marginally by 1.54E-07 and 8.34E-07 units, respectively.

Being resident in North-east, North-west, South-east and South-south increased the probability of being food insecure relative to residing in South-west. However, being resident in the North-central zone had a negative relationship with food insecurity relative to being in the South-west. Ashagidigbi (2012) also posited that households in North-east geo-political zone of Nigeria had higher probability of being food insecure. This could be linked to the incessant insurgency in the region, which has made most rural households internally displaced without livelihood activities and income.

Table 5. Determinants of food insecurity in rural Nigeria

Variables	Coefficient	Marginal Effects			
		Food Secure	Mildly Food Insecure	Moderately Food Insecure	Severely Food Insecure
Age	0.0058** (0.0029)	-0.0014** (0.0007)	0.0000 (0.0000)	0.0002* (0.0001)	0.0012** (0.0006)
Gender	-0.2296* (0.1387)	0.0557* (0.0336)	0.0011 (0.0009)	-0.0089 (0.0054)	-0.0479* (0.0289)
Marital Status	-0.0228 (0.0265)	0.0055 (0.0064)	0.0001 (0.0001)	-0.0009 (0.0010)	-0.0048 (0.0055)
Occupation	0.0165 (0.1361)	-0.0040 (0.0329)	-0.0000 (0.0006)	0.0007 (0.0053)	0.0034 (0.0284)
No formal education	-0.3906*** (0.1082)	0.0947*** (0.0262)	0.0018 (0.0011)	-0.1507*** (0.0044)	-0.0815*** (0.0226)
Primary education	-0.1956 (0.1217)	0.0474 (0.0295)	0.0009 (0.0008)	-0.0075 (0.0047)	-0.0408 (0.0254)
Secondary education	(-0.2512)* (0.1319)	0.0609 (0.0319)	0.0012 (0.0009)	-0.0097* (0.0052)	-0.0524 (0.0275)
Household size	-0.0415*** (0.0145)	0.0101*** (0.0035)	0.0002 (0.0001)	-0.0016** (0.0006)	-0.0087*** (0.0030)
Remittances	0.3843** (0.1888)	-0.0932** (0.0458)	-0.0018 (0.0013)	0.0148** (0.0074)	0.0802** (0.0394)
Farm size	1.77E-06 (5.13E-06)	-4.03E-07 (1,24E-06)	-8.34E-09 (2.46E-08)	6.84E-08 (1.98E-07)	3.70E-07 (1.07E-06)
Membership into cooperatives	-0.0517 (0.0476)	0.0125 (0.0115)	0.0002 (0.0003)	-0.0019 (0.0018)	-0.0108 (0.0099)
Access to loans	0.0153 (0.0955)	-0.0037 (0.0232)	-0.0000 (0.0005)	0.0006 (0.0037)	0.0032 (0.0199)

continues on following page

Table 5. Continued

Variables	Coefficient	Marginal Effects			
		Food Secure	Mildly Food Insecure	Moderately Food Insecure	Severely Food Insecure
Access to extension	-0.2293 (0.1563)	0.0556 (0.0379)	0.0011 (0.0009)	-0.0089 (0.0061)	-0.0478 (0.0326)
Land ownership	-0.2049 (0.3326)	0.0497 (0.0806)	0.0009 (0.0017)	-0.0079 (0.0129)	-0.0427 (0.0694)
Per capita income	-4.00E-06*** (1.27E-06)	-9.69E-07*** (3.09E-07)	-1.88E-08 (1.18E-08)	1.54E-07*** (5.09E-08)	8.34E-07*** (2.66E-07)
North-central	-0.4508*** (0.1675)	0.1093*** (0.0406)	0.0021 (0.0014)	-0.0174*** (0.0066)	-0.0940*** (0.0349)
North-east	0.6313*** (0.1679)	-0.1530*** (0.0407)	-0.0029* (0.0018)	0.0244*** (0.0068)	0.1317*** (0.0350)
North-west	0.3725** (0.1674)	-0.0903** (0.0406)	-0.0018 (0.0012)	0.0144** (0.0066)	0.0777** (0.0349)
South-east	1.1617*** (0.1581)	-0.2816*** (0.0383)	-0.0055* (0.0031)	0.0448*** (0.0072)	0.2423*** (0.0331)
South-south	0.8041*** (0.1732)	-0.1949*** (0.0419)	-0.0038* (0.0022)	0.0310*** (0.0072)	0.1677*** (0.0362)
Constant cut 1	-0.7245 (0.4741)				
Constant cut 2	-0.1590 (0.4740)				
Constant cut 3	0.4910 (0.4739)				

LR chi2(20)= 303.11
Prob>chi2= 0.0000
Log likelihood= -3404.1197
No of observation = 2809
*** Significant at 1%, ** significant at 5% and * significant at 10%.
Figures in parenthesis are the standard errors.

Effect of Remittances on Food Insecurity Categories Among Rural Households in Nigeria

The Nested logit model was used to determine the conditional effects of remittances on food insecurity categories among rural households that received foreign remittances in Nigeria (Table 6). In order to isolate the effect of location differences in food insecurity status, Model I is the nested model without the effect of the geopolitical zones while model II includes the effects of the geopolitical zones. One of the assumptions of nested logit is that the dissimilarity parameter should not exceed one. The dissimilarity parameter measures the degree of correlation of random shock within each of the food insecurity categories. This conforms to the dissimilarity parameters gotten for food secure and food insecure categories. In model I, severely food insecure households with remittances was positive and significant for categories suggesting that increasing in household heads' access to foreign remittances will the increase probability of the households being severely food insecure relative to being food secure. Castaldo and Reilly (2007) found from Albanian households' expenditure pattern that households with access to foreign remittances spent more on health and education but less on food, and they were food insecure.

Age had a positive relationship with food insecurity status of households with remittances suggesting that as household head age increases, the more food insecure the household becomes. The global food-insecure population has risen since 2014, reaching 821 million in 2017 with rural women amongst the worst affected (Botreau & Cohen, 2019). Female headed households in rural Nigeria were more food insecure than male counterparts, which is similar to the findings of a Felker-Kantor and Wood (2012) that the odds of food insecurity are higher among female-headed households compared to male-headed households in Brazil. Women are often marginalized in their families and have less access to access productive inputs, services and markets required for rural livelihoods (SIDA, 2015). Household heads with secondary occupation also had a negative relationship with food insecurity which implies that household head with remittances and secondary education had lower probability of being food insecure. Thus, being literate and having access to foreign remittances reduce the chance of becoming food insecure in the sample households. Maxwell (2003) observed that education also has other important components of human capital that is the purchasing efficiency and food knowledge of the household head. Household heads will also be able to adopt more modern farm technologies on their farms thus improving their productivity. This is line with the findings of Ojogho (2010) that education of household heads reduces the probability of being food secure.

Membership of household heads into cooperatives had a negative relationship with food insecurity status of households with remittances implying that household heads in cooperatives are less food insecure than household heads who do not belong to cooperatives. This is line with the findings of Ahmed et al., (2015) that level of participation of households in cooperatives increases the probability of being food secure. He opined that active participation in cooperatives activities tend to attract benefits in terms of helping members in mobilizing resources within society for agricultural operations and marketing, enables to take advantage economies of scale. Access to extension was significant and had a negative relationship with food insecurity which means that household heads that had access to extension are less food insecure. Access to extension agents by households exposes them to new farm techniques and adoption of innovation on their diverse income activities. This is line with Olaoye et al., (2016) that households that had access to extensions services tends to be more food secure. Further, household size also had a negative relationship with food insecurity which implies that increased in number of households decreases the likelihood of them being food insecure. This may be so if the households had more members that are working which reduces the number of dependents in a household and thus increasing their income and making them to be less food insecure. This is line with an earlier finding in this study that household size reduces food insecurity.

Farm size was significant and had a negative relationship with food insecurity with remittances suggesting household with remittances and large farm sizes were less likely to be food insecure. Reddy et al., (2004) observed that greater efficiencies in the use of resources are associated with the large farms than the small farms. They pointed out that the smallness of holdings deters the use of mechanization and does not allow the use of modern inputs due to lack of purchasing power in the hands of small farmers. This may result in low productivity and hence low food security status. Income of household heads had a negative relationship with food insecurity status of households with remittances. This implies that as income of household head increases, the less food insecure the household becomes. Household income is important as it determines how much can be spent on various needs of the households. The quality and quantity of a household's expenditure patterns are highly correlated with the purchasing power which is expected to improve their food security. This is line with the findings of Bashir et al., (2010) that income improves the probability of household being food secure.

Table 6. Determinants of interaction of food insecurity status and remittances among rural households in Nigeria

Choice-Specific Variables	Model I		Model II	
	Coefficient	Standard Error	Coefficient	Standard Error
Remittances (Mildly food insecure category)	0.2196	0.3535	0.1969	0.3678
Remittances (Moderately food insecure category)	0.1601	0.4044	0.1673	0.4230
Remittances (Severely food insecure category)	0.7512*	0.3745	0.4618	0.3920
Type-specific variables				
Insecure category				
Age of the household head	0.0132***	0.0031	0.0085***	0.0033
Gender	-0.2843*	0.1465	-0.0166	0.1570
Marital status	-0.0005	0.0286	-0.0084	0.0297
Occupation	-0.1536	0.1493	-0.0530	0.1536
No formal education	-0.3597***	0.1185	-0.3499***	0.1212
Primary education	0.0369	0.1266	-0.1247	0.1389
Secondary education	-0.3737***	0.1358	-0.3889***	0.1438
Household size	-0.0344**	0.0154	-0.0314*	0.0161
Farm size	-0.0000*	5.76E-06	5.79E-07	5.77E-06
Membership of cooperatives	-0.0357	0.0507	-0.0447	0.0527
Access to credit	0.0909	0.1042	0.0335	0.1082
Access to extension	-0.5034**	0.1616	-0.3655**	0.1660
Landownership	0.2428	0.3445	-0.0081	0.3602
Income	-0.0046***	0.0014	-0.0039***	0.0014
North-central			-0.4198**	0.1796
North-east			0.4835***	0.1837
North-west			0.2898	0.1824
South-east			1.1564***	0.1788
South-south			0.5413***	0.1902
Dissimilarity parameters				
Food secure_tau	1		1	
Food insecure_tau	0.4152	0.4392	0.2068	0.4728
LR test for IIA (tau=1): chi2(1) 1.76			2.80	
Wald chi2 (17) = 105.56			Wald chi2 (22) =240.06	
Number of observations = 11236				
Number of cases = 2809				

The food secure category was used as a baseline. ***, ** and * represent 1%, 5% 10% levels of significance.

Being resident in the North-central geo-political zone with remittances had a negative relationship with food insecurity status which implies that households with remittances in this zone were less likely to be food insecure relative to other households in other zones. North-east, North-west, South-east and South-south had positive and significant relationship with food insecurity status which implies that households residing in these zones had higher probability of being food insecure. This result conforms with an earlier findings in this study that North-central had the least severity of food insecurity.

CONCLUSION AND RECOMMENDATIONS

The study revealed that only minority of the rural households used for this study had access to remittances and that more than half of the rural households falls into the food insecure categories. Increased age of household heads enhanced household food security. Female headed households were more food insecure than male headed households. Thus, increasing access to production assets by female-headed households will enhance their livelihood activities and food security status. Primarily farming households tend to be more food insecure status than non-farming households. Farmers should therefore be encouraged to diversify their economic activities to non-farm activities in order to increase their risk variance, earn short-run income and enhance their food security status. Also the effect of education household food insecurity confirms that education plays a significant role in ensuring better living condition. Therefore, education in form of formal, informal or vocational training should been encouraged and promoted among the rural households which can guarantee them more income sources to meet food needs so as to ensure food security in the study area. Increase in farm size decreases food insecurity which implies that food insecurity seems to reduce with larger farm size. Thus, agricultural policy should place more emphasis on increasing agricultural productivity rather than increasing farm size. Finally, the results revealed that level of remittance to rural households is very low hence it does not contribute meaningfully to food security among the respondents.

REFERENCES

Adeyemo, T. A., & Olajide, O. A. (2013). *Urban Rural Pattern of Remittances and Effects on Food Security of Rural Households in Nigeria*. In *Conference on International Research on Food Security, Natural Resource Management and Rural Development organised by the University of Hohenheim*, Stuttgart, Germany.

Ahmed, F. F., Eugene, C. E., & Abah, P. O. (2015). Analysis of food security among farming households in Borno State, Nigeria. *Journal of Agricultural Economics, Environment and Social Sciences*, *1*(1), 130–141.

Babatunde, R. O. (2018). *Impact of Remittances on Food Security and Nutrition of Migrant's Household: Evidence from Nigeria*. Paper presented at the 30th International Association of Agricultural Economist, Vancouver, Canada.

Bashir, M. K., Schilizzi, S., & Pandit, R. (2012). *The determinants of rural household food security: The case of landless households of the Punjab, Pakistan.* Working Paper 1208, School of Agricultural and Resource Economics, University of Western Australia, Crawley, Australia.

Botreau, H., & Cohen, M. J. (2019). *Gender Inequalities and Food Insecurity. Oxfam Briefing Paper 2019.* Oxford, UK: Oxfam International.

Castaldo, A., & Reilly, B. (2007). Do Migrant Remittances Affect the Consumption Patterns of Albanian Households? Southern-Eastern Europe. *Journal of Economics*, *1*, 25–54.

Coastes. J., Swindale, A., & Bilinsky, P. (2007). *Household food Insecurity Access Scale (HFIAS) for measurement of food access: indicator guide* (vol. 3). Washington, DC: FHI 360/FANTA.

Combes, J. L., Ebeke, C. H., Etoundi, S. M. N., & Yogo, T. U. (2014). Are remittances and foreign aid a hedge against food price shocks in developing countries? *World Development*, *54*, 81–98. doi:10.1016/j.worlddev.2013.07.011

Ehebhamen, O. G., Obayelu, A. E., Vaughan, I. O., & Afolabi, W. A. O. (2017). Rural households' food security status and coping strategies in Edo State Nigeria. *International Food Research Journal*, *24*(1), 333–340.

Etim, N. A., Okon, S., & Akpabio, I. A. (2011). Determinants of deprivation among part-time cassava farming households in the humid tropics. *International Journal of Agricultural Management and Development*, *1*(1), 45–51.

Etowa, E. B. (2015). *Effects of Migrant Remittances on Farm Household Welfare in Nigeria* (Unpublished Ph.D thesis). Department of Agricultural Economics, University of Nigeria.

Felker-Kantor, E., & Wood, C. H. (2012). Female-headed households and food insecurity in Brazil. *Food Security*, *4*(4), 607–617. doi:10.100712571-012-0215-y

Fonta, M. W., Onyukwu, E. O., & Nwosu, O. E. (2011). International Remittance Inflow and Household Welfare: Empirical Evidence from Nigeria. *Research Journal of Finance and Accounting*, *2*(3), 140–149.

Food and Agriculture Organization. (2009). *Measuring Food Security Using Respondents Perception of Food Consumption Adequacy*. Rome, Italy: Food and Agriculture Organization of the United Nations.

Gebre, G. G. (2012). Determinants of food insecurity among households in Addis Ababa City, Ethiopia. *Interdisciplinary Description of Complex Systems*, *10*(2), 159–173. doi:10.7906/indecs.10.2.9

Hedeker, D., Siddiqui, O., & Hu, F. B. (2000). Random-effects regression analysis of correlated grouped-time survival data. *Statistical Methods in Medical Research*, *9*(00), 161–179. doi:10.1177/096228020000900206 PMID:10946432

IFAD. (2007). *Sending money home: Worldwide remittance flows to developing countries. International Fund for Agricultural Development.* Rome, Italy: IFAD.

IFAD. (2009). *Report on migration, remittances and rural development.* International Fund for Agricultural Development. Rome, Italy: IFAD. Retrieved from https://www.un.org/en/development/desa/population/events/pdf/8/P14_IFAD.pdf

Jabo, M. S. M., Mansor, I. S., Nasir, S. M., Mahir, A. A., & Maikasuwa, A. M. (2014). Examining the Incidence, Depth and Severity of Food Insecurity among rural Households in Nigeria. *Journal of Economics and Sustainable Development, 5*(26).

Kabunga, N.S., Dubois, T., & Qaim, M. (2014). Impact of tissue culture banana technology on farm household income and food security in Kenya. *Journal of Food policy, 45*, 25-34.

Kuuire, V. Z., Mkandawire, P., Arku, G., & Luginaah, I. (2013). Abandoning farms in search of food. Food remittance and household food security in Ghana. *African Geographical Review, 32*(2), 125–139. doi:10.1080/19376812.2013.791630

Langworthy, M., Frankenberger, T., Bonnard, P., Onwwubah, C., & Green, H. (2003). Food Access Indicator Review: A Food and Nutrition Technical Assistance Project. Academic Press.

Luginaah, I., Weis, T., Galaa, S., Nkrumah, K. M., Bezner-Kerr, R., & Bagah, D. (2009). Enviroment, migration and food security in the Upper West Region of Ghana. In *Environment and health in Sub-Saharan African: Managing an emerging crises* (pp. 25–38). Springer. doi:10.1007/978-1-4020-9382-1_2

Mango, N., Zamasiya, B., Makate, C., Nyikahadzoi, K., & Siziba, S. (2014). Factors influencing household food security among smallholder farmers in the Mudzi district of Zimbabwe. *Development Southern Africa, 31*(4), 625–640. doi:10.1080/0376835X.2014.911694

Maphosa, F. (2007). Remittances and development: The impact of migration to South Africa on rural livelihood in Southern Zimbabwe. *Development Southern Africa, 24*(1), 123–136. doi:10.1080/03768350601165942

Martin, M. (2005). *Africa's external constraints: What developed countries should do. Africa in the World Economy. The National, Regional and International Challenges*. Retrieved from w.w.w.fondad.org

Maziya, M., Mudhara, M., & Chitja, J. (2017). What factors determine household food security among smallholder farmers? Insights from Msinga, KwaZulu-Natal, South Africa. *Agrekon, 56*(1), 40–52. doi:10.1080/03031853.2017.1283240

Momodu, A.S., Akinbami, C.A.O., & Obisanya, J.F. (2011). Achieving food security and climate change mitigation through development in rural Nigeria: Gender perspective. *African Journal of Environmental Science and Technology, 5*(10), 834–854.

Nguyen, M. C., & Winters, P. (2011). The Impact of migration on food consumption patterns: The case of Vietnam. *Food Policy, 36*(1), 71–87. doi:10.1016/j.foodpol.2010.11.001

Obayelu, O. A., & Idowu, O. O. (2019). Dietary diversity status of rural households in Nigeria: A gendered perspective. *Economia Agro-Alimentare, 21*(3). doi:10.3280/ECAG2019-003003

Odozi, J., Awoyemi, T. T., & Omonona, B. T. (2010). Household poverty and inequality: The implication of migrants. *Journal of Economic Policy Reform, 13*(2), 191–199. doi:10.1080/17487871003700788

Ojogho, O. (2010). Determinants of food insecurity among arable farmers in Edo State, Nigeria. *Agricultural Journal, 5*(3), 151–156. doi:10.3923/aj.2010.151.156

Olaoye, T. A., Kayode, A. O., & Oladiran, J. O. (2016). Determinants of food security in Ogbomoso Agricultural zone. *International Journal of Research and Development Organization*, *2*(1).

Olowa, O. W., Awoyemi, T. T., Shittu, M. A., & Olowa, O. A. (2013). Effects of remittances on poverty among rural households in Nigeria. *European Journal of Sustainable Development*, *2*(4), 263–284.

Randazzo, T., & Piracha, M. (2014). *Remittances and Household Expenditure Behaviour in Senegal*. IZA Discussion Paper No. 8106.

Reddy, S. S., Ram, P. R., Sastry, T. V. N., & Devi, I. B. (2004). *Agricultural Economics*. New Delhi: Oxford and IBH Publishing Co. Pvt Ltd.

Shaw, W. (2007). *Migration in Africa: A review of the economic literature on international migration in 10 countries*. Washington, DC: Development Prospects Group, The World Bank.

SIDA. (2015). *Women and food security. Gender tool Box (Brief)*. https://www.sida.se/contentassets/3a820dbd152f4fca98bacde8a8101e15/women-and-food-security.pdf

Sule, A. M., Ogunwale, S. A., & Atala, T. K. (2002). Factors affecting adoption of fishing innovation among fishing entrepreneurs in Jebba Lake Community. *Journal of Agricultural Extension, 6, 48-60.*

The Economist. (2017). *The Global Food Security Index 2017: Measuring food security and the impact of resource risks is the sixth edition of an Economist Intelligence Unit study*. Retrieved from http://foodsecurityindex.eiu.com/Home/DownloadResource?fileName=EIU%20Global%20Food%20Security%20Index%20-%202017%20Findings%20%26%20Methodology.pdf

Tijani, B. A., & Ndukwe, O. K. (2012). Assessment of household food security situation in Marte Local Government Area of Borno State, Nigeria. *Maiduguri Journal of Arts and Social Science*, *10*(1), 47–60.

Tsegai, D. (2007). Migration as a household decision: What are the roles of income differences? Insights from the Volta Basin of Ghana. *European Journal of Development Research*, *19*(2), 305–326. doi:10.1080/09578810701289212

Vision. (2010). *Agriculture 1997 Final Report of the Sub Committee of Agricultural Economic Sector*. Federal Government of Nigeria.

World Bank. (2012). *Nigeria Agricultural Production: World Bank Indicators*. Washington, DC: World Bank.

World Bank. (2016). *Migration and Remittances. Factbook 2016, Global Knowledge Partnership on Migration and Development*. KNOMAD.

ENDNOTE

1 It is complex to interpret the coefficient estimate of ordered logit but it possible to interpret both the magnitude and the sign of marginal effects. Further, the sum of the marginal effects of a variable in the four categories is zero, owing to the fact that a high likelihood in one category results in less likelihood in the other categories.

APPENDIX

The HFIAS consists of two types of related questions. The first question type is called an occurrence question. There are nine occurrence questions that ask whether a specific condition associated with the experience of food insecurity ever occurred during the previous four weeks (30 days). Each severity question is followed by a frequency-of-occurrence question, which asks how often a reported condition occurred during the previous four weeks. The occurrence question is asked whether the condition in the question happened at all in the past four weeks (yes or no). If the respondent answers "yes" to an occurrence question, a frequency-of-occurrence question is asked to determine whether the condition happened rarely (once or twice), sometimes (three to ten times) or often (more than ten times) in the past four weeks. There are three response options representing a range of frequencies (1 = rarely, 2 = sometimes, 3 = often).

Also a score known as the HFIA score is given to the household in the past four weeks (30 days) which is a continuous measure of the degree of food insecurity (access) in the household. First, a HFIAS score variable is calculated for each household by summing the codes for each frequency-of-occurrence question. The maximum score for a household is 27 (the household response to all nine frequency-of-occurrence questions was "often", coded with response code of 3); the minimum score is 0. The higher the score, the more food insecurity (access) the household experienced. The lower the score, the less food insecurity (access) a household experienced.

Lastly, Food Insecurity is categorized into categories using the HFIAP (Household Food Insecurity Access Prevalence indicator status indicator. The HFIAP indicator categorizes households into four levels of household food insecurity (access): food secure and mild, moderately and severely food insecure. Households are categorized as increasingly food insecure as they respond affirmatively to more severe conditions and/or experience those conditions more frequently.

Household Food Insecurity Access category (HFIA) includes:

1=Food Secure, 2=Mildly Food Insecure Access, 3=Moderately Food Insecure Access, 4=Severely Food Insecure Access

HFIA category = 1 if [(Q1a=0 or Q1a=1) and Q2=0 and Q3=0 and Q4=0 and Q5=0 and Q6=0 and Q7=0 and Q8=0 and Q9=0]

HFIA category = 2 if [(Q1a=2 or Q1a=3 or Q2a=1 or Q2a=2 or Q2a=3 or Q3a=1 or Q4a=1) and Q5=0 and Q6=0 and Q7=0 and Q8=0 and Q9=0]

HFIA category = 3 if [(Q3a=2 or Q3a=3 or Q4a=2 or Q4a=3 or Q5a=1 or Q5a=2 or Q6a=1 or Q6a=2) and Q7=0 and Q8=0 and Q9=0]

HFIA category = 4 if [Q5a=3 or Q6a=3 or Q7a=1 or Q7a=2 or Q7a=3 or Q8a=1 or Q8a=2 or Q8a=3 or Q9a=1 or Q9a=2 or Q9a=3]

Table 7. Description of variables

Independent Variables	Description of Variables
Age of the household head	In years
Gender	(1= male, 0 if otherwise)
female	
Marital status	(1= married, 0 if otherwise)
Occupation	(1= farming, 0 if otherwise)
Educational level of household head	Years of formal education
Household size	Number of household members
Remittances	(1= received remittances, 0 if otherwise)
Farm size	Farm size in hectares
Membership of cooperatives	(1= if the head is a member, 0 if otherwise)
Access to credit	(1= yes, 0 if otherwise)
Access to extension	(1= yes, 0 if otherwise)
Landownership	(1= yes, 0 if otherwise)
Income	in monetary terms

Chapter 13
Impact of Rural-Urban Migration on the Food Consumption Pattern of Farming Households in Ibadan/Ibarapa Agricultural Zone of Oyo State, Nigeria

Fatai Abiola Sowunmi
https://orcid.org/0000-0003-4108-8830
Department of Agricultural Economics, University of Ibadan, Ibadan, Nigeria

Funmi Lydia Adeduntan
Department of Agricultural Economics, University of Ibadan, Ibadan, Nigeria

ABSTRACT

The study examined the impact of rural-urban migration on the food consumption pattern of farming households. The study revealed that 73.8% of the households had migrants, while 80.2% of the migrants were male. The highest level of education of most of the migrants was secondary school (71.4%). The study showed that the major reason (63.3%) for migration was for job. The average remittance sent per year was ₦108,119.14. The study revealed that household expenditure on carbohydrate food group accounted for 54.4% of the total households' expenditure on food. The average dietary diversity indices for the migrant (0.345) and non-migrant (0.346) households were low. The study revealed that migration (short and long term) positively influenced per capita food expenditure of respondent. Despite the remittance from some of the migrants, the need to develop the rural areas in terms of provision of basic infrastructures by government is imperative in order to reduce rural-urban migration.

DOI: 10.4018/978-1-7998-2599-9.ch013

INTRODUCTION

High poverty rate and lack of basic infrastructures in the rural areas have contributed to shortage of labour for farming activities in Nigeria (Ogun, 2010). The neglect of agriculture for the crude oil compels rural households to look for opportunities away from the agricultural sector as a way to escape from poverty. The migration of youths from rural to urban areas has contributed significantly to the decline in agricultural production (Maharjan *et al.,* 2013; Olayide, 2009). However, temporary migration to urban centre is undertaken by farmers during the dry season which affords farming households incomes that are invested in variety of production and consumption uses. Adewale (2005) affirmed that migration occurs as a response to economic development as well as social, cultural, environmental and political factors and effects on areas of origin as well as destination. Studies (Lewis, 2004; Osondu and Ibezim, 2001; Mbah *et al.,* 2016) have shown that rural-urban migration is associated with heavy drain on the supply of rural family labour and also pulls out the individuals who are vital for agricultural production. The studies highlighted low agriculture productivity, reduction in agricultural labor force, farm work mostly done by the aged, food insecurity in households, poor standard of living, high cost of labor, among others as the negative effects of rural-urban migration. Through migration, able bodied youth relocate to urban areas making the old farmers left behind to overwork themselves with attendant decline in health. According to Amrevurayire (2016), migration places a greater burden on farmers by compelling them to cultivate the same farm size as when he had enough farm labour, working much longer hours thus, depriving farmer's time for leisure or participation in various social activities. Rural-urban migration brings about rapid deterioration of rural economy leading to chronic poverty and food insecurity that arises mainly due to excessive drain of youth from the rural populace (Mini 2000)

Poverty has a strong relationship with poor nutrition, thus compelling households to consume mainly carbohydrate based foods (Harrington *et al.* (2009). Carbohydrates, fat and protein comprise the three principal sources of balanced diet in human nutrition. Individual foods contain different proportions of these three principal macronutrients. Animal products (meat and dairy) are rich sources of protein and fat, while cereals, fruits and vegetables contain a large proportion of carbohydrate (Kennedy, 2001; National Institutes of Health, 2019). The need for human dietary diversity cannot be overemphasized. According to Ruel (2003) and International Dietary Data Expansion Project (2015), dietary diversity is the number of individual food items or food groups consumed over a given period of time. Dietary diversity is an essential element of diet quality; consuming a variety of foods across and within food groups is associated with adequate intake of essential nutrients and promotes good health (Waswa et al., 2015; Arimond et al., 2010). However, limited access to variety of foods is a predominant problem among rural households. Their diets consist mainly of starchy staples with few or no animal products, fresh fruits and vegetables (Ruel, 2003; Torheim et al., 2004; Henjum et al., 2015). Consumption of poor quality diets and general lack of access to wide food diversity has been acknowledged as the major predisposing factors for low productivity among farming households in the rural areas (Moursi et al., 2008; Arimond et al., 2010; Headey & Ecker, 2013). A study reported by the National Bureau of Statistics (2007) showed that the farming households in Southwest, Nigeria consumed more of "Eko/Agidi", bread, yam flour, yam tuber, and garri.., consuming a nutritious diet can be a challenge for those struggling with poverty for a variety of reasons, including limited finances and resources, competing priorities, and stress (Braveman, 2011; Liu et al., 2014; Stahre et al., 2015; King, 2016). This is attributable to poverty caused by several factors including rural-urban migration.

Rural-urban migration is a serious problem in the southwest Nigeria most especially the agrarian communities because of their proximity to major towns and cities. Movements of people from these communities to urban areas have been identified to be one of the reasons agricultural production is still low in the region despite the vast resources that abound (Lawal & Okeowo, 2014). Most farmers in these communities are forced to reduce their farm size while the few that can afford hired labour (for land clearing, weeding, harvesting) paid exorbitantly for labour whose availability may not be timely. While some households benefits from the remittance sent by their migrant member(s), other households with migrant member do not receive any support either in cash or kind after years of departure from village to urban area. Rural–urban migration is known to have adverse effect on farmer's productivity, which predisposes farmers to poverty and consequently poor consumption pattern.

Various reasons have been adduced for rural-urban migration. Vargas-Lundius et al. (2008) opined that migration may be prompted by major economic, demographic and social disparities, as well as conflicts, environmental degradation or natural disasters. Regardless of their origin and the causes of the relocation, the migrant's productivity and earnings constitute a powerful force for poverty reduction. They affirmed that remittances play an essential role in ensuring food for many rural poor households and thus constitute an efficient strategy for facing adversities such as low agricultural productivity and the inherent risks and instability of farming activities. Although, it is not all that migrated to urban areas that are gainfully employed (paid job or self employed), majority of those who engaged in economic activities often send money home (Jørgen, 2008; Hagen-Zanker, 2015). According to Ajaero and Onokala (2013), households that receive remittances tend to use the proceeds primarily for current consumption (food, clothing) investments in children's education, health care, improvement in water and sanitation. Nevertheless, the ability of remittances to compensate for the labour shortage in rural areas is still a function of the amounts and value of remittances received by migrants' households at home (World Bank, 2005).

Previous studies on rural-urban migration examined its effects on agricultural production (Adaku, 2013), food security (FAO, 2018; Ayanwu et al., 2018), farm labour (Lawal & Okeowo, 2014), rural development (Ajaero & Onokala, 2013) and poverty (Skeldon, 1997; Qiu et al., 2011). Most of the previous studies (Chandrasekhar et al., 2014; Zezza et al., 2011; Karamba et al., 2011; Beegle et al., 2011; Meeske, 2018) on the effect of migration on food consumption pattern were carried out outside the shore of Nigeria. In Nigeria however; much has not been done in the area of linking migration with food consumption pattern bearing in mind the importance of food in the wellbeing and by extension productivity of farmer. This study is aimed at addressing the scanty literature on the impact of rural-urban migration and food consumption pattern in Nigeria. The study is also an attempt to fill the gap in literature by generating evidence on variations in food consumption pattern of households with and without migrants. The following research questions were raised and answered in the course of this study: What proportion of the farming households had migrants? What are the reasons for the migration of farming household member? What proportion of the migrant households received remittance, food and clothing periodically? Is there significant difference in the average per capita expenditure on carbohydrate, protein, vitamins and mineral food groups of migrants and non-migrants households? What is the extent of food diversity among the migrants and non-migrants households? Does migration affects per capita food expenditure of farming households in the study area?

THEORETICAL FRAMEWORK AND LITERATURE REVIEW

The study is supported by dual labour market theory.The theory states that migration is mainly caused by pull factors. The theory assumes that the labour markets consist of two segments: the primary markets, which require high skilled labour, and the secondary markets, which is a very labour intensive requiring low skilled worker (Dickens & Lang, 1984). It posits a bifurcated occupational structure and a dual pattern of economic organization in advanced economics (Kurekova et al., 2011). Better standard of living and good working environment in the urban areas are like "pull factors" that attract able bodied youths from rural to urban areas. As a result of these, youth are able to work and send home items (money, food and clothing) that improve the food consumption patterns and general wellbeing of the households left behind in the rural areas.

Several analytical tools have been used to analyse the impact of migration on some key variables (such as poverty, food security, rural development, farm labour and food consumption pattern) in the rural areas. Such include propensity score matching (Meeske, 2018; Caliendo & Kopeinig, 2005) ordinary least squares (Chandrasekhar et al., 2014), Binary logistic regression (Dheeraj et al., 2010) difference in difference (Angrist & Pischke, 2009; Blundell & Coastal Dias, 2009), and discontinuity designs (Angrist, 1998). However, apart from the peculiar short comings associated with each of these tools which include inconsistency, heavy data requirement, and complex computation among others; endogeniety is a common occurrence in migration study which none of the aforementioned tools can identify (test) and proffer necessary solution to, hence the use of instrumental variable (two-stage least squares) which addresses this problem. According to Antonakis et al. (2014), a coefficient in a regression output may appear to adequately reflect the hypothesized relationship—for example, it is the right direction and the effect is highly significant—but with the presence of endogeneity it will be inconsistent and will not reflect the true population parameter and often render parameter estimates uninterpretable. They affirmed that changes in x (independent variable) produce changes in y (dependent variable) holding all other things equal. This is clearly the case if x varies randomly and independently from the system of variables under study. However, if x depends on some unmodeled causes that also drive other variables in the model, then x would be said to be endogenous—hence the problem of endogeneity.

Food consumption has been used extensively to measure human nutrition in literature (Hoddinott & Yohannes, 2002; Mishra & Ray, 2006; Nguyen & Winters, 2011). The following approaches are commonly used: per capita food expenditures, per capita calorie consumption, and food consumption diversity. However, this study adopted per capita food expenditure and food consumption diversity approach. Food count, Dietary Diversity Score (DDS) developed by Kant et al. (1993), Simpson-Weiner index and entropy have been used to estimate dietary diversity in literature. However, this study utilized Shannon index. The measure reflects food richness (count) and evenness (abundance). A definitive cut-off point does not bind the Shannon-Weiner index score; thus an increase in the score reflects greater dietary diversity in a household (Magurran & McGill, 2011; McArt et al., 2012).

Analytical Framework of Two - Stage Least Squares

Two-stage least squares (2SLS) method is used to handle model with endogenous explanatory variables in a linear regression framework. An endogenous variable is a variable which is correlated with the error term in the regression model. Using endogenous variable is in contradiction with the linear regression assumptions. This kind of variable can be encountered when variable are measured with error (XLSTAT, 2019). Following NCSS approach, 2SLS model is comprised of the following two linear regression models.

$$y = X_{ex}\beta_{ex} + V_{en}\beta_{en} + e = X\beta + e \quad (1)$$

$$Ven = X_{en}\Gamma_{ex} + X_{iv}\Gamma_{iv} = Z\Gamma + E \quad (2)$$

Where:

y: $n \times 1$ vector of dependent variable
X_{ex}: $n \times k_{ex}$ matrix of exogenous regressor variables
X_{iv}: $n \times k_{iv}$ matrix of instrumental variables
V_{en}: $n \times k_{en}$ matrix of endogenous regressor variables
$\beta e_{n:}$ $ke_n \times 1$vector of endogenous regressor parameters
$\beta e_{x:}$ $ke_x \times 1$vector of included exogenous parameters

$$\beta = \begin{bmatrix} \beta_{ex} \\ \beta_{en} \end{bmatrix} : \left(k_{ex} + k_{en}\right) \times 1 \text{ vector of parameters}$$

$X = [X_{ex} | V_{en}]$
$Z = [X_{ex} | X_{iv}]$
Γ_{ex}: $k_{ex} \times k_{en}$ matrix of parameters
Γ_{iv}: $k_{iv} \times k_{en}$ matrix of parameters
e: $n \times 1$vector of errors
E: $n \times k_{en}$ vector of errors

The 2SLS estimator of $\beta$is b given by:

$$b = \left\{X'Z\left(Z'Z\right)^{-1} Z'X\right\}^{-1} Z'y \quad (3)$$

$$Var\left(b\right) = s^2 \left\{X'Z\left(Z'Z\right)^{-1} Z'X\right\}^{-1} \quad (4)$$

Where:

$$s^2 = \frac{E_{ss}}{\left[n - \left(k_{ex} + k_{en}\right)\right]} : \text{mean square error}$$

$$E_{ss} = \sum_{i=1}^{n} u_i^2$$

Hausman's Test of Endogeneity

Cameron and Trivedi (2010) present a special version of Hausman's test that may be used to test whether one or more explanatory variables are endogenous. For a single explanatory variable, the test is:

$$T_{H,1} = \frac{b_{2SLS} - b_{OLS}}{Var\left(b_{2SLS}\right) - Var\left(b_{OLS}\right)} \tag{5}$$

The test statistic is distributed as a chi-square with one degree of freedom under the null hypothesis that the regressor is exogenous. NCSS also provides an overall test of endogeneity of all the designated endogenous variables. This is calculated as:

$$T_{H,k_{en}} = \left(b_{2SLS} - b_{OLS}\right)' \left(V\left(b_{2SLS}\right) - V\left(b_{OLS}\right)\right)^{-1} \left(b_{2SLS} - b_{OLS}\right) \tag{6}$$

The test statistic is distributed as a chi-square with ken degrees of freedom under the null hypothesis that the regressors are exogenous. Note that b_{2SLS} and b_{OLS} represent only those regression coefficients corresponding to endogenous variables.

METHODOLOGY

Description of the Study Area

The study was conducted in Ibadan/ Ibarapa agricultural development zone of Oyo state. The study area fell within theequatorial climatewith distinct dry and wet seasons and relatively high humidity. The vegetation of the zone is mostly tropical rain forest, found in the southern part of the state where lumbering, plantation farming of cocoa, Oil palm and Cashew is practiced (Adeola et al., 2008). Ibarapa area falls within latitudes $70^{0}15^{1}$N and $70^{0}55^{1}$N and Longitude 30^{0}E and $30^{0}30^{I}$E. It is located approximately 100km North of the coast of Lagos, and about 95km west of Oyo State capital and neighbouring city of Ibadan. Ibadan/ Ibarapa zone is one of the four (4) Agricultural Development Programme zones in Oyo State, Nigeria. The four zones are Ibadan/Ibarapa zone, Oyo zones, Saki zone and Ogbomosho zone. Ibadan/ Ibarapa zone is made up of nine (9) Local Government Areas (LGAs). Oyo zone is made up of six (6) LGAs which Saki zone is eight (8) LGAs and Ogbomosho zone which is made up five (5) LGAs. All the LGAs in the four (4) ADP zones are twenty-eight (28). The remaining five local governments out of the thirty three local governments in Oyo state are urban areas. The choice of the study area was based on agricultural activities of the residents. The main occupation of the residents in the study area (Ibadan/Ibarapa zone) is farming. The zone is known as the food basket of Oyo State and Southwest in general (Momoh & Kormawa, 2002). Food crops such as maize, cassava, cowpea, melon, yam,rice and vegetables (fruit and leafy) are cultivated mostly by small scale farmers who cultivate less than one hectare of farm land.

Figure 1. Map of Oyo state showing the study area in green boundary.
Source: Google Maps (2019)

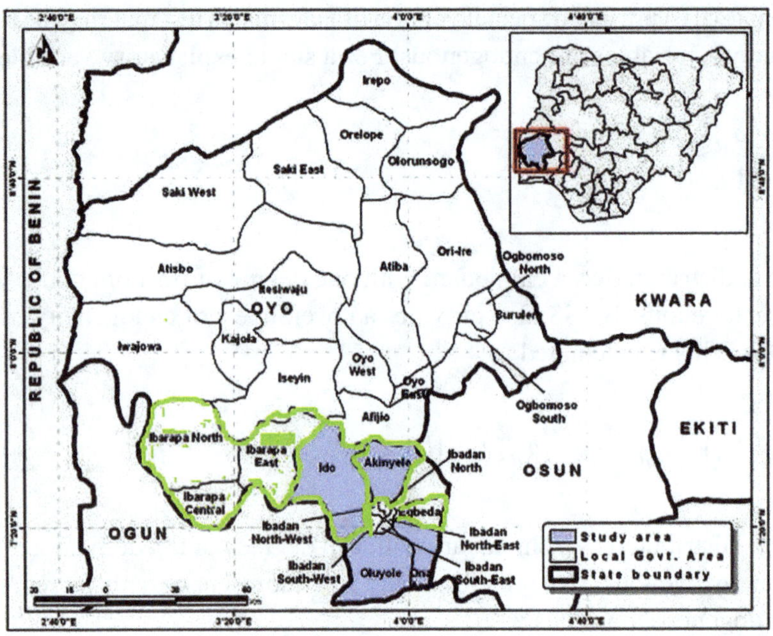

Sample Selection and Data Collection

A four-stage sampling procedure was used to select the respondents. The first stage involved the purposive selection of Ibadan/Ibarapa zone of Agricultural Development Project which is made up of rural communities notable for production of various food crops in Oyo State. The second stage was the random selection of five (5) local government areas out of nine (9) local government areas in Ibadan/Ibarapa zones. These are Akinyele, Egbeda, Ibarapa east, Ibarapa central, Ibarapa North. The third stage involved purposive selection of contiguous villages in the selected LGAs. In the fourth stage, households were randomly chosen proportionate to the household size of the selected villages. The number of villages and the household sizes from each of the local government areas are contained in the parenthesis: Akinyele (14, 114), Egbeda (11, 33), Ibarapa North (21, 38) Ibarapa East (21, 42) and Ibarapa Central (15, 43). Two hundred and seventy households were used for this study.

Data Collected and Utilized

Structured questionnaire was used to collect primary data. Data collected included socioeconomic characteristics (age of household head, sex, marital status, household size, educational status, weekly/ monthly income and occupation of the household head), the basic foodstuffs (group into carbohydrate, protein, vitamins and minerals), amount spent on basic foodstuffs per week, household with (at least one person migrated) /without migrants, short (1-2years) and long term (more than 2years) migration, age of migrants at the time of departure, reasons for migration, number of migrants, commodity sent home by migrant (money, food and clothing), money (₦) sent home, frequency of sending money, type of labour used, number of hired labour engaged by migrant and non-migrant households and amount paid to hired labour per cropping season among others. A total of 270 questionnaires were administered while 260 were returned to time and 240 were good for analysis

Data Analysis

Data were analysed using descriptive statistics to profile the characteristics of the migrants and non-migrants households. The descriptive statistics included measure of central tendency (mean and mode), measure of dispersion (standard deviation/variance and skewness) and frequency distribution. Equality test was used to explain whether there were significant differences in the per capita expenditure on carbohydrate, protein, vitamins and mineral food group of migrants and non migrant households. Per capital food expenditure for each food group was calculated, while Shannon-Weiner dietary diversity index was used to estimate the extent of dietary diversification of migrants and non-migrants households. The study utilized two-stage least squares to determine the impact of migration on food consumption patterns of respondents as well as to test for endogeneity. Two-stage least squares (Instrumental Variable) involved the use of two equations. The first equation had migration status as the dependent variable. The predicted migration status in the first equation was used as one of the independent variables in the second equation. Per capita food expenditure (a proxy for food consumption pattern) is the dependent variable in the second equation. The explicit equations are shown below:

First Equation (Short and Long Term Migration):

$$Mig_{st} = \alpha_0 + \alpha_1 Pd + \alpha_2 Ag_{28-37} + \alpha_3 Ag_{38-47} + \alpha_4 Ag_{48-57} + \alpha_5 Cl_{b1ha} + \alpha_6 Cl_{1-2ha} + \alpha_7 Cl_{2.1-3ha} + \alpha_8 Hs_{1-4} + \alpha_9 Hs_{5-8} + \alpha_{10} Sx + \alpha_{11} Ms + \alpha_{12} Lb_{fa} + \alpha_{13} Lb_{hi} + \alpha_{14} Lb_{fahi} + \alpha_{15} Ye + \alpha_{16} Oc + U_0 \quad (7)$$

Second Equation (Short and Long Term Migration):

$$LnPCE_{st} = \alpha_0 + \alpha_1 Migst + \alpha_2 Ag_{28-37} + \alpha_3 Ag_{38-47} + \alpha_4 Ag_{48-57} + \alpha_5 Cl_{b1ha} + \alpha_6 Cl_{1-2ha} + \alpha_7 Cl_{2.1-3ha} + \alpha_8 Hs_{1-4} + \alpha_9 Hs_{5-8} + \alpha_{10} Sx + \alpha_{11} Ms + \alpha_{12} Lb_{fa} + \alpha_{13} Lb_{hi} + \alpha_{14} Lb_{fahi} + \alpha_{15} Ye + \alpha_{16} Oc + U_0 \quad (8)$$

Where:

Mig_{st} is the dependent variable which represents the migration status for short (Yes =1, No = 0) and long (Yes =1, No) migrant household;

Pd represents the migration density which served as instrument in the first equation;

Ag_{28-37} represents the migrant household member in the age bracket of 28-37(Yes =1, No)

Ag_{38-47} represents the migrant of household member in the age bracket of 38-47 (Yes =1, No)

Ag_{48-57} represents the migrant of household member in the age bracket of 48-57(Yes =1, No)

Cl_{b1ha} represents the households that cultivated less than 1ha(Yes =1, No)

Cl_{1-2ha} represents the households that cultivated 1-2ha (Yes =1, No)

$Cl_{2.1-3ha}$ represents the households that cultivated 2.1-3ha (Yes =1, No)

Hs_{1-4} represents the households with 1-4 household members (Yes =1, No)

Hs_{5-8} represents the household with 5-8 household members (Yes =1, No)

Sx represents the sex of the migrant (short/long term migration) (male = 1, female = 0)

Ms represents the marital status of the migrant (married = 1, others =0)
Lb_{fa} represents the household that used family labour (Yes =1, No)
Lb_{hi} represents the household that used hired labour (Yes =1, No)
Lb_{fahi} represents the household that used family labour and hired labour (Yes =1, No)
Ye represents respondent's year spent in school
Oc represents respondents major economic activities (crop farming =1, others = 0)
Mig_{st} represents the predicted values of migration status (Yes =1, No) serving as independent variable in the ***second equation***
Uo represents residual error
α_i reprecentsthe coefficienti = 0-16

Note that the first stage of the instrumental variable estimation includes a discrete variable, being a migrant household, and is estimated as linear probability model.

The dietary diversity index of the migrants and non-migrants households was estimated using Shannon-Weiner index. The formula for the index is:

$$H = -\sum_{i=1}^{s} P_I \ln\left(P_I\right) \tag{9}$$

Where:

H represents Shannon index
S represents number of food groups
P_i represents proportion of each food group from source *i*.
Ln represents natural log

RESULTS AND DISCUSSION

Characteristics of Respondents

Distribution of migration status among the households shows that 26.3% of the households had no migrant while 34.2% had short-term migrants and 39.6% long-term migrants (see Figure 1). Majority of the non-migrant households (25.4%), short-term migrants (31.7%) and long term migrants' (33.7%) households heads were within the age brackets of 38 - 42, 58 - 62 and 58 - 62 years respectively (see Table 1). Also, 67.8% of the migrants were at most 37years old. This is in agreement with the United Nations Population Fund (2014) report that the typical profile of migrants comprises young women and men from 15 to 35 years of age, generally belonging to medium and low socioeconomic groups.

The averages of household size for non-migrant and migrant households were 5.0 and 7.1 respectively (see Table 1). The average ages of the short-term and long term migrants when they left for urban areas were 34.7years and 31.4 years respectively, while 80.2% of the migrants were male (see Table 2). Most of the migrants (83.6%) were single before they left for urban areas. The highest level of education of most of the migrants was secondary school (71.4%). This confirms the finding of Amuakwa-Mensah et al. (2016) that migration from rural to urban areas is common among secondary school leavers. The study revealed that the major reason (63.3%) for migration was to search for job, 28.8% migrated for schooling while 6.8% migrated for job and schooling. The study showed that 69.1% of the migrants sent money and other useful items home since they migrated while 30.9% of the migrants did not. This confirms the findings by Vargas-Lundius (2008) that not all migrants send remittances. The breakdown of the migrants that sent money and other items revealed that 61.8% sent money occasionally, 28.5% sent money and foodstuffs regularly while 8.9% and 0.8% sent money and clothing, and clothing respectively (see Table 2).

Majority of the migrants (68.3%) sent remittance (₦) in the range of ₦5501-79500 per annum and the average remittance per year was ₦108119.14. The average amount spent on hired labour by the non-migrant farming households (₦12016.89) was higher than that of the migrant farming households (₦11714.79). This is contrary to expectation bearing in mind the access of migrant farming households to remittances. Maybe migrant farming households used most of their remittances for food and other non-agricultural ventures at the detriment of farming business (see Table 2).

Table 3 shows that non-migrant households used larger percentage of the family labour while migrant households engaged more of hired labour. The significant difference between the average amounts paid for hired labour may be attributed to the bargaining ability for charges by the farmer when engaging hired labour. The breakdown of wages paid by the migrants and non-migrants farming households is shown in Table 4. The long term migrant households had the highest average paid to hired labour. The positive skweness shows that most of the long term migrant households paid below the average amount for the hired labour.

Distribution of Food Expenditure of Farming Households (Migrant and Non-Migrant)

Carbohydrate food group (rice, eba, amala, fufu, bread and pap) had the highest average annual per capita food expenditure for the migrant (₦499,523.40) and non-migrant (₦427,683.84) households (Table 5). The amount spent on carbohydrate food group corroborates the submission of National Bureau of Statistics (2007) farming households in Southwestern, Nigeria consumed more of "Eko/Agidi", bread, yam flour, yam tuber, and garri. The study revealed that expenditure on carbohydrate food group accounted for 54.4% of the total households' expenditure on food. The breakdown of the total food expenditure showed that non-migrants spent 52.8% on carbohydrate food groups while migrant households spent 55.0% on carbohydrate. Protein food group (bean, meat, fish and egg) had ₦352,332.62 and ₦375,495.57 as the averages of annual per capita expenditure for non-migrant and migrant households respectively. The average food expenditure of migrant households was ₦908,012.63 while that of non-migrant household was ₦810,354.48

Table 6 shows that there were significant differences in the averages of total food expenditure ($p<0.01$), expenditure on carbohydrate ($p<0.01$) and vitamins and minerals ($p<0.01$)food groups between the non-migrant and migrant households. The higher expenditure on food by migrant households may be attributed to money that some of the migrants sent home. According to Olowa *et al.* (2013), some migrants send something in form of cash, clothings, farming materials among others (Olowa, 2009). He submitted that remittances help to alleviate poverty through good feeding that invariably enhances productivity which brings about a healthy nutrition. This may also be confirming Crush (2012) submission that over 80 percent of recipient households used remittances to cover, on average, half of their expenditures on food.

Figure 2. Migration status of households

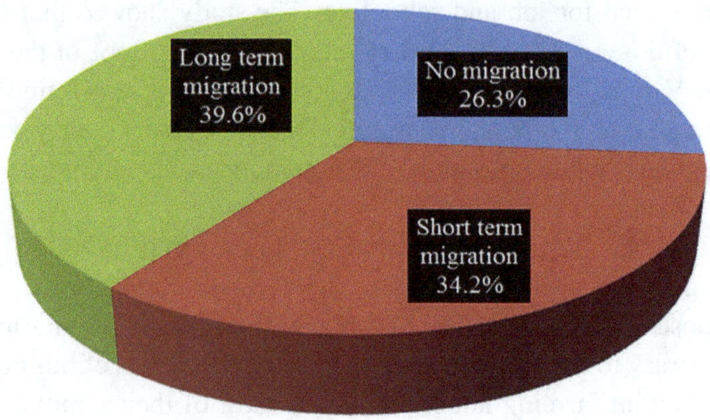

Table 1. Socioeconomic characteristics of respondents

		No Migration	%	STM	%	LTM	%
Age	28 - 32	2	3.2	1	1.2	0	0.0
	33 - 37	5	7.9	2	2.4	0	0.0
	38 - 42	16	25.4	3	3.7	1	1.1
	43 - 47	8	12.7	16	19.5	10	10.5
	48 - 52	7	11.1	13	15.9	19	20.0
	53 - 57	11	17.5	20	24.4	25	26.3
	58 - 62	14	22.2	26	31.7	32	33.7
	63 & above	0	0.0	1	1.2	8	8.4
Average age (year)		**48.1**		**55.5**		**52.4**	
Household size	1 - 4	45	71.4	26	31.7	19	20.0
	5 - 8	17	27.0	49	59.8	63	66.3
	9 - 12	1	1.6	7	8.5	13	13.7
Average		**6.1**		**6.9**		**7.1**	
Farm size (ha)	less than 1	9	14.3	6	7.4	3	3.2
	1 - 2	39	61.9	47	58.0	43	45.3
	2.1 - 3	15	23.8	28	34.6	49	51.6
Average farm size (ha)		**1.6**		**1.9**		**2.0**	
Other Occupation	Artisanal	11	17.5	28	35.4	22	23.7
	Trading	33	52.4	36	45.6	61	65.6
	Civil/public service	10	15.9	10	12.7	3	3.2
	Private service	8	12.7	4	5.1	4	4.3
	Retiree	1	1.6	1	1.3	3	3.2

NB: STM means short term migrant, LTM means long term migrant

Table 2. Characteristics of migrants

		Frequency	Percentage
Migrant Age (Year) Interval	14 - 21	31	17.5
	22 - 29	37	20.9
	30 - 37	52	29.4
	38 - 45	41	23.2
	46 & above	16	9.0
Pooled average age		32.3	
Short term migrant average age		34.7	
Long term migrant average age		31.4	
Sex of migrant	Male	142	80.2
	Female	35	19.8
Educational status of migrant	No Formal education	31	35.4
	Prim School	59	65.5
	Secondary School	63	70.9
	NCE/OND	18	20.9
	HND/BSc	6	7.2
Reasons for Migration	School	51	28.8
	Job opportunity	112	63.3
	Visit	2	1.1
	School and Job opportunity	12	6.8
Remittance	Remittance sent	123	75.9
	No Remittance sent	39	24.1
What was sent home by migrant	Clothing	1	0.8
	Money	76	61.8
	Cloth and Money	11	8.9
	Money and Foodstuffs	35	28.5
Annual remittance (₦)	5501 - 79500	82	68.3
	79501 - 153500	10	8.3
	153501 - 227500	14	11.7
	227501 - 301500	1	0.8
	301501 - 375500	8	6.7
	375501 - 449500	0	0
	449501 - 523500	0	0
	523501 - 597500	0	0
	597501 - 671500	5	4.2
Average		108119.14	
Skewness		2.267086	

Table 3. Distribution of labour type engaged by non-migrant and migrant households

Labour Type Used for Farming	Non- migrant (%)	Migrant (%)
Family Labour	11.3	3.4
Hired labour	30.6	37.9
Family and Hired Labour	58.1	58.6

Table 4. Wage distribution among labour used in migrant and non-migrants households

Wage (₦) Interval	No Migration		Short Term Migration		Long Term Migration	
	Frequency	Percentage	Frequency	Percentage	Frequency	Percentage
501 - 5500	6	9.8	7	8.8	2	2.1
5501 - 10500	25	41.0	36	45.0	41	43.2
10501 - 15500	5	8.2	21	26.3	18	18.9
15501 - 20000	25	41.0	16	20.0	34	35.8
Average	12016.89		10875.5		12421.55	
Skewness	-0.069		0.236		0.0742	

Determination of Dietary Diversity Index

The dietary diversity in the study area ranged from 0.249 – 0.379. Majority of the migrant (62.7%) and non-migrant (55.6%) households were in the dietary diversity range of 0.336 – 0.357. The average dietary diversity indices for the migrant and non-migrant households were 0.345 and 0.346 respectively (see Table 7). Generally, there was low dietary diversity in the study area regardless of the migratory status. The low dietary diversity confirms the report of the survey by NBS (2007) that the farming households in Southwest, Nigeria consumed more of carbohydrate food group ("Eko/Agidi", bread, yam flour, yam tuber, and garri). This is affirmed by households spending 54.4% of their total food expenditure on carbohydrate food group at the detriment of protein, vitamins and mineral food groups. Even the migrant farming households that had access to money sent home by migrant (remittance), spent the substantial part that went into feeding on carbohydrate food group (see Table 5). Lack of food diversity has been acknowledged among the major predisposing factors for low productivity among farming households in rural areas (Moursi et al., 2008; Arimond et al., 2010; Headey & Ecker, 2013).

Table 5. Average annual per capita food expenditure (₦) of food groups

Migration Status	Carbohydrate	Protein	Vitamins and Mineral
Non-migrant households	427683.84	352332.62	30338.13
Migrant households	499523.40	375495.57	32993.79

Table 6. Equality test results for food groups annual expenditure based on migration status

Parameter	Non-Migrant Households (n = 63)		Migrant Households (n = 177)		Z-Value	p-Value
	Mean	Sd	Mean	Sd		
Total food expenditure (₦)	810354.48	177285.60	908012.63	205839.36	3.59	0.000***
Expenditure (₦) on food group						
(i) Carbohydrate	427683.81	131926.34	499523.39	135654.61	3.64	0.000***
(ii) Protein	352332.57	132184.49	375495.46	129901.09	1.20	0.233
(iii) Vitamins and minerals	30338.10	26739170.51	32993.79	28080472.52	3.48	0.000***

Source: Author's Compilation (2019)

Impact of Migration on Food Consumption Pattern of Respondents

Diagnostic Tests for Two-Stage Least Squares

Test of Endogeneity

This shows the correlation between the instrument and the endogenous variables. The Durbin and Wu-Husman were statistically significant, thus, we reject the null that the instrument is not exogenous but endogenous for both short-term and long-term migration (see Table 8). Hence, the justification for the use of two-stage least squares to determine the effect of migration on the per capita food expenditure of households.

Test of Validity

The rule of thumb for the validity of instrument states that the partial R-square should be large and the F value should be larger than all the critical values. The results below show that the partial R-squares for short-term and long-term migration are large and the F-values are greater than all the critical values. Thus, we reject the null hypothesis that the instrument is not weak. Column 1 in Table 7 presents the estimated first-stage relationship between the short term migration and household migration density. Migration density is significant and positively related to short ($p<0.01$) and long ($p<0.01$) term migrations (see Tables 7 and 8). The explanatory variables explain about 68 and 77 percent of the variations in short and long term migrations respectively (see Table 9). About 77 percent of the explanatory variables explain the variation in long term migration. The strong relationship means there is no problem of a weak instrument. Thus, migration density is a strong and valid instrument.

Table 10 shows that migratory density ($p<0.01$), households using family labour positively influenced the probability of household member embarking on short term migration. Also, households with the family size of 1-4 ($p<0.05$) reduced the probability of household member embarking on short term migration. The need to migrate is often high among the large households because of the inadequacy of resources to cater adequately for the needs of family. The table also shows that pe capita food expenditure is influenced by short term migration (instrumental variable). The significance of short term migration ($p<0.01$) implies that an increase in short-term migration will increase per capita household food expenditure by 53 percent. Nguyen and Winter (2010) opined that short-term migration is a mechanism by which households maintain food security and increase food expenditure (see Table 10 column 4).

Table 7. Shannon food diversity index range of respondents

Dietary Diversity Index	Pooled Households		Non-Migrant Households		Migrant Households	
	Frequency	%	Frequency	%	Frequency	%
0.249 - 0.269	1	0.4	0	0.00	1	0.5
0.270 - 0.291	3	1.3	1	1.6	2	1.1
0.292 - 0.313	17	7.1	5	7.9	12	6.8
0.314 - 0.335	21	8.8	6	9.5	15	8.5
0.336 - 0.357	146	61.1	35	55.6	111	62.7
0.358 - 0.379	51	21.3	16	25.4	36	20.3
Total	**240**	**100**	**63**	**100**	**177**	**100**

Source: Author's compilation, (2019)

Table 8. Test of endogeneity

		Statistic	p-Value
Short Term Migration	Durbin (score) chi2 (1)	36.85	0.0000
	Wu-Hausman F(1,144)	40.45	0.0000
Long Term Migration	Durbin (score) chi2 (1)	22.73	0.0000
	Wu-Hausman F(1,144)	23.02	0.0000

Source: Authors Computation (2019)

Table 9. Validation of instrument

Variable	R^2	Adjusted R^2	Partial R^2	F(1, 224)	Prob>F
Short term migration	0.679	0.657	0.655	425.607	0.000
Long term migration	0.768	0.749	0.737	618.15	0.000

Column 2 Table 11 shows that migratory density and households that used hired labour were positively related to the probability of household embarking on long-term migration. Households in the age brackets of 48-57 and 58-67, sex, marital status and years of education were negatively related to the probability of household embarking on long-term migration. The significance of the two age brackets means that at advanced ages, the probability of embarking on long-term migration deceases. This is partially due to their inability to withstand the stress involved in migration which often takes long time to settle down. Also, the negative relationships of marital status of migrant and being a female implied that a reduction in the probability of long term migration among household members. This finding is consistent with Dodson et al. (2008) reported that female migrants maintain strong ties to family members in their home countries and as such stay closer to family members. Moreover, the coefficient of years of education implies that an increase in the years of education reduce the probability of long term migration. This is in agreement with Bockerman and Haapanen (2013) that there is a propensity to

Table 10. Instrumental variable estimate of short-term migration and food expenditure

		Dependent Variable: Short Term Migration	**Dependent Variable: Per Capita Food Expenditure**	
		OLS (First-Stage) (1)	**OLS (2)**	**IV-2SLS (Second-Stage) (3)**
Intercept		0.577 (0.339)	11.659*** (0.539)	11.577*** (0.533)
Migration		-	0.093 (0.063)	0.329*** (0.077)
Migration density		1.829*** (0.090)	-	-
Age	*28-37*	-0.201 (0.303)	0.302 (0.482)	0.604 (0.488)
	38-47	-0.064 (0.290)	-0.000 (0.461)	0.245 (0.467)
	48-57	-0.086 (0.287)	0.775 (0.457)	0.250 (0.463)
	58-67	-0.094 (0.288)	0.024 (0.458)	0.176 (0.463)
Cultivated land	*<1 ha*	-0.038	-0.164* (0.090)	-0.109 (0.089)
	1-2ha	0.011 (0.036)	0.181*** (0.0568)	0.181*** (0.057)
	2.1-3ha	-0.004 (0.038)	0.128** (0.059)	0.096 (0.059)
Household size	*1-4*	-0.161** (0.074)	-	-
	5-8	-0.070 (0.068)	-	-
Sex		0.005 (0.048)	0.051 (0.076)	0.055 (0.077)
Marital status		0.001 (0.020)	-0.049 (0.032)	-0.047 (0.033)
Labour	*Family*	0.329** (0.138)	-0.065 (0.220)	-0.115 (0.223)
	Hired	0.245* (0.133)	0.069 (0.213)	-0.049 (0.216)
	Family and hired	0.198 (0.133)	-0.208 (0.212)	-0.116 (0.215)
Years of education		-0.002 (0.019)	-0.018 (0.030)	-0.019 (0.030)
Farming primary occupation		0.015 (0.017)	-0.015 (0.028)	-0.015 (0.028)
No. of observation		240	240	240

Source: Authors Compilation (2019)

Table 11. Instrumental variable estimate of long-term migration and food expenditure

		Dependent Variable: Short Term Migration	Dependent Variable: Per Capita Food Expenditure	
		OLS (First-Stage)	OLS	IV-2SLS
Intercept		0.642** (0.292)	11.560*** (0.552)	11.437*** (0.593)
Migration		-	0.081 (0.065)	0.259*** (0.074)
Migration density		2.215*** (0.089)	-	-
Age	*28-37*	-0.588 (0.458)	0.364 (0.488)	0.491 (0.477)
	38-47	-0.527 (0.246)	0.120 (0.465)	0.221 (0.454)
	48-57	-0.560** (0.245)	0.223 (0.463)	0.316 (0.452)
	58-67	-0.505** (0.245)	0.191 (0.463)	0.285 (0.452)
Cultivated land	*<1 ha*	0.009 (0.049)	-0.028 (0.092)	-0.022 (0.090)
	1-2ha	-0.012 (0.031)	-0.009 (0.058)	-0.009 (0.057)
	2.1-3ha	-0.007 (0.033)	0.024 (0.062)	0.025 (0.060)
Household size	*1-4*	-0.138** (0.063)	-	-
	5-8	-0.088 (0.059)	-	-
Sex		-0.056** (0.023)	0.115 (0.801)	0.111 (0.079)
Marital status		-0.027* (0.014)	0.028 (0.036)	0.036 (0.035)
Labour	*Family*	0.159 (0.119)	0.255 (0.224)	0.202 (0.219)
	Hired	0.216* (0.114)	0.208 (0.216)	0.138 (0.211)
	Family and hired	0.113 (0.114)	0.251 (0.215)	0.204 (0.210)
Years of education		-0.030* (0.018)	-0.026 (0.034)	0.049** (0.024)
Farming primary occupation		-0.004 (0.014)	-0.024 (0.029)	-0.020 (0.028)
Observation		240	240	240

Source: Authors Compilation (2019)

migrate in the short run when individuals are young and in the long run these effects dissipate. Column 4 table 11 shows that long-term migration and years of education positively influenced household food expenditure. The significance of long-term migration is consistent with Nguyen et al. (2017) that in the long run migration increase food expenditure through remittances.

CONCLUSION AND RECOMMENDATIONS

The study affirmed the positive influence of migration on the per capital food expenditure of households. However, with the remittance received by some migrant households the old diet of high consumption of carbohydrate food group is still maintained. Generally, the dietary diversity was low with carbohydrate food group accounting for the larger percentage of food expenditure. While substantial percentage of the migrants did not send remittance; others sent irregularly. This does not allow proper planning of farm activities. Moreover, the average amount of annual remittance available for farming households was small to make significant impact on farm production required for the upliftment of farming households in terms of food consumption. Despite the irregular remittance from some of the migrants, the need to develop the rural areas in terms of provision of basic infrastructures by government and community based associations will address the major reason for migration in farming communities.

REFERENCES

Adaku, A. (2013). The effect of rural-urban migration on agricultural production in the northern region of Ghana. *Journal of Agricultural Science and Applications*, *02*(04), 193–201. doi:10.14511/jasa.2013.020402

Adeola, R. G., Adewale, J. G., & Adebayo, O. O. (2008). Information Needs of Cowpea Farmers in Ibadan/Ibarapa Agricultural Zone of Oyo State, Nigeria. *International Journal of Agricultural Economics & Rural Development*, *1*(2), 78–83.

Adewale, J. G. (2005). Socio-economic factors associated with urban-rural migration in Nigeria: A case study of Oyo State, Nigeria. *Journal of Human Ecology (Delhi, India)*, *17*(1), 13–16. doi:10.1080/09709274.2005.11905752

Ajaero, C. K., & Onokala, P. C. (2013). The effcts of rural-urban migration on rural communities of Southeastern Nigeria. *International Journal of Population Research*, *2013*, 1–10. doi:10.1155/2013/610193

Amrevurayire, E. O. (2016). Consequences of rural-urban migration on the source region of Ughievwen clan delta state Nigeria. *European Journal of Geography*, *7*(3), 42–57.

Amuakwa-Mensah, F., Boakye-Yiadom, L., & Baah-Boateng, W. (2016). Effect of education on migration decisions in Ghana: A rural-urban perspective. *Journal of Economic Studies (Glasgow, Scotland)*, *43*(2), 336–356. doi:10.1108/JES-09-2013-0138

Angrist, J. D. (1998). Estimating the Labor Market Impact of Voluntary Military Service Using Social Security Data on Military Applicants. *Econometrica*, *66*(2), 249–288. doi:10.2307/2998558

Angrist, J. D., & Pischke, J. (2009). *Mostly Harmless Econometrics: An Empiricist's Companion*. Princeton University Press. doi:10.1515/9781400829828

Antonakis, J., Bendahan, S., Jacquart, P., & Lalive, R. (2014). Causality and endogeneity: Problems and solutions. In D. V. Day (Ed.), *The Oxford Handbook of Leadership and Organizations* (pp. 93–117). New York: Oxford University Press.

Anyanwu, C. I., Njoku, M. E., Agwu, N. M., & Iwunze, Q. E. (2018). Effect of rural-urban migration on household food security in Umuahia South Local Government Area of Abia State, Nigeria. *Nigerian Agricultural Journal*, *49*(1), 20–31.

Arimond, M., Wiesmann, D., Becquey, E., Carriquiry, A., Daniels, M. C., Deitchler, M., ... Torheim, L. E. (2010). Simple Food Group Diversity Indicators Predict Micronutrient Adequacy of Women's Dietsin Diverse, Resource-Poor Settings. *The Journal of Nutrition*, *140*(11), 2059S–2069S. doi:10.3945/jn.110.123414 PMID:20881077

Beegle, K., De Weerdt, J., & Dercon, S. (2011). Migration and economic mobility in Tanzania: Evidence from a tracking survey. *The Review of Economics and Statistics*, *93*(3), 1010–1033. doi:10.1162/REST_a_00105

Blundell, R., & Dias, M. C. (2009). Alternative Approaches to Evaluation in Empirical Microeconomics. *The Journal of Human Resources*, *44*(3), 565–640. doi:10.3368/jhr.44.3.565

Böckerman, P., & Haapanen, M. (2013). The effect of polytechnic reform on migration. *Journal of Population Economics*, *26*(2), 593–617. doi:10.100700148-012-0454-4

Braveman, P., Dekker, M., Egerter, S., Sadegh-Nobari, T., & Pollack, C. (2011). *Housing and Health*. Prepared for the Robert Wood Johnson Foundation.

Caliendo, M., & Kopeining, S. (2005). *Some Practical Guidance for the Implementation of Propensity Score Matching*. IZA Discussion Paper, 1588. Bonn, Germany: Institute of Labor Economics (IZA).

Cameron, A. C., & Trivedi, P. K. (2010). *Microeconometrics Using Stata, Revised Edition*. Stata Press Books, StataCorp LP. Available on http://cameron.econ.ucdavis.edu/musbook/mus.html

Carletto, C., De Brauw, A., & Banerjee, R. (2012). Measuring migration in multi-topic household surveys. Handbook of Research Methods in Migration, 207.

Chandrasekhar, S., Das, M., & Sharma, A. (2014). Short-term Migration and Consumption Expenditure of Households in Rural India. *Oxford Development Studies*, *43*(1), 105–122. doi:10.1080/13600818.2014.964194

Crush, J. (2012). *Migration, development and urban food security*. Urban Food Security Series No. 9. Kingston and Cape Town, Queen's University and African Food Security Network (AFSUN).

Deaton, A., & Paxson, C. (1998). Economies of scale, household size, and the demand for food. *Journal of Political Economy*, *106*(5), 897–930. doi:10.1086/250035

Dehejia, R. H., & Wahba, S. (2002). Propensity score-matching methods for non experimental causal studies. *The Review of Economics and Statistics*, *84*(1), 151–161. doi:10.1162/003465302317331982

Dheeraj, B., Satija, A., Khandpur, N., Bowen, L., Kinra, S., Prabhakaran, D., ... Shah, E. (2010). Effects of migration on food consumption patterns in a sample of Indian factory workers and their families. *Public Health Nutrition*, *13*(12), 1982–1989. doi:10.1017/S1368980010001254 PMID:20507672

Dickens, W. T., & Lang, K. (1984). *A Test of Dual Labor Market Theory*. NBER Working Paper No. 1314. Retrieved from https://www.nber.org/papers/w1314

Dodson, B., Simelane, H., Tvera, D., Green. T., Chikanda, A., & de Vletter, F. (2008). Gender, Migration and Remittances in Southern Africa. Southern African Migration Project (SAMP), in Migration Policy Series, 29.

Echebiri, R. N. (2005). *Characteristics and Determinants of Urban Youth Unemployment in Umuahia, Nigeria: Implications for Rural Development and Alternative Labor Market Variables*. Paper presented at the ISSER/Cornell/World Bank conference on "Shared Growth in Africa", Accra, Ghana.

Food and Agricultural Organization. (2018). *The linkages between migration, agriculture, food security and rural development*. Technical report by the Food and Agriculture Organization of the United Nations, the International Fund for Agricultural Development, the International Organization for Migration and the World Food Programme. Retrieved from http://www.fao.org/3/CA0922EN/CA0922EN.pdf

Hagen-Zanker, J. (2015). *Effects of remittances and migration on migrant sending countries, communities and households*. Retrieved from https://assets.publishing.service.gov.uk/media/57a08999ed915d3cfd000326/Effects_of_remittances_and_migration_56.pdf

Hagen-Zanker, J., & Azzarri, C. (2010). Are internal migrants in Albania leaving for the better? *Eastern European Economics*, *48*(6), 57–84. doi:10.2753/EEE0012-8775480604

Harrington, J., Lutomski, J., Molcho, M., & Perry, I. J. (2009). Food poverty and dietary quality: Is there a relationship? *Journal of Epidemiology and Community Health*, *63*(Suppl II), A1–A36. doi:10.1136/jech.2009.096701p

Headey D., & Ecker O. (2013). Rethinking the measurement of food security: from first principles to best practice. *Food Security, 5*, 327–343. Doi:10.1007%2Fs12571-013-0253-0

Heckman, J. J., Ichimura, H., & Todd, P. (1998). Matching as an econometric evaluation estimator. *The Review of Economic Studies*, *65*(2), 261–294. doi:10.1111/1467-937X.00044

Heckman, J. J., & Smith, J. A. (1999). The pre-programme earnings dip and the determinants of participation in a social programme. Implications for simple programme evaluation strategies. *Economic Journal (London)*, *109*(457), 313–348. doi:10.1111/1468-0297.00451

Henjum, S., Torheim, L. E., Thorne-Lyman, A. L., Chandyo, R., Fawzi, W. W., Shrestha, P. S., & Strand, T. A. (2015). Low dietary diversity and micronutrient adequacy among lactating women in a peri-urban area of Nepal. *Public Health Nutrition*, *18*(17), 3201–3210. doi:10.1017/S1368980015000671 PMID:25824344

Hoddinott, J., & Yohannes, Y. (2002). *Dietary diversity as a households' food security*. Washington, DC: Food and Nutrition Technical Assistance Project, Academy for Educational Development. Available on https://pdfs.semanticscholar.org/d2b3/ce8d0ad79364f0b5583f2c6e8b77e6575297.pdf

International Dietary Data Expansion Project. (2015). *Dietary diversity*. Available on https://inddex.nutrition.tufts.edu/data4diets/data-source/dietary-diversity

Jørgen, C. (2008). The determinants of migrant remittances. *Oxford Review of Economic Policy*, *24*(3), 581–598. doi:10.1093/oxrep/grn022

Karamba, W. R., Quiñones, E. J., & Winters, P. (2011). Migration and food consumption patterns in Ghana. *Food Policy*, *36*(1), 41–53. doi:10.1016/j.foodpol.2010.11.003

Kennedy, G. L. (2009). *Evaluation of dietary diversity scores for assessment of micronutrient intake and food security in developing countries*. Wageningen University.

King, C. (2016). *Food insecurity and housing instability in vulnerable families. Review of Economics of the Household, published online ahead of print.* Lincoln, NE: Nutrition and Health Sciences - Faculty Publications.

Kurekova, L. (2011). *Theories of migration: Conceptual review and empirical testing in the context of the EU East-West flows*. Academic Press.

Lawal, A. S., & Okeowo, T. A. (2014). Effects of rural urban migration on labour supply in cocoa production in ondo east Local Government Area of Ondo State. *Academic Journal of Interdisciplinary Studies*, *3*(7), 25–29.

Liu, Y., Njai, R. S., Greenlund, K. J., Chapman, D. P., & Croft, J. B. (2014). Relationships between Housingand food insecurity, frequent mental distress, and insufficient sleep amongadults in12 US states, 2009. *Preventing Chronic Disease*, *11*, 130334. doi:10.5888/pcd11.130334 PMID:24625361

Magurran, A. E., & McGill, B. J. (2011). *Biological diversity: frontiers in measurement and assessment*. Oxford: Oxford University Press.

Maharjan, A., Bauer, S., & Knerr, B. (2013). International migration, remittances and subsistence farming: Evidence from Nepal. *International Migration (Geneva, Switzerland)*, *5*(1), 249–263. doi:10.1111/j.1468-2435.2012.00767.x

Mbah, E. N., Ezeano, C. I., & Agada, M. O. (2016). Effects of rural-urban youth migration on farm families in Benue state, Nigeria. *International Journal of Agricultural Research, Innovation and Technology*, *6*(1), 14–20. doi:10.3329/ijarit.v6i1.29207

McArt, S. H., Cook-Patton, S. C., & Thaler, J. S. (2012). Relationships between arthropod richness, evenness, and diversity are altered by complementarity among plant genotypes. *Oecologia*, *168*(4), 1013–1021. doi:10.100700442-011-2150-6 PMID:22002039

Meerza, S. I. A. (2010). Rural Urban Migration and its Consequences on Rural Children. In Mini The impact of Rural-Urban Migration and Rural Economy in Rural Village. Academic Press.

Meeske, M. (2018). *Rural-Urban migration, food consumption pattern and trend in the Local food system: A case study for Vietnam*. Wageningen University, Department of Social Sciences.

Mini, S. E. (2001). *The impact of rural-urban migration on rural economy in rural village*. Retrieved from www. geofileonline.com

Mishra, V., & Ray, R. (2006). *Dietary pattern, calorie intake and undernourishment: the Vietnamese experience.* Retrieved from https://www.researchgate.net/publication/237123306

Momoh, P., & Kormawa, N. (2002). Need assessment study for market-driven agricultural technology transfer. *IITA*. Retrieved from www.https://books.google.com.ng/books?

Moursi, M. M., Arimond, M., Dewey, K. G., Trèche, S., Ruel, M. T., & Delpeuch, F. (2008). Dietary diversity is a good predictor of the micronutrient density of the diet of 6-to 23-month-old children in Madagascar. *The Journal of Nutrition*, *138*(12), 2448–2453. doi:10.3945/jn.108.093971 PMID:19022971

National Bureau of Statistics. (2007). *Agricultural Survey Report 1994/95– 2005/06*. Abuja: National Bureau of Statistics.

National Institutes of Health. (2019). *Important Nutrients to Know: Proteins, Carbohydrates, and Fats.* Retrieved from https://www.nia.nih.gov/health/important-nutrients-know-proteins-carbohydrates-and-fats

NCSS Statistical Software NCSS. (2019). *Two-stage least squares*. Retrieved from https://www.ncss-wpengine.netdna-ssl.com/wp content/themes/ncss/pdf/Procedures/NCSS/Two Stage_Least_Squares.pdf

Nguyen, D. L., Grote, U., & Nguyen, T. T. (2017). Migration and rural household expenditures: A case study from Vietnam. *Economic Analysis and Policy*, *56*, 163–175. doi:10.1016/j.eap.2017.09.001

Nguyen, M. C., & Winters, P. (2011). The impact of migration on food consumption patterns: The case of Vietnam. *Food Policy*, *36*(1), 71–87. doi:10.1016/j.foodpol.2010.11.001

Olayide, D. W. (2009). Rural-urban migration and agriculture development in Nigeria. *Art Soc. Sci. International Research Journal*, *1*(30), 44–57.

Olowa, O. W. (2009). The Impact of migrant remittances on household welfare in rural Nigeria. *Social Sciences*, *4*(1), 59–64.

Olowa, O. W., Awoyemi, T. T., Shittu, M. A., & Olowa, O. A. (2013). Effects of Remittances on Poverty among Rural Households in Nigeria European. *Journal of Sustainable Development*, *2*(4), 263–284.

Osondu, C. K., & Ibezim, C. O. (2001). The impact of rural-urban migration on plantation Agriculture in the Niger-Delta Region, Nigeria. *Journal of Rural Social Sciences*, *1*(20).

Qui, P., Yang, Y., Zhang, J., & Ma, X. (2011). Rural-to-urban migration and its implication for new cooperative medical scheme coverage and utilization in China. *BMC Public Health*, *11*(1), 520. doi:10.1186/1471-2458-11-520 PMID:21718506

Ruel, M. T. (2003). Operationalizing dietary diversity: A review of measurement issues and research priorities. *The Journal of Nutrition*, *133*(11), 3911S–3926S. doi:10.1093/jn/133.11.3911S PMID:14672290

Skeldon, R. (1997). Rural-to-urban migration and its implications for poverty alleviation. *Asia-Pacific Population Journal*, *12*(1), 3–16. doi:10.18356/cd2c964e-en PMID:12292421

Stahre, M., VanEenwyk, J., Siegel, P., & Njai, R. (2015). Housing insecurity and the association with-health outcomes and unhealthy behaviors, Washington State, 2011. *Preventing Chronic Disease*, *12*, 140511. doi:10.5888/pcd12.140511 PMID:26160295

Torheim, L. E., Ouattara, F., Diarra, M. M., Thiam, F. D., Barikmo, I., Hatløy, A., & Oshaug, A. (2004). Nutrient adequacy and dietary diversity in rural Mali: Association and determinants. *European Journal of Clinical Nutrition*, *58*(4), 594–601. doi:10.1038j.ejcn.1601853 PMID:15042127

United Nations Population Fund. (2014). *Adolescents, youth and the transformation of the future*. Retrieved from https://www.unfpa.org/sites/default/files/pub-pdf/EN-SWOP14-Report_FINAL-web.pdf

Vargas-Lundius, R., Guillaume, L., Villarreal, M., & Osorio, M. (2008). *International migration, remittances and rural development*. Retrieved from http://www.fao.org/3/a-ak405e.pdf

Waswa, L. M., Jordan, I., Herrmann, J., Krawinkel, M. B., & Keding, G. B. (2015). Community-based educational intervention improved the diversity of complementary diets in western Kenya: Results from a randomized controlled trial. *Public Health Nutrition*, *18*(18), 3406–3419. doi:10.1017/S1368980015000920 PMID:25857703

World Bank Report. (2005). *Empowering People by Transforming Institutions, Social Development in World Bank Operations*. Retrieved from http://documents. worldbank.org/curated/en/563751468780547056/Empowering-people-by-transforminginstitutions-social-development-in-World-Bank-operations.

World Bank Report. (2005). *Empowering People by Transforming Institutions, Social Development in World Bank Operations*. Retrieved from http://documents.worldbank.org/curated/en/563751468780547056/Empowering-people-by-transforming-institutions-social-development-in-World-Bank-operations

XLSTAT. (2019). *Two-Stage Least Squares Regression*. Retrieved from https://www.xlstat.com/en/solutions/features/two-stage-least-squares-regression

Zezza, A., Carletto, C., Davis, B., & Winters, P. (2011). Assessing the impact of migration on food and nutrition security. *Food Policy*, *36*(1), 1–6. doi:10.1016/j.foodpol.2010.11.005

Compilation of References

Ababouch, L. (2015). *Fisheries and Aquaculture in the Context of Blue Economy*. UN Economic Commission for Africa. Retrieved from https://www.afdb.org/fileadmin/uploads/afdb/Documents/Events/DakAgri2015/Fisheries_and_Aquaculture_in_the_Context_of_Blue_Economy.pdf

Abbas, I. I. (2009). An overview of land cover changes in Nigeria, 1975–2005. *Journal of Geography and Regional Planning*, *2*, 62–65.

Abiodun, B. J., Salami, A. T. & Tadross, M. (2011). *Climate change scenarios for Nigeria: understanding biophysical impacts.* A report published by Building Nigeria's Responses to Climate Change (BNRCC).

Abubakar, K. (2015). *Land use land cover changes and their implications in Nigeria using geospatial technique* (MSc. Dissertation). Ahmadu Bello University, Zaria, Nigeria.

AC-Chukwuocha, N. B. (2015). Analysis of structural changes in land use and land cover of Otamiri watershed in Owerri, south east Nigeria. *Journal of Environment and Earth Science*, *5*, 51–59.

Accordino, J., & Johnson, G. T. (2000). Addressing the Vacant and Abandoned Property Problem. *Journal of Urban Affairs*, *22*(3), 301–315. doi:10.1111/0735-2166.00058

Action Aid International. (2008). *Food, Farmers and Fuel: Balancing Global Grain and Energy policies with Sustainable Land Use*. Durban, South Africa: Action Aid International Secretariat.

ActionAid International., CARE., Christian Aid., Concern Worldwide., Find Your Feet., Oxfam Practical Action., … University of East Anglia. (2012). What works for women: Proven approaches for empowering women smallholders and achieving food security. *Oxfam Policy and Practice*: *Agriculture*. *Food and Land*, *12*(1), 113–140.

ActionAid. (2014). *The Great Land Heist: How the world is paving the way for corporate land grabs*. Retrieved from www.actionaid.org/publications/great-land-heist

Adaku, A. (2013). The effect of rural-urban migration on agricultural production in the northern region of Ghana. *Journal of Agricultural Science and Applications*, *02*(04), 193–201. doi:10.14511/jasa.2013.020402

Adam, A. G., Bidoli, T. D., Ammani, A. A., & Oduehie, T. C (2018). Gender Participation in Rice Processing Value Chain in kebbiand Sokoto States, Nigeria. *Global Journal of Science Frontier Research: Agriculture and Veterinary, 18*(2).

Adebayo, K. (2009). Dynamics of technology adoption in rural-based cassava processing enterprises in South-West Nigeria. *International Journal of Agricultural Economics and Rural Development*, *2*(1), 15–24.

Adebayo, O. O., & Adeola, R. G. (2008). Sources and uses of agricultural credit by small scale farmers in Surulere Local Government Area of Oyo State. *Anthropologist*, *10*(4), 313–314. doi:10.1080/09720073.2008.11891069

Adegboye, O. R., Smith, C., Anang, D., & Musa, H. (2016). Comparing and Contrasting Three Cultural Food Customs from Nigeria and Analyzing the Nutrient Content of Diets from These Cultures with the Aim of Proffering Nutritional Intervention. *Critical Reviews in Food Science and Nutrition*, *56*(15), 2483–2494. doi:10.1080/10408398.2013.862201 PMID:25830896

Adegboye, R. A. (1993). Agricultural land: its use, tenure and conservation in Nigeria. In E. A. Aiyelani, E. O. Lucas, M. O. Abatan, & O. A. Akinboade (Eds.), *Fundamentals of Agriculture* (pp. 9–13). Ibadan: Afrika-Link Books.

Adejo, P. E., Okwu, J. O., & Edoka, M. H. (2015). Assessment of Gender Participation in Post-harvest Management Practices of Maize in Kogi State, Nigeria. *Proceedings of the 49th Annual National Conference of the Agricultural Society of Nigeria "Agriculture: The Nigerian Economy Beyond Oil"*.

Adejo, P. E., Okwu, J. O., & Saliu, O. J. (2016). Assessment of Postharvest Information Needs ofMaize Farmers in Kogi State, Nigeria. *Journal of Agricultural & Food Information*, *17*(4), 260–273. doi:10.1080/10496505.2016.1221766

Ademola, A. (1994). Women farmers in Ondo State, Nigeria. *Journal of Anthropological Research*, *50*(3), 311–326.

Adeola, R. G., Adewale, J. G., & Adebayo, O. O. (2008). Information Needs of Cowpea Farmers in Ibadan/Ibarapa Agricultural Zone of Oyo State, Nigeria. *International Journal of Agricultural Economics & Rural Development*, *1*(2), 78–83.

Adepoju, A. O., Ewolor, S., & Obayelu, O. A. (2018). Welfare implications of Domestic land grabs among rural households in Delta State, Nigeria. *Rural Sustainability Research*, *39*(334), 19–25. doi:10.2478/plua-2018-0003

Adesina, H. A. (2003). *The gender variable in agricultural research*. Macmillan Publishers.

Adewale, J. G. (2005). Socio-economic factors associated with urban-rural migration in Nigeria: A case study of Oyo State, Nigeria. *Journal of Human Ecology (Delhi, India)*, *17*(1), 13–16. doi:10.1080/09709274.2005.11905752

Adeyemo, T. A., & Olajide, O. A. (2013). *Urban Rural Pattern of Remittances and Effects on Food Security of Rural Households in Nigeria*. In *Conference on International Research on Food Security, Natural Resource Management and Rural Development organised by the University of Hohenheim*, Stuttgart, Germany.

Adi, D. N. (2009). *Critical Mass Representation in Uganda. Independent Study Project (ISP) Collection Paper 674*. Retrieved from http://digitalcollections.sit.edu/isp_collection/674

Afolabi, M. M. (2008). *Women as pillars of national economy in Nigeria: A study of economic activities of rural women in six local government areas of Ondo State*. IAFFE summer conference, International Association for Feminist Economics, Toronto, Italy.

Africa Check. (2015). *Who owns the land in Nigeria? Africa Check factsheet*. Retrieved from https://africacheck.org/factsheets/factsheet-who-owns-the-land-in-nigeria/

Agbarevo, M. N. B., & Okeke, S. O. (2015). The Effect of Adoption of Cassava Value added Technologies on Farmers' Production in Abia State, Nigeria. *European Journal of Physical and Agricultural Science*, *3*(1), 48–53.

Agbarevo, M. N., & Obinne, C. P. O. (2008). Evaluation of the effect of Agricultural extensiondelivery on cassava production in Nigeria. A case study of Cross-River state of Nigeria. *Nigerian Agricultural Journal*, *39*(1&2), 16–21.

Agboola, T. A. (2000). Water management in Fadama: a report on Fadama implementation. Academic Press.

Aguilar, A., Carranza, E., Goldstein, E., Kilic, T., & Oseni, G. (2015). Decomposition of gender differentials in agricultural productivity in Ethiopia. *Agricultural Economics*, *46*(3), 311–334. doi:10.1111/agec.12167

Agwu, K., Amasiatu, O. G., & Onuoha, O. U. (2010). Land Rights Characteristics and Access to Land: Implications on Food Security in Nigeria. *Journal of Agriculture and Environment For Developing Countries*, 2(2&3), 146–152.

Ahmed, F. F., Eugene, C. E., & Abah, P. O. (2015). Analysis of food security among farming households in Borno State, Nigeria. *Journal of Agricultural Economics, Environment and Social Sciences*, *1*(1), 130–141.

Ajaero, C. K., & Onokala, P. C. (2013). The effcts of rural-urban migration on rural communities of Southeastern Nigeria. *International Journal of Population Research*, *2013*, 1–10. doi:10.1155/2013/610193

Ajala, A. S. & Gana, A. (2015). Analysis of Challenges Facing Rice Processing in Nigeria. *Journal of Food Processing*.

Akers, J. (2013). Making markets: Think tank legislation and private property in Detroit. *Urban Geography*, *34*(8), 1070–1095. doi:10.1080/02723638.2013.814272

Akter, S., Rutsaert, P., Luis, J., Htwe, N. M., San, S. S., Raharjo, B., & Pustika, A. (2017). Women's empowerment and gender equity in agriculture: A different perspective from Southeast Asia. *Food Policy*, *69*, 270–279. doi:10.1016/j.foodpol.2017.05.003

Aladjadjiyan, A. (2012). *Food Production: Approaches, Challenges and Tasks*. InTech. Retrieved from https://www.intechopen.com/books/food-production-approaches-challenges-and-tasks/physical-factors-for-plant-growth-stimulation-improve-food-quality

Alexander, S., Panayotis, K., & Luc, C. (2007). The Stated Benefits From Commodity Price and Weather. In C. Luc & S. Alexander (Eds.), *Rural Household Vulnerability and Insurance Against Commodity Risks: Evidence From the United Republic of Tanzania*. Food and Agriculture Organization of the United Nations.

Ali, D., Bowen, D., Deininger, K., & Duponchel, M. (2016). Investigating the gender gap in agricultural productivity: Evidence from Uganda. *World Development*, *87*, 152–170. doi:10.1016/j.worlddev.2016.06.006

Alker, S., Joy, V., Roberts, P., & Smith, N. (2000). The definition of brownfield. *Journal of Environmental Planning and Management*, *43*(1), 49–69. doi:10.1080/09640560010766

Allison, E. H. (2011, Dec.). *Aquaculture, Fisheries, Poverty and Food Security*. Working Paper 2011-65. The WorldFish Center.

Aluko, B. T., & Amudu, A. (2006, March). *Women and Land Rights Reforms in Nigeria*. [Paper presentation]. The 5th FIG Regional Conference, Accra, Ghana. Retrieved from https://www.fig.net/resources/proceedings/fig_proceedings/accra/papers/ts09/ts09_04_aluko_amidu.pdf

Aluko, B. T. (2001). *Relationship between family and chieftancy land in Ibokun, LLB long essay*. Ibadan: Faculty of Law, University of Ibadan.

Amanor, S. K. (2012). Global Resource Grabs. Agribusiness Concentration and Smallholder; Two West African Case Studies. *The Journal of Peasant Studies*, *39*(3-4), 731–749. doi:10.1080/03066150.2012.676543

Amrevurayire, E. O. (2016). Consequences of rural-urban migration on the source region of Ughievwen clan delta state Nigeria. *European Journal of Geography*, *7*(3), 42–57.

Amuakwa-Mensah, F., Boakye-Yiadom, L., & Baah-Boateng, W. (2016). Effect of education on migration decisions in Ghana: A rural-urban perspective. *Journal of Economic Studies (Glasgow, Scotland)*, *43*(2), 336–356. doi:10.1108/JES-09-2013-0138

Anderson, E. C., & Minor, E. S. (2017, January 1). Vacant lots: An underexplored resource for ecological and social benefits in cities. In *Urban Forestry and Urban Greening*. Elsevier GmbH. Doi:10.1016/j.ufug.2016.11.015

Andersson, E. (2006). Urban landscapes and sustainable cities. *Ecology and Society*, *11*(1), 34. doi:10.5751/ES-01639-110134

Angrist, J. D. (1998). Estimating the Labor Market Impact of Voluntary Military Service Using Social Security Data on Military Applicants. *Econometrica, 66*(2), 249–288. doi:10.2307/2998558

Angrist, J. D., & Pischke, J. (2009). *Mostly Harmless Econometrics: An Empiricist's Companion*. Princeton University Press. doi:10.1515/9781400829828

Anseeuw, W. L., Alden, W., Cotula, L., & Taylor, M. (2012a). *Land Rights and the Rush for Land: Findings of the Global Commercial Pressures on Land Research Project*. Rome: International Land Coalition.

Anseeuw, W., Boche, M., Breu, T., Giger, M., Lay, J., & Messerli, P. (2012b). *Transnational Land Deals for Agriculture in the Global South: Analytical Report based on the Land Matrix Database*. Hamburg: The Land Matrix Partnership.

Antonakis, J., Bendahan, S., Jacquart, P., & Lalive, R. (2014). Causality and endogeneity: Problems and solutions. In D. V. Day (Ed.), *The Oxford Handbook of Leadership and Organizations* (pp. 93–117). New York: Oxford University Press.

Antwi-Agyakwa, K. T. (2014). *Assessing the Effect of Land Use-Land Cover Change on Weija Catchment* (MSc Thesis). Kwame Nkrumah University of Science and Technology (KNUST), Kumasi, Ghana.

Anyanwu, C. I., Njoku, M. E., Agwu, N. M., & Iwunze, Q. E. (2018). Effect of rural-urban migration on household food security in Umuahia South Local Government Area of Abia State, Nigeria. *Nigerian Agricultural Journal, 49*(1), 20–31.

Anyiro, C. O., & Onyemachi, A. D. (2014). *Adoption of Cassava Value Added Innovation and Its Implication on Rural Livelihood: A Case of Rural Women in Abia State, Nigeria*. Retrieved from https://www.researchgate.net/publication/296183628

Apata, T. G. (2015). *Analysis of Cassava Value Chain in Nigeria, From a Pro-poor and Gender Perspective of Farming Households in Southwest, Nigeria*. Retrieved from www.agrodep.org RV_Apata_T_final

Aquaculture Resources. (2019). *Aquaculture resources*. Retrieved from http://www.fao.org/fishery/resources/aquaculture/en

Arezki, R., Deininger, K., & Selod, H. (2011). *What Drives the Global Land Rush?* World Bank Policy Research Working Paper, No.WPS5864.

Arimond, M., Wiesmann, D., Becquey, E., Carriquiry, A., Daniels, M. C., Deitchler, M., ... Torheim, L. E. (2010). Simple Food Group Diversity Indicators Predict Micronutrient Adequacy of Women's Dietsin Diverse, Resource-Poor Settings. *The Journal of Nutrition, 140*(11), 2059S–2069S. doi:10.3945/jn.110.123414 PMID:20881077

Ariyo, J. A., & Mortimore, M. (2011). *Land Deals and Commercial Agriculture in Nigeria: The New Nigerian Farms in Shonga District, Kwara State.* Paper Presented at the International Conference on Global Land Grabbing Organised by the Land Deals Politics Initiatives (LDPI) in Collaboration with the *Journal of Peasant Studies* and hosted by the Future Agricultural Consortium at the Institute of Development Studies, University of Sussex, UK.

Arowolo, A. O., Bhowmik, A. K., Qi, W., & Deng, X. (2017). Comparison of spatial interpolation techniques to generate high-resolution climate surfaces for Nigeria. *International Journal of Climatology, 37*, 179–192. doi:10.1002/joc.4990

Arowolo, A. O., & Deng, X. (2018). Land-use/land-cover change and statistical modelling of cultivated land change drivers in Nigeria. *Regional Environmental Change, 18*(1), 247–259. doi:10.100710113-017-1186-5

Arowolo, A. O., Deng, X., Olatunji, O. A., & Obayelu, A. E. (2018). Assessing changes in the value of ecosystem services in response to land-use/land-cover dynamics in Nigeria. *The Science of the Total Environment, 636*, 597–609. doi:10.1016/j.scitotenv.2018.04.277 PMID:29723833

Arts, B., Leroy, P., & Tatenhove, J. (2006). Political Modernisation and Policy Arrangements: A Framework for Understanding Environmental Policy Change. *Public Organization Review. A Global Journal, 6*(2), 93–106.

Aruwa, A. (2011). *Kogi State Governor Launches Bio-fuels Refinery in "Itobe"*. Graphic Newspaper.

Asche, F., Bellemare, M. F., Roheim, C., Smith, M. D., & Tveteras, S. (2015). Fair Enough? Food Security and the International Trade of Seafood. *World Development*, *67*, 151–160. doi:10.1016/j.worlddev.2014.10.013

Asian Development Bank. (2013). *Gender Equality and Food Security. Women's Empowerment as a Tool against Hunger.* Retrieved from https://www.adb.org/sites/default/files/publication/30315/gender-equality-and-food-security.pdf

Audet, R., Lefèvre, S., Brisebois, É., & El-Jed, M. (2017). Structuring tensions and key relations of Montreal seasonal food Markets in the Sustainability Transition of the Agri-food sector. *Sustainability*, *9*(3), 320. doi:10.3390u9030320

Ayanlade, A. (2012). Evaluating Environmental Change Impacts on Ecological Services in the Niger Delta of Nigeria. *Ife Research Publications in Geography*, *11*, 111–125.

Ayanlade, A., & Proske, U. (2016). Assessing wetland degradation and loss of ecosystem services in the Niger Delta, Nigeria. *Marine and Freshwater Research*, *67*, 828–836. doi:10.1071/MF15066

Ayodele, D., Udah, A., Elechi, N., Oriuwa, C., Tijani, G., & Sanni, L. (2011). Cassava Value Chain Analysis in the Niger Delta. A Report by Foundation for Partnership Initiatives in the Niger Delta (PIND).

Azadi, H., Barati, A. A., Raðaani, P., Taheri, F., Gebrehiwot, K., Witlox, F., & Lebailly, P. (2017). Evolution of land use-change modeling: Routes of different schools of knowledge. *Landscape and Ecological Engineering*, *13*(2), 319–332. doi:10.100711355-016-0311-9

Baan, L., Alkemade, R., & Koellner, T. (2012). Land use impacts on biodiversity in LCA: A global approach. *The International Journal of Life Cycle Assessment*, *18*(6), 1216–1230. doi:10.100711367-012-0412-0

Babatunde, R. O. (2018). *Impact of Remittances on Food Security and Nutrition of Migrant's Household: Evidence from Nigeria*. Paper presented at the 30th International Association of Agricultural Economist, Vancouver, Canada.

Ball, R. M. (1998). *The use and re-use of vacant industrial buildings*. EPSRC Research Grant GR=K93464, Final Project Report, EPSRC.

Ball, R.M. & Bord, D. (1994, Sept.). *Vacant industrial premises in Stoke-on-Trent: a survey and analysis of the situation in August 1994*. Consultancy Report submitted to Stoke-on-Trent City Council and the Staffordshire Training and Enterprise Council.

Ball, R. M. (1996). *The use and re-use of vacant industrial buildings, Proposal for funding under the EPSRC Sustainable Cities Managed Programme*. Division of Geography, Staffordshire University.

Ball, R. M. (2002). Re use potential and vacant industrial premises: Revisiting the regeneration issue in Stoke-on-Trent. *Journal of Property Research*, *19*(2), 93–110. doi:10.1080/09599910210125223

Balmori, H. H. (2003). *Gender and Budgets Overview Reports. BRIDGE (Development – Gender)*. Brighton: Institute of Development Studies, University of Sussex.

Bashir, M. K., Schilizzi, S., & Pandit, R. (2012). *The determinants of rural household food security: The case of landless households of the Punjab, Pakistan.* Working Paper 1208, School of Agricultural and Resource Economics, University of Western Australia, Crawley, Australia.

Battersby, J. (2011). *The state of urban food insecurity in Cape Town: Urban food security series No 11*. Retrieved from http://www.afsun.org/wp-content/uploads/2013/09/AFSUN_11.pdf.

Bature, B. N. (2013). The G20 and the Nigerian Vision 20:2020. *IOSR Journal Of Humanities And Social Science, 15*, 53-60.

Bearre, C. D. (1976, April 8). Farm-a-Lot: Detroiters turn vacant lots into gardens. *Ann Arbor Sun*. Retrieved from http://oldnews.aadl.org/node/201103

Beegle, K., De Weerdt, J., & Dercon, S. (2011). Migration and economic mobility in Tanzania: Evidence from a tracking survey. *The Review of Economics and Statistics*, *93*(3), 1010–1033. doi:10.1162/REST_a_00105

Bello, R.S., Bello, M.B., Essien, B.A., & Saidu, M.J. (2015). Economic Potentials of Oil Palm Production in UDI, Enugu State. *Science Journal of Business and Management*. Special Issue: *Sustainable Entrepreneurial Developments in Agribusiness, 3*(5-1), 16-20. doi:10.11648/j.sjbm.s.2015030501.14

Bene, C., Prager, S. D., Achicanoy, H. A., Toro, P. A., Lamotte, L., Cedrez, C. B., & Mapes, B. R. (2019). Understanding food systems drivers: A critical review of the literature. *Global Food Security*, *23*(1), 149–159. doi:10.1016/j.gfs.2019.04.009

Benjamin, C. A., & Joseph, U. C. (2012). *Food insecurity determinants among rural farm households in Nigeria.* International conference on ecology, agriculture and chemical engineering, Thailand.

Bennett, A., Patil, P., Kleisner, K., Rader, D., Virdin, J., & Basurto, X. (2018). *Contribution of Fisheries to Food and Nutrition Security: Current Knowledge, Policy, and Research.* NI Report 18-02. Durham, NC: Duke University.

Biyi, D. (2005). *Government Policies and Competitiveness of Nigerian Rice Economy*. Paper presented at the `Workshop on Rice Policy & Food Security in Sub-SaharanAfrica' organized by WARDA, Cotonou, Benin.

Blundell, R., & Dias, M. C. (2009). Alternative Approaches to Evaluation in Empirical Microeconomics. *The Journal of Human Resources*, *44*(3), 565–640. doi:10.3368/jhr.44.3.565

Böckerman, P., & Haapanen, M. (2013). The effect of polytechnic reform on migration. *Journal of Population Economics*, *26*(2), 593–617. doi:10.100700148-012-0454-4

Bogale, A., & Shimelis, A. (2009). Household level determinants of food security in rural areas of Dire Dawa, Eastern Ethiopia. *African Journal of Food, Agriculture, Nutrition and Development*, *9*(9) 1914-1926.

Bohn, K., & Viljoen, A. (2011). The edible city: Envisioning the continuous productive urban landscape. *Field J*, *4*(1), 149–161.

Boland, M. (2009). *What is Value added Agriculture?* Retrieved from http://www.agmrc.org/business_developm ent/getting_prepared/valueadded_agriculture/articles/index.cfm

Bontje, M. (2004). Facing the challenge of shrinking cities in East Germany: The case of Leipzig. *GeoJournal*, *61*(1), 13–21. doi:10.100710708-005-0843-2

Borras, S. M. Jr, Franco, J., & van der Ploeg, J. D. (2013). Land concentration, land grabbing and people's struggles in Europe: introduction to the collection of studies. In J. Franco & S. Borras Jr., (Eds.), *Land Concentration, Land Grabbing and People's Struggles in Europe* (pp. 6–30). Amsterdam: Transnational Institute, European Coordination Via Campesina.

Borras, S. M. Jr, Hall, R., Scoones, I., White, B., & Wolford, W. (2011). Towards a better understanding of global land grabbing: An editorial introduction. *The Journal of Peasant Studies*, *38*(2), 209–216. doi:10.1080/03066150.2011.559005

Borras, S. M. Jr, McMichael, P., & Scoones, I. (2010). the politics of biofuel, land and agrarian change: Editors' introduction. *The Journal of Peasant Studies*, *27*(4), 575–592. doi:10.1080/03066150.2010.512448 PMID:20873025

Bostancı, S. H., & Yıldırım, S. (2018). *From Sustainable Communities to Ecological Citizenship: Searching For Ecological Environment-Sensitive Life*. ICOMEP'18-Spring, International Congress of Management, Economy and Policy, Turkey.

Botreau, H., & Cohen, M. J. (2019). *Gender Inequalities and Food Insecurity. Oxfam Briefing Paper 2019*. Oxford, UK: Oxfam International.

Bowman, A. O. M., & Pagano, M. A. (2000). *Vacant Land in Cities: And Urban Resource*. Brookings Institution, Center on Urban and Metropolitan Policy.

Branas, C. C., Cheney, R. A., MacDonald, J. M., Tam, V. W., Jackson, T. D., & Ten Have, T. R. (2011). A difference-in-differences analysis of health, safety, and greening vacant urban space. *American Journal of Epidemiology*, *174*(11), 1296–1306. doi:10.1093/aje/kwr273 PMID:22079788

Brandspurng. (2018, Feb 13). Agriculture in Nigeria: Food Production and Consumption Rates. *Brandspurng*. Retrieved from https//brandspurng.com/category/agro-news/

Brauw, A., & Herskowitz, S. (2019). *Consumer demand and processed foods in Nigeria: a complex, evolving picture*. Retrieved from https://a4nh.cgiar.org/2019/02/28/consumer-demand-and-processed-foods-in-nigeria-a-complex-evolving-picture/

Braveman, P., Dekker, M., Egerter, S., Sadegh-Nobari, T., & Pollack, C. (2011). *Housing and Health*. Prepared for the Robert Wood Johnson Foundation.

Bravo-Baumann, H. (2000). *Capitalisation of experiences on the contribution of livestock projects to gender issues. Working Document*. Bern: Swiss Agency for Development and Cooperation.

Breeman, G., Dijkman, J., & Termeer, C. (2015). Enhancing food security through a multi-stakeholder process: The global agenda for sustainable livestock. *Food Security*, *7*(2), 425–435. doi:10.100712571-015-0430-4

Brooks, J. (2016, Apr 6). *Insights, New Approaches to Economic Challenges Food Security and the Sustainable Development Goals*. Academic Press.

Brown, K. H., & Jameton, A. L. (2000). Public health implications of urban agriculture. *Journal of Public Health Policy*, *21*(1), 20–39. doi:10.2307/3343472 PMID:10754796

Brulle, R. J., & Pellow, D. N. (2005). Environmental justice: Human health and environmental inequalities. *Annual Review of Public Health*, *27*(1), 103–124. doi:10.1146/annurev.publhealth.27.021405.102124 PMID:16533111

Brüntrup, M. (2011). *Detrimental Land Grabbing or Growth Poles? Determinants and Potential Development Effects of Foreign Direct Land Investments. Technikfolgenabschätzung – Theorie und Praxis*. Bonn: German Development Institute. doi:10.14512/tatup.20.2.8

Bureau, J. C., & Swinnen, J. (2018). EU policies and global food security. *Global Food Security*, *16*, 106–115. doi:10.1016/j.gfs.2017.12.001

Burkholder, S. (2012). *The New Ecology of Vacancy: Rethinking Land Use in Shrinking Cities*. Academic Press.

Burkholder, S. (2012). The new ecology of vacancy: Rethinking land use in shrinking cities. *Sustainability*, *4*(6), 1154–1172. doi:10.3390u4061154

Burnham, B. O. (1973). Markov intertemporal land use simulation model. *Journal of Agricultural and Applied Economics*, *5*(1), 253–258. doi:10.1017/S0081305200011110

Butler, R. A. (2006). *Why is palm oil replacing tropical rainforests?* Retrieved from https://news.mongabay.com/2006/04/why-is-palm-oil-replacing-tropical-rainforests/

C2A Notes. (2010). *Land grabbing, a threat to food security*. Agriculture and food in question, Issue No. 3.

Cadena, P. M. (2017). *Stimulating cassava growth in Nigeria and Tanzania*. Retrieved from https://agroecology.ifas.ufl.edu/2017/09/06/august-30-2017-simulatin-cassava-growth- in- nigeria-and-tanzania/

Caliendo, M., & Kopeining, S. (2005). *Some Practical Guidance for the Implementation of Propensity Score Matching*. IZA Discussion Paper, 1588. Bonn, Germany: Institute of Labor Economics (IZA).

Camacho Olmedo, M. T., Paegelow, M., Mas, J. F., & Escobar, F. (2018). Geomatic Approaches for Modeling Land Change Scenarios. An Introduction. In M. T. Camacho Olmedo, M. Paegelow, J.F., Mas, & F. Escobar (Eds.), Geomatic Approaches for Modeling Land Change Scenarios. Lecture Notes in Geoinformation and Cartography. Cham: Springer.

Cameron, A. C., & Trivedi, P. K. (2010). *Microeconometrics Using Stata, Revised Edition*. Stata Press Books, StataCorp LP. Available on http://cameron.econ.ucdavis.edu/musbook/mus.html

Candel, J. J. L., & Biesbroek, R. (2018). Policy integration in the EU governance of global food security. *Food Security*, *10*(1), 195–209. doi:10.100712571-017-0752-5

Caponio, T., & Correa, M. (2017). Theorising migration policy in multilevel states: The multilevel governance perspective. *Journal of Ethnic and Migration Studies*, *44*(12), 1995–2010. doi:10.1080/1369183X.2017.1341705

Capture Fisheries Resources. (2019). *Capture fisheries resources*. Retrieved from http://www.fao.org/fishery/resources/capture/en

Carletto, C., De Brauw, A., & Banerjee, R. (2012). Measuring migration in multi-topic household surveys. Handbook of Research Methods in Migration, 207.

Castaldo, A., & Reilly, B. (2007). Do Migrant Remittances Affect the Consumption Patterns of Albanian Households? Southern-Eastern Europe. *Journal of Economics*, *1*, 25–54.

Caswel, J., Bredahl, M., & Hooker, N. (1998). How Quality Management Metasystems Are Affecting the Food Industry. *Review of Agricultural Economics, 20*(2), 547-557.

CBN. (2017). *CBN: Local Rice Production Records 70% Increase.* Retrieved from https://s3west1.amazonaws.com/leadersandco/wpcontent/uploads/2016/07/25051645/CentralBank-of-Nigeria-CBN.jpg

CFS (Committee on World Food Security). (2015a). *Framework for Action for Food Security and Nutrition In Protracted Crises*. Retrieved from http://www.fao.org/3/a-bc852e.pdf

CFS. (2009). *Reform of the committee on world food security. Final version*. Committee on world food security. Thirty-fifth Session. Agenda Item III. Retrieved from: http://www.fao.org/3/a-k7197e.pdf

CFS. (2015b). *Water for Food Security and Nutrition. Policy recommendations*. CFS 42 Final Report. Retrieved from http://www.fao.org/3/a-av046e.pdf

CFS. (2017). *Global Strategic Framework for Food Security and Nutrition (GSF)*. Retrieved from http://www.fao.org/3/MR173EN/mr173en.pdf

C-GIDD. (2016). *GDP and income distribution data*. Available at https://www.cgidd.com/

Chamorro, J., & Katz, E. (2002). *Gender, Land Rights and the Household Economy in Rural Nicaragua and Honduras, 11*. Paper prepared for the Regional Workshop on Land Issues in Latin America and the Caribbean (USAID).

Chandrasekhar, S., Das, M., & Sharma, A. (2014). Short-term Migration and Consumption Expenditure of Households in Rural India. *Oxford Development Studies*, *43*(1), 105–122. doi:10.1080/13600818.2014.964194

Chaudhuri, G., & Clarke, K. C. (2013). The SLEUTH Land Use Change Model: A Review. *The International Journal of Environmental Resources Research*, *1*, 88–104.

Chen, J., Chen, J., Liao, A., Cao, X., Chen, L., Chen, X., ... Mills, J. (2015). Global land cover mapping at 30 m resolution: A POK-based operational approach. *ISPRS Journal of Photogrammetry and Remote Sensing*, *103*, 7–27. doi:10.1016/j.isprsjprs.2014.09.002

Chodur, G. M., Zhao, X., Biehl, E., Mitrani-Reiser, J., & Neff, R. (2018, July 3). Assessing food system vulnerabilities: A fault tree modeling approach. *BMC Public Health*, *18*(1), 817. doi:10.118612889-018-5563-x PMID:29970077

Chuenpagdee, R., McConney, P., Munro, G., Ferreira, B., Marschoff, E., Rice, J., & Rosenberg, A. (2016). *Social and Economic Aspects of Sea-Based Food and Fisheries*. United Nations.

CIESIN/ITOS. (2013). Global roads open access dataset, Version 1 (gROADSv1). Palisades, NY: NASA Socioeconomic Data and Applications Center (SEDAC).

City of Portland. (2007). *Voices from the community: The vision PDX community input summary* (Abridged Version). Retrieved from https://www.visionpdx.com/downloads/visionPDX%20CIS%20Abrdg%20Versi.pdf

Coastes. J., Swindale, A., & Bilinsky, P. (2007). *Household food Insecurity Access Scale (HFIAS) for measurement of food access: indicator guide* (vol. 3). Washington, DC: FHI 360/FANTA.

Cohen, B. (2006). Urbanization in Developing Countries: Current Trends, Future Projections, and Key Challenges for Sustainability. *Technology in Society*, *28*(1), 63–80. doi:10.1016/j.techsoc.2005.10.005

Colasanti, K., Hamm, M., & Litjens, C. (2012). The city as an "agricultural powerhouse?" Perspectives on expanding urban agriculture from Detroit, Michigan. *Urban Geography*, *33*(3), 348–369. doi:10.2747/0272-3638.33.3.348

Coleman, A. (1982). Dead Space in the Dying Inner City. *The International Journal of Environmental Studies*, *19*(2), 103–107. doi:10.1080/00207238208709976

Collard, S. J., & Zammit, C. (2006). Effects of land-use intensification on soil carbon and ecosystem services in Brigalow (Acacia harpophylla) landscapes of Southeast Queensland, Australia. *Agriculture, Ecosystems & Environment*, *117*(2-3), 185–194. doi:10.1016/j.agee.2006.04.004

Combes, J. L., Ebeke, C. H., Etoundi, S. M. N., & Yogo, T. U. (2014). Are remittances and foreign aid a hedge against food price shocks in developing countries? *World Development*, *54*, 81–98. doi:10.1016/j.worlddev.2013.07.011

Committee on World Food Security. (2012). *Coming to terms with terminology: Food security, nutrition security, food security and nutrition, food and nutrition security*. Retrieved online on 10 Nov 2019 at http://www.fao.org/fsnforum/sites/default/files/file/Terminology/MD776(CFS___Coming_to_terms_with_Terminology).pdf

Communauté Urbaine de Montréal. (1996). [Montreal Urban Community]. *Carte d'Occupation du Sol* [Land Occupancy Card]. Retrieved from http://cmm.qc.ca/biotopes/docs/volets_1_et_2.pdf

Costanza, R., d'Arge, R., de Groot, R., Farber, S., Grasso, M., Hannon, B., ... van den Belt, M. (1997). The value of the world's ecosystem services and natural capital. *Nature*, *387*(6630), 253–260. doi:10.1038/387253a0

Costanza, R., de Groot, R., Sutton, P., van der Ploeg, S., Anderson, S. J., Kubiszewski, I., ... Turner, R. K. (2014). Changes in the global value of ecosystem services. *Global Environmental Change*, *26*, 152–158. doi:10.1016/j.gloenvcha.2014.04.002

Cotula, L., Dyer, N., & Vermeulen, S. (2008). Fuelling Exclusion? The Biofuels Boom and Poor People's Access to Land. FAO/IIED.

Cotula, L., Toulmin, C., & Hesse, C. (2004). *Land Tenure and Administration in Africa; Lessons of Experience and Emerging Issues* (p. 50). London: International Institute for Environmental and Development.

Cotula, L., & Vermeulen, S. (2009b). Deal or no Deal: The Outlook of Agricultural Land Investment in Africa. *International Affairs*, *85*(6), 1233–1247. doi:10.1111/j.1468-2346.2009.00859.x

Cotula, L., Vermeulen, S., Leonard, R., & Keeley, J. (2009a). *Land Grab or Development Opportunity?* London: Agricultural Investment and International Land Deals in Africa. IIED/FAO/IFAD.

CPLP. (2011). *Estratégia de Segurança Alimentar e Nutricional ESAN-CPLP. Parte 1: Enquadramento* [Food and Nutritional Security Strategy ESAN-CPLP. Part 1: Background]. Retrieved from: https://www.cplp.org/Admin/Public/Download.aspx?file=Files%2FFiler%2Fcplp%2FEstrategia%2FI-_Enquadramento_final.pdf

CPLP. (2015). *CPLP em números* [CPLP in numbers]. Retrieved from: https://www.CPLP.org/Default.aspx?ID=4447&Action=1&NewsId=3984&M=NewsV2&PID=10872

Crane, A., Viswanathan, L., & Whitelaw, G. (2012). Sustainability through intervention: A case study of guerrilla gardening in Kingston, Ontario. *Local Environment*. doi:10.1080/13549839.2012.716413

Crouch, P. (2012, October 23). Evolution or gentrification: do urban farms lead to higher rents? *Grist.*

Crush, J. (2012). *Migration, development and urban food security*. Urban Food Security Series No. 9. Kingston and Cape Town, Queen's University and African Food Security Network (AFSUN).

D'Odorico, P., & Rulli, M. C. (2013). The fourth food revolution. *Nature Geoscience*, *6*(6), 417–418. doi:10.1038/ngeo1842

Dadhich, P. N., & Hanaoka, S. (2010). Remote sensing, GIS and Markov's method for land use change detection and prediction of Jaipur district. *Journal of Genomics*, *4*, 9–15.

Daniel, S., & Mittal, A. (2009). *The great land grab. Rush for world's farmland threatens food security for the poor.* Oakland, CA: The Oakland Institute.

Davies, A. R. (2014). Co-creating sustainable eating futures: Technology, ICT and citizen-consumer ambivalence. *Futures*, *62*, 181–193. doi:10.1016/j.futures.2014.04.006

Davis, K. F., D'Odorico, P., & Rulli, M. C. (2014). Land grabbing: A preliminary quantification of economic impacts on rural livelihoods. *Population and Environment*, *36*(2), 36. doi:10.100711111-014-0215-2 PMID:25400309

Dayo, P., Olumuyiwa, O. J., Yarma, N., & Fatunbi, O. (2018). Innovation Opportunities in the Rice Value Chain in Nigeria. *FARA Results, 2*(3). Retrieved from https://research4agrinnovation.org/wpcontent/uploads/2018/03/Volume-2-No- 3_2018.pdf

De Groot, R., Brander, L., van der Ploeg, S., Costanza, R., Bernard, F., Braat, L., . . . McVittie, A. (2012). Global estimates of the value of ecosystems and their services in monetary units. Ecosystem Services, 1, 50-61.

De Muro, P., & Burci, F. (2007). *Education for Rural People and Food Security A Cross Country Analysis.* Rome: Food and Agriculture Organization of the United Nations.

De Schutter, O. (2009). *Large-scale land acquisitions and leases: A set of core principles and measures to address the human rights challenges*. Retrieved from http://www.oecd.org/site/swacmali2010/44031283.pdf

De Schutter, O. (2011). How not to think of land-grabbing: Three critiques of large-scale investments in farmland. *The Journal of Peasant Studies*, *38*(2), 249–279. doi:10.1080/03066150.2011.559008

de Sherbinin, A. (2002). *Land-Use and Land-Cover Change, A CIESIN Thematic Guide*. Palisades, NY: Center for International Earth Science Information Network of Columbia University.

De Zoysa, R. (2013). *The Implications of Large Scale Land Acquisition on Smallholder's Food Security*. Development Planning Unity. Working Paper No. 156, p. 34.

Deaton, A., & Paxson, C. (1998). Economies of scale, household size, and the demand for food. *Journal of Political Economy*, *106*(5), 897–930. doi:10.1086/250035

Dehejia, R. H., & Wahba, S. (2002). Propensity score-matching methods for non experimental causal studies. *The Review of Economics and Statistics*, *84*(1), 151–161. doi:10.1162/003465302317331982

Deininger, K., & Byerlee, D. (2011). *Rising global interest in farmland: Can it yield sustainable and equitable benefits?* Washington, DC: World Bank. doi:10.1596/978-0-8213-8591-3

Deng, X., Jiang, Q. O., Zhan, J., He, S. & Lin, Y. (2010). Simulation on the dynamics of forest area changes in Northeast China. *J. Geogr. Sci.*, *20*, 495-509.

Deng, X., Su, H., & Zhan, J. (2008). Integration of Multiple Data Sources to Simulate the Dynamics of Land Systems. *Sensors (Basel)*, *8*(2), 620–634. doi:10.33908020620 PMID:27879726

Department for Environment, Food and Rural Affairs (DEFRA). (2006). *Food security and the UK: An evidence and analysis paper.* Retrieved from http://archive.defra.gov.uk/evidence/economics/foodfarm/reports/documents/foodsecurity.pdf

DeSousa, C. A. (2004). The greening of brownfields in American cities. *Environmental Planning and Management*, *47*(4), 579–600. doi:10.1080/0964056042000243249

DETR. (1998). Sustainable Development: Opportunities for Change. Department of the Environment, Transport and the Regions. Consultation Paper on a Revised UK Strategy, DETR, London.

DETR. (2000). Our Towns and Cities: The Future. CM 4911, Department of the Environment Transport and the Regions. London: DETR.

Detroit Works Project. (2014). *Detroit future city: Detroit strategic framework plan*. Retrieved from http://detroitfuturecity.com/wp-content/uploads/2014/12/DFC_Full_2nd.pdf

Dheeraj, B., Satija, A., Khandpur, N., Bowen, L., Kinra, S., Prabhakaran, D., ... Shah, E. (2010). Effects of migration on food consumption patterns in a sample of Indian factory workers and their families. *Public Health Nutrition*, *13*(12), 1982–1989. doi:10.1017/S1368980010001254 PMID:20507672

Diaz-Bonilla, E., & Hepburn, J. (2016). Trade, Food Security, and the 2030 Agenda. Geneva, Switzerland: International Centre for Trade and Sustainable Development (ICTSD), International Environment House 2.

Dickens, W. T., & Lang, K. (1984). *A Test of Dual Labor Market Theory*. NBER Working Paper No. 1314. Retrieved from https://www.nber.org/papers/w1314

Dietary Guidelines Advisory Committee (DGAC). (2015). Dietary Guidelines Advisory Committee: Advisory Report to the Secretary of Health and Human Services and the Secretary of Agriculture.. *Scientific Reports*, 2015.

DIVA-GIS. (2016). *Free spatial data*. Author.

Dodson, B., Simelane, H., Tvera, D., Green. T., Chikanda, A., & de Vletter, F. (2008). Gender, Migration and Remittances in Southern Africa. Southern African Migration Project (SAMP), in Migration Policy Series, 29.

Doss, C. (2014). If women hold up half the sky, how much of the world's food do they produce? In A. R. Quisumbing, R. Meinzen-Dick, T. L. Raney, A. Croppenstedt, J. A. Behrman, & A. Peterman (Eds.), *Gender in Agriculture*. Springer. doi:10.1007/978-94-017-8616-4_4

Doss, C. R. (2017). Women and agricultural productivity: Reframing the issues. *Development Policy Review*, *36*(1), 35–50. doi:10.1111/dpr.12243 PMID:29263585

Douglas, J. R. (1975). *Hydroponics: The Bengal System* (5th ed.). Madison, NY: Oxford University Press.

Draus, P. J., Roddy, J., & McDuffie, A. (2014). We don't have no neighbourhood: Advanced marginality and urban agriculture in Detroit. *Urban Studies (Edinburgh, Scotland)*, *51*(12), 2523–2538. doi:10.1177/0042098013506044

Dubbeling, M., Bracalenti, L., & Lagorio, L. (2009b). Participatory design of public spaces for urban agriculture, Rosario, Argentina. *Open House International*, *34*, 36–49.

Duncan, J. (2017). Global Food Security Governance: Civil society engagement in the reformed Committee on World Food Security. London: Routledge.

EBRD (European Bank for Reconstruction and Development). (2011). *Private Sector for Food Security Initiative*. Retrieved from https://www.ebrd.com/downloads/research/factsheets/foodsecurity.pdf

Echebiri, R. N. (2005). *Characteristics and Determinants of Urban Youth Unemployment in Umuahia, Nigeria: Implications for Rural Development and Alternative Labor Market Variables*. Paper presented at the ISSER/Cornell/World Bank conference on "Shared Growth in Africa", Accra, Ghana.

Echebiri, R. N., & Edaba, M. E. I. (2008). Production and Utilization of Cassava in Nigeria: Prospects for Food Securit and Infant Nutrition. *Production Agriculture and Technology*, *4*(1), 38–52.

Edelman, M. (2013). Messy hectares: Questions about the epistemology of land grabbing data. *The Journal of Peasant Studies*, *40*(3), 485–501. doi:10.1080/03066150.2013.801340

Effiong, J. B., Ijioma, J. C., & Okolo, L. C. (2015). Participation of women farmers in rice production in bende local government area, abia state. *International Journal of Agricultural Extension and Rural Development Studies*, *2*(2), 1–9.

Ehebhamen, O. G., Obayelu, A. E., Vaughan, I. O., & Afolabi, W. A. O. (2017). Rural households' food security status and coping strategies in Edo State Nigeria. *International Food Research Journal*, *24*(1), 333–340.

Eigenbrod, F., Armsworth, P. R., Anderson, B. J., Heinemeyer, A., Gillings, S., Roy, D. B., ... Gaston, K. J. (2010). Error propagation associated with benefits transfer-based mapping of ecosystem services. *Biological Conservation*, *143*(11), 2487–2493. doi:10.1016/j.biocon.2010.06.015

Ejaro, S. P., & Abdullahi, U. (2013). Spatiotemporal analyses of land use and land cover changes in Suleja local government area, Niger state, Nigeria. *Journal of Environment and Earth Science*, *3*, 72–83.

Eke-okoro, O. N., & Njoku, D. N. (2012). A Review of Cassava Development in Nigeria from 1940-2010. *Journal of Agricultural and Biological Science*, *7*(1), 59–65.

Ekong, E. E. (2010) Diffusion and Adoption of Agricultural Innovations in Nigeria: Rural Sociology – Third Edition. Uyo: DOVE Educational Publishers.

El Sheikha, A. F., & Xu, J. (2017). Traceability as a Key of Seafood Safety: Reassessment and Possible Applications. *Reviews in Fisheries Science & Aquaculture*, *25*(2), 158–170. doi:10.1080/23308249.2016.1254158

Ellis, F. (2000). *Rural livelihood diversity in developing countries: Evidence and policy implications*. London: Overseas Development Institute; Retrieved from http://hdl.handle.net/10535/4486

Elmqvist, T., Zipperer, W., & Güneralp, B. (2016). Urbanization, habitat loss, biodiversity decline: solution pathways to break the cycle. In K. Seta, W. Solecki, & C. Griffith (Eds.), *Routledge handbook of urbanization and global environmental change*. London, UK: Routledge.

Eludoyin, O. M., Adelekan, I. O., Webster, R. E., & Eludoyin, A. O. (2013). Air temperature, relative humidity, climate regionalizationand thermal comfort of Nigeria. *International Journal of Climatology*, *34*(6), 2000–2018. doi:10.1002/joc.3817

Elver, H. (2015). *Special Rapporteur on the Right to Food.* Interim Rep. of the Special Rapporteuron the Right to Food, U.N. Doc. A/70/287. Retrieved from https://www.un.org/en/ga/search/

Ely, A., Geall, S., & Song, Y. (2016). Sustainable maize production and consumption in China: Practices and politics in transition. *Journal of Cleaner Production, 134*(Part A), 259–268. doi:10.1016/j.jclepro.2015.12.001

Enderlein, H., Walti, S., & Zurn, M. (Eds.). (2010). *Handbook on Multi-level Governance*. Cheltenham: Edward Elgar Publishing. doi:10.4337/9781849809047

Eneh, O. C. (2011). Nigeria's Vision 20:2020-Issues, Challenges and Implications for Development Management. *Asian Journal of Rural Development*, *1*(1), 21–40. doi:10.3923/ajrd.2011.21.40

Erhie, E., Iwelurio, M., Agbeyi, E., Oladipo, O., Oyaniran, T., Akinbiyi, A., & Adegunle, E. (2018). *Boosting Rice Production through increased Mechanization*. Retrieved from http://www.pwc.com.ng

Ericksen, P. J. (2008). Conceptualizing food systems for global environmental change research. *Global Environmental Change*, *18*(1), 234–245. doi:10.1016/j.gloenvcha.2007.09.002

Ericksen, P. J. (2008). What is the vulnerability of a food system to global environmental change? *Ecology and Society*, *13*(2), 14. doi:10.5751/ES-02475-130214

Ericksen, P., Stewart, B., Dixon, J., Barling, D., Loring, P., Anderson, M., & Ingram, J. (2010). The value of a food system approach. In *Food Security and Global Environmental Change* (pp. 25–45). London, UK: Earthscan.

Erle, E., & Pontius, R. (2007). Land-use and land-cover change. In C. J. Cleveland (Ed.), *Encyclopaedia of Earth*. Washington, DC: Environmental Information Coalition, National Council for Science and the Environment.

Etim, N. A., Okon, S., & Akpabio, I. A. (2011). Determinants of deprivation among part-time cassava farming households in the humid tropics. *International Journal of Agricultural Management and Development*, *1*(1), 45–51.

Etowa, E. B. (2015). *Effects of Migrant Remittances on Farm Household Welfare in Nigeria* (Unpublished Ph.D thesis). Department of Agricultural Economics, University of Nigeria.

European Coordination Via Campesina (ECVC). (2013). *Land Concentration and Land Grabbing, and People's Struggles in Europe*. Via Campesina.

Ewepu, G. (2018, April 11). Nigeria Rice Sub-sector to attract N250bn Investment. *Vanguard Newspaper*. Retrieved from https://www.vanguardngr.com/category/agric/

Fan, F., Wang, Y., & Wang, Z. (2008). Temporal and spatial change detecting (1998-2003) and predicting of land use and land cover in core corridor of Pearl River Delta (China) by using TM and ETM + images. *Environmental Monitoring and Assessment*, *137*(1-3), 127–147. doi:10.100710661-007-9734-y PMID:17564805

Fankun, D. S., & Evbuomwan, G. O. (2017). An evaluation of agricultural financing, policies and initiatives for sustainable development in Nigeria, in the 21st Century: 1990-2014. *IOSR Journal of Economics and Finance*, *8*(3), 32–38. doi:10.9790/5933-0803033238

Fanzo, J., Cogill, B., & Mattei, F. (2012). *Metrics of sustainable diets and food systems*. Technical Brief - Biodiversity International. Retrieved from https://www.bioversityinternational.org/uploads/tx_news/Metrics_of_sustainable_diets_and_food_systems_1572.pdf

FAO, WFP, & IFAD. (2013). The State of Food Insecurity in the World 2013: The Multiple Dimensions of Food Security. Rome: Author.

FAO. (1996). Fisheries and Aquaculture In Sub-Saharan Africa: Situation And Outlook In 1996. FAO Fisheries Circular No. 922 FIPP/C922. Rome: FAO.

FAO. (2002). Small Scale Palm Oil Processing in Africa. FAO Agricultural Services Bulletin, 148.

FAO. (2002a). Reducing Poverty and Hunger: The Critical Role of Financing for Food. *Agriculture and Rural Development. International Conference on Financing for Development*, Monterrey, Mexico.

FAO. (2002b). *Trade and Food Security: Conceptualizing the Linkages Expert Consultation*. Rome: FAO.

FAO. (2006). *Food Security, Policy Brief, June 2006, Issue 2*. FAO.

FAO. (2008a). *An Introduction to the Basic Concepts of Food Security*. Retrieved from http://www.fao.org/3/al936e/al936e00.pdf

FAO. (2008b). Food Security Information for Action, Availability Assessment and Analysis, Lesson, Assessing Availability, Learners'. *Notes*.

FAO. (2010). *Private Food Safety Standards: Their Role in Food Safety Regulation and their Impact*. Retrieved from: http://www.fao.org/3/ap236e/ap236e.pdf

FAO. (2012). *Decent rural employment for food security: A case for action. Knowledge Platform. Rural Employment. Gender, Equity and Rural Employment Division*. Rome, Italy: Economic and Social Development Department.

FAO. (2013). *Save and Grow: Casssava "Harvest, Post-harvest and Value addition"*. Retrieved from http//www.fao.org/ag/save-and-grow/cassava/index_en.html

FAO. (2016). FAO/INFOODS Global Food Composition Database for Fish and Shellfish Version 1.0- uFiSh1.0. Rome, Italy: FAO.

FAO. (2017). *Country fact sheet on Food and Agricultural Policy Trends*. Article prepared by Food and Agriculture Policy Decision Analysis (FAPDA). Retrieved from www.fao.org/in-action/fapda/tool

FAO. (2017). *Food and Agriculture: Driving action across the 2030 Agenda for Sustainable Development*. Rome: Food and Agriculture Organization of the United Nations (FAO). Retrieved from http://www.fao.org/3/ai7454e.pdf

FAO. (2017). *Strengthening Sector Policies for Better Food Security and Nutrition Results*. Policy Guidance Series.

FAO. (2018). *Nigeria at a glance*. Retrieved from www.fao.org/nigeria/programmes-and-projects/en/

FAO. (2018a). Transforming Food and Agriculture to Achieve the SDGs. 20 interconnected actions to guide decision-makers. Rome: FAO.

FAO. (2018b). *Achieving Blue Growth Building Vibrant Fisheries and Aquaculture Communities*. Retrieved from http://www.fao.org/3/CA0268EN/ca0268en.pdf

FAOSTAT. (2018). *Cassava Production (tons) for Nigeria from 1960 – 2018*. Retrieved from www.factfish.com/country/Nigeria

Farnworth, C. R., & Colverson, K. E. (2015). Building a Gender Transformation Extension and Advisory Facilitation System in sub-Saharan Africa. *Journal of Gender, Agriculture & Food Security*, *1*(1), 20–39.

Feenstra, G. (2002). Creating space for sustainable food systems: Lessons from the field. *Agriculture and Human Values*, *19*(99).

Felker-Kantor, E., & Wood, C. H. (2012). Female-headed households and food insecurity in Brazil. *Food Security*, *4*(4), 607–617. doi:10.100712571-012-0215-y

Fiquepron, J., Garcia, S., & Stenger, A. (2013). Land use impact on water quality: Valuing forest services in terms of the water supply sector. *Journal of Environmental Management*, *126*, 113–121. doi:10.1016/j.jenvman.2013.04.002 PMID:23681358

Fisseha, M. (2011). *Commercial Pressure on Land in Africa: A regional overview of opportunities challenges and Impacts*. International Land Coalition.

Fitz Gerald, E., & Winston, N. (2005). Housing, equality and inequality. In M. Norris M & D. Redmond (Eds.), Housing Contemporary Ireland: Economy, Society, Space and Shelter (pp. 224-244). Institute of Public Administration.

Folbre, N. (2006). Measuring Care: Gender, Empowerment, and the Care Economy. *Journal of Human Development*, *7*(2), 183–199. doi:10.1080/14649880600768512

Foley, J., DeFries, R., Asner, G., Barford, C., Bonan, G., Carpenter, S., ... Snyder, P. (2005). Global consequences of land use. *Science*, *309*(5734), 570–574. doi:10.1126cience.1111772 PMID:16040698

Fonta, M. W., Onyukwu, E. O., & Nwosu, O. E. (2011). International Remittance Inflow and Household Welfare: Empirical Evidence from Nigeria. *Research Journal of Finance and Accounting*, *2*(3), 140–149.

Food and Agricultural Organization. (2010). *Gender and land rights database*. Rome: Food and Agriculture Organization of the United Nations. Retrieved from http://www.fao.org/gender/landrights/en/

Food and Agricultural Organization. (2011). *The State of food and agriculture: Women in agriculture closing the gender gap for development*. Rome: Food and Agriculture Organization of the United Nations; doi:10.1109/PVSC.2008.4922754

Food and Agricultural Organization. (2012). *Investing in agriculture for a better future*. Rome: Food and Agriculture Organization of the United Nations. Retrieved from https://doi.org/10.18356/3e6e633a-en

Food and Agricultural Organization. (2013). *FAO Policy on Gender Equality. Attaining Food Security Goals in Agriculture and Rural Development*. Rome: Food and Agriculture Organization of the United Nations. Retrieved from http://www.fao.org/3/a-i3205e.pdf

Food and Agricultural Organization. (2016). *Food Security Indicators*. Rome: Food and Agriculture Organization of the United Nations. Retrieved from http://www.fao.org/economic/ess/ess-fs/ess-fadata/en/#.WKOLFhIrLR0

Food and Agricultural Organization. (2018). *The linkages between migration, agriculture, food security and rural development*. Technical report by the Food and Agriculture Organization of the United Nations, the International Fund for Agricultural Development, the International Organization for Migration and the World Food Programme. Retrieved from http://www.fao.org/3/CA0922EN/CA0922EN.pdf

Food and Agriculture Organisation – FAO. (1997). *Gender: the key to sustainability and food security*. SD Dimensions, May 1997. Retrieved from www.fao.org/sd

Food and Agriculture Organisation - FAO. (2018). *Sustainable Food Systems: Concepts and Framework*. Retrieved from http://www.fao.org/3/ca2079en/CA2079EN.pdf

Food and Agriculture Organisation. (2017). *Leaving no one behind: Achieving gender equality for food security, nutrition and sustainable agriculture*. FAO Conference 40th Conference, Rome, Italy.

Food and Agriculture Organization of the United Nations - FAO. (2011). *The State of Food and Agriculture 2010–11: Women in Agriculture: Closing the Gender Gap for Development.* Rome: FAO. Retrieved from www.fao.org/docrep/013/i2050e/i2050e00.htm

Food and Agriculture Organization. (2002). *Gender and access to land. FAO Land Tenure Studies 4.* Rome: Food and Agriculture Organization of the United Nations. Retrieved from ftp://ftp.fao.org/docrep/fao/005/y4308e/y4308e00.pdf

Food and Agriculture Organization. (2009). *Measuring Food Security Using Respondents Perception of Food Consumption Adequacy.* Rome, Italy: Food and Agriculture Organization of the United Nations.

Förster, J., Barkmann, J., Fricke, R., Hotes, S., Kleyer, M., Kobbe, S., ... Wittmer, H. (2015). Assessing ecosystem services for informing land-use decisions: A problem-oriented approach. *Ecology and Society, 20*(3), 31. doi:10.5751/ES-07804-200331

Frederic, L., Olaf, E., Akande, S. O., Titilola, S. O., Akpaokodje, G., & Ogundele, O. O. (2003). The Nigerian Rice Economy in a Competitive World. Constraints, Opportunities Strategic Choice. Imported Rice Retailing and Purchasing in Nigeria. A Report of a Survey funded by USAID and jointly Implemented by WARDA and NISER.

Freibauer, A., Mathijs, E., Brunori, G., Damianova, Z., Faroult, E., Gomis, J. G., & Treyer, S. (2011). *Sustainable food consumption and production in a resource-constrained world. European Commission - Standing Committee on Agricultural Research (SCAR).* The Third SCAR Foresight Exercise. doi:10.2777/49719

Frenken, K. (2009). *Irrigation in The Middle East Region in Figures AQUASTAT Survey – 2008.* FAO Water Reports 34. Rome: FAO.

Friis, C., & Reenberg, A. (2010). *Land grab in Africa: Emerging land system drivers in a teleconnected world.* GLP Report No.1. GLP-IPO.

Fuhrich, M., & Goderbauer, E. (2011). Urban restructuring: Making 'more' from 'less. In B. Müller (Ed.), *Ur-ban regional resilience: How do cities and regions deal with change* (pp. 49–58). Berlin: Springer.

Garnett, T. (2014). Three perspectives on sustainable food security: Efficiency, demand restraint, food system transformation. What role for life cycle assessment? *Journal of Cleaner Production, 73*, 10–18. doi:10.1016/j.jclepro.2013.07.045

Garvin, E. C., Cannuscio, C. C., & Branas, C. C. (2013). Greening vacant lots to reduce violent crime: A randomised controlled trial. *Injury Prevention, 19*(3), 198–203. doi:10.1136/injuryprev-2012-040439 PMID:22871378

Gashaw, T. A. (2015). Women's Land Use Right Policy and Household Food Security in Ethiopia. *International Journal of African and Asian Studies, 2*, 2409–6938.

Geary, K. (2012). *Our Land Our Lives, Time out on the Global Land Rush.* Oxford, UK: OXFAM GB for Oxfam International.

Gebre, G. G. (2012). Determinants of food insecurity among households in Addis Ababa City, Ethiopia. *Interdisciplinary Description of Complex Systems, 10*(2), 159–173. doi:10.7906/indecs.10.2.9

Ge, L., Anten, N. P. R., van Dixhoorn, I. D. E., Feindt, P. H., Kramer, K., Leemans, R., ... Sukkel, W. (2016). Why we need resilience thinking to meet societal challenges in biobased production systems. *Current Opinion in Environmental Sustainability, 23*, 17–27. doi:10.1016/j.cosust.2016.11.009

German, L., Schoneveld, G., & Mwangi, E. (2011). *Contemporary Processes of Large Scale Land Acquisition by Investors; Case Studies from Sub-Saharan Africa.* Occasional Paper 68, p. 56. CIFOR, Bagor, Indonesia.

Gillett, R., & Lightfoot, C. (2002). *The Contribution of Fisheries to the Economies of Pacific Island Countries*. A report prepared for the Asian Development Bank, the Forum Fisheries Agency, and the World Bank, Asian Development Bank.

Gladek, E., Fraser, M., Roemers, G., Sabag, M. O., Hirsch, P., & Kennedy, E. (2016). *The global food system: An analysis*. Retrieved from https://www.metabolic.nl/publications/global-food-system-analysis

Global Panel on Agriculture and Food Systems for Nutrition. (2014). *How can Agriculture and food system policies improve nutrition?* Technical brief, London, UK: Global Panel on Agriculture and Food Systems for Nutrition. Retrieved from http://www.glopan.org/sites/default/files/Global%20Panel%20Technical%20Brief%20Final.pdf

Global Panel on Agriculture and Food Systems for Nutrition. (2016). Food systems and diets: Facing the challenges of the 21st century. London, UK: Author.

Gobena, J. M. (2010). *Effects of Large-scale Land Acquisition in Rural Ethiopia; The Case of Bako-Tibe Woreda, Uppsala* (Master's dissertation). Swedish University of Agricultural Science.

Gough, M. Z., & Accordino, J. (2013). Public Gardens as Sustainable Community Development Partners: Motivations, Perceived Benefits, and Challenges. *Urban Affairs Review*, *49*(6), 851–887. doi:10.1177/1078087413477634

Grabar, H. (2013). *Could Earth's Population Peak in 2050?* Retrieved from https://www.citylab.com/life/2013/04/could-earths-population-peak-2050/5208/

GRAIN. (2008). *Seized! The 2008 land grab for food and financial security*. GRAIN Briefing. Available at: https://www.grain.org/article/93-seized-the-2008-landgrab-for-food-and-financial-security

GRAIN. (2015). *Dominion Farm's land grab in Nigeria*. Available at: https://www.grain.org/article/entries/5126-dominionfarm-s-land-grab-in-nigeria

Green, M. M. (2003). *Igbo village Affair* (2nd ed.). London: Frank Cass.

Greenstein, R., & Sungu-Eryilmaz, Y. (2004). *Recycling the City: The Use and Reuse of Urban Land*. Cambridge, MA: Lincoln Inst of Land Policy.

Groverman, V., & van der Wees, C. (2016). How to Green Food Systems in a Gender Smart Way: A matter of Insight and Smart Interventions. Academic Press.

Guan, D., Li, H., Inohae, T., Su, W., Nagaie, T., & Hokao, K. (2011). Modeling urban land use change by the integration of cellular automaton and Markov model. *Ecological Modelling*, *222*(20-22), 3761–3772. doi:10.1016/j.ecolmodel.2011.09.009

Guerena, A., & Burgos, S. (2014). *Smallholder at risk, monoculture expansion, land, food and livelihoods in Latin America*. Oxford, UK: Oxfam International.

Gujarati, D. N. (2005). *Basic Econometrics*. Tat McGraw-Hill Publishing Company Ltd.

Hagen-Zanker, J. (2015). *Effects of remittances and migration on migrant sending countries, communities and households*. Retrieved from https://assets.publishing.service.gov.uk/media/57a08999ed915d3cfd000326/Effects_of_remittances_and_migration_56.pdf

Hagen-Zanker, J., & Azzarri, C. (2010). Are internal migrants in Albania leaving for the better? *Eastern European Economics*, *48*(6), 57–84. doi:10.2753/EEE0012-8775480604

Hagman, J., Chuma, E., Murwira, K., & Connolly, M. (2000). Learning Together through Participatory Extension. *African Journal of Marketing Management*, *3*(10), 253–260.

Haile, H. K., Alemu, Z. G., & Kuhlande, G. (2005). *Causes of household food insecurity in Koredegaga pleasant association, Oromiya zone, Ethiopia*. Bloemfontein: University of Free State.

Hall, P. (1998, May 28) Vacant looks. *Guardian*.

Hall, P. G. (1998). *Cities on Civilisation*. London: Wiedenfeld and Nicolson.

Halmy, M., Gessler, P., Hicke, J., & Salem, B. (2015). Land use/land cover change detection and prediction in the north-western coastal desert of Egypt using Markov-CA. *Applied Geography (Sevenoaks, England)*, *63*, 101–112. doi:10.1016/j.apgeog.2015.06.015

Halseth, G., & Ryser, L. (2006). Trends in Service Delivery: Examples from rural and small town Canada, 1998 to 2005. *Journal of Rural and Community Development*, *14*, 69–90.

Hamilton, A. J., Burry, K., Mok, H. F., Barker, F., Grove, J. R., & Williamson, V. G. (2013). Give peas a chance? Urban agriculture in developing countries. A review. *Agronomy for Sustainable Development*, *34*(1), 45–73. doi:10.100713593-013-0155-8

Hammelman, C., & Hayes-Conroy, A. (2015). Understanding cultural acceptability in urban food policy. *Journal of Planning Literature*, *30*(1), 37–48. doi:10.1177/0885412214555433

Hanemann, W. M., & Kanninen, B. (1998). *The Statistical Analysis of Discrete-Response Data*. Working Paper No.798, Department of Agricultural and Resource Economics and Policy, University of California.

Hannah, H., & Jost, C. (2011). Public Health Participatory Epidemiology. *Introductory Training Module, manual for trainees*. Available at: https://cgspace.cgiar.org/bitstream/handle/10568/24715/OneHealthManual.pdf?sequence=1

Haralambous, S., Liversage, H., & Romano, M. (2009). *The Growing Demand for Land – Risks and Opportunities for Smallholder Farmers*. Rome: IFAD. Available at: www.ifad.org/events/gc/32/roundtables/2.pdf

Harrington, J., Lutomski, J., Molcho, M., & Perry, I. J. (2009). Food poverty and dietary quality: Is there a relationship? *Journal of Epidemiology and Community Health*, *63*(Suppl II), A1–A36. doi:10.1136/jech.2009.096701p

Hasan, S. S., Deng, X., Li, Z., & Chen, D. (2017). Projections of future land use in Bangladesh under the background of baseline, ecological protection and economic development. *Sustainability*, *9*(4), 505. doi:10.3390u9040505

Haug, R. (2018). *Food Security Indicators: How To Measure And Communicate Results*. Noragric Report No. 83 January 2018. Department of International Environment and Development Studies, Noragric Faculty of Landscape and Society, Norwegian University of Life Sciences.

Havinga, T., Waarden, F., & Casey, D. (Eds.). (2015). *The Changing Landscape of Food Governance—Public and Private Encounters*. Cheltenham, UK: Edward Elgar Publishing Limited.

Headey D., & Ecker O. (2013). Rethinking the measurement of food security: from first principles to best practice. *Food Security, 5*, 327–343. Doi:10.1007%2Fs12571-013-0253-0

Heckman, J. J., Ichimura, H., & Todd, P. (1998). Matching as an econometric evaluation estimator. *The Review of Economic Studies*, *65*(2), 261–294. doi:10.1111/1467-937X.00044

Heckman, J. J., & Smith, J. A. (1999). The pre-programme earnings dip and the determinants of participation in a social programme. Implications for simple programme evaluation strategies. *Economic Journal (London)*, *109*(457), 313–348. doi:10.1111/1468-0297.00451

Hedeker, D., Siddiqui, O., & Hu, F. B. (2000). Random-effects regression analysis of correlated grouped-time survival data. *Statistical Methods in Medical Research*, *9*(00), 161–179. doi:10.1177/096228020000900206 PMID:10946432

Hengl, T., Heuvelink, G. B. M., Kempen, B., Leenaars, J. G. B., Walsh, M. G., Shepherd, K. D., … Tondoh, J. E. (2015). *Mapping soil properties of Africa at 250 m resolution: random forests significantly improve current predictions*. Academic Press.

Henjum, S., Torheim, L. E., Thorne-Lyman, A. L., Chandyo, R., Fawzi, W. W., Shrestha, P. S., & Strand, T. A. (2015). Low dietary diversity and micronutrient adequacy among lactating women in a peri-urban area of Nepal. *Public Health Nutrition*, *18*(17), 3201–3210. doi:10.1017/S1368980015000671 PMID:25824344

Herzog, K. (2011, Sept 19). Milwaukee recognized for urban farms, aquaponics. *Journal Sentinel*.

Hettig, E., Lay, J., & Sipangule, K. (2016). Drivers of Households' Land-Use Decisions: A Critical Review of Micro-Level Studies in Tropical Regions. *Land (Basel)*, *5*(4), 32. doi:10.3390/land5040032

Heynen, N., Perkins, H. A., & Roy, P. (2006). The political ecology of uneven urban green space. *Urban Affairs Review*, *42*(1), 3–25. doi:10.1177/1078087406290729

High Level Panel of Experts on Food Security and Nutrition - HLPE. (2014). Food losses and waste in the context of sustainable food systems. A report by the High Level Panel of Experts on Food Security and Nutrition of the Committee on World Food Security, Rome.

Hilhorst, T., & Zomers, A. (2011). *How Can Large Scale Transnational Land Acquisition Contribute to Inclusive and Sustainable Growth?* University of Utrecht.

HLPE (High Level Panel of Experts). (2012). Food security and climate change. A report by the High Level Panel of Experts on Food Security and Nutrition of the Committee on World Food Security.

HLPE. (2014). Food losses and waste in the context of sustainable food systems. A report by the High Level of Experts on Food Security and Nutrition of the Committee on World Food Security.

HLPE. (2017). Nutrition and food systems. A report by the High Level Panel of Experts on Food Security and Nutrition of the Committee on World Food Security, Rome.

HLPE. (2017). *Sustainable Forestry for Food Security and Nutrition*. A report by The High Level Panel of Experts. HLPE Report 11.

HLPE. (2018). *Multi-stakeholder partnerships to finance and improve food security and nutrition in the framework of the 2030 Agenda*. A report by the High-Level Panel of Experts on Food Security and Nutrition of the Committee on World Food Security. Retrieved from: http://www.fao.org/3/CA0156EN/CA0156en.pdf

Hoddinott, J., & Yohannes, Y. (2002). *Dietary diversity as a households' food security*. Washington, DC: Food and Nutrition Technical Assistance Project, Academy for Educational Development. Available on https://pdfs.semanticscholar.org/d2b3/ce8d0ad79364f0b5583f2c6e8b77e6575297.pdf

Hodgson, K., Caton Campbell, M., & Bailkey, M. (2011). *Urban agriculture: growing healthy, sustainable places*. Washington, DC: American Planning Association.

Hoffman, M., Westermann, J. R., Kowarik, I., & van der Meer, E. (2012). Perceptions of parks and urban derelict land by landscape planners and residents. *Urban Forestry & Urban Greening*, *11*(3), 303–312. doi:10.1016/j.ufug.2012.04.001

Horemans, B. & Kébé, M. (2006). Enhancing The Economic Contribution Of Fisheries To West And Central African Nations. *IIFET 2006 Portsmouth Proceedings*.

Hospes, O., & Brons, A. (2016). Food system governance: A systematic literature review. In Food Systems Governance: Challenges for Justice, Equality and Human Rights (pp. 13-42). Abingdon-on-Thames: Routledge.

Houet, T., Loveland, T., Hubert-Moy, L., Gaucherel, C., Napton, D., Barnes, C., & Sayler, K. (2010). Exploring subtle land use and land cover changes: A framework for future landscape studies. *Landscape Ecology*, *25*(2), 249–266. doi:10.100710980-009-9362-8

Hough, M., & Barrett, S. (1987). *People and City Landscapes: a study of people and open space in metropolitan areas of Ontario*. Toronto: Conservation Council of Ontario.

HUD. (2015). *HUD Aggregated USPS Administrative Data on Address Vacancies.* U.S. Department of Housing and Urban Development. Retrieved from http://www.huduser.org/portal/datasets/usps.html

Hueston, W., & McLeod, A. (2012). Overview of the global food system: changes over time/space and lessons for future food safety. In *Institute of Medicine (US). Improving Food Safety Through a One Health Approach: Workshop Summary.* Washington, DC: National Academies Press (US). Retrieved from https://www.ncbi.nlm.nih.gov/books/NBK114491/

Huggins, C. (2012). *Consolidating Land, Consolidating Control: What Future for Smallholder Farming in Rwanda's Green Revolution.* Carleton University.

Hu, H., Liu, W., & Cao, M. (2008). Impact of land use and land cover changes on ecosystem services in Menglun, Xishuangbanna, Southwest China. *Environmental Monitoring and Assessment*, *146*(1-3), 147–156. doi:10.100710661-007-0067-7 PMID:18157650

Hui, S. C. M. (2011). *Green roof urban farming for buildings in high-density urban cities.* Invited paper for the Hainan China World Green Roof Conference 2011, Hainan, China.

Hutchinson, M. F. (2006). *Anusplin version 4.36 user guide*. Canberra, Australia: Centre for Resource and Environmental Studies.

Hu, Y., Zheng, Y., & Zheng, X. (2013). Simulation of land-use scenarios for Beijing using CLUE-S and Markov composite models. *Chinese Geographical Science*, *23*(1), 92–100. doi:10.100711769-013-0594-9

Iacono, M., Levinson, D., El-Geneidy, A., & Wasfi, R. (2015). A Markov chain model of land use change. *TeMA Journal of Land Use, Mobility and Environment*, *8*, 263–276.

Ibitoye, O. O., Akinsorotan, A. O., Meludu, N. T., & Ibitoye, B. O. (2011). Factors Affecting Oil Palm Production in Ondo State of Nigeria. *Journal of Agriculture and Social Research*, 2.

IFAD (International Fund for Agricultural Development). (2013). *Gender and Rural Development Brief Southeast Asia.* Retrieved from www.ifad.org/gender/policy/ gender_e.pdf

IFAD. (2007). *Sending money home: Worldwide remittance flows to developing countries. International Fund for Agricultural Development.* Rome, Italy: IFAD.

IFAD. (2009). *Report on migration, remittances and rural development.* International Fund for Agricultural Development. Rome, Italy: IFAD. Retrieved from https://www.un.org/en/development/desa/population/events/pdf/8/P14_IFAD.pdf

IISD (International Institute for Sustainable Development). (2012). *Sustainable Development Timeline.* Retrieved from https://www.iisd.org/sites/default/files/publications/sd_timeline_2012.pdf

Ijere, M. O. (2000). *Microfinance and Economic Development.* Ibadan: University of Ibadan Press.

Ijewere, A. A., & Odia, E. O. (2012). Cultural Influences on Product Choice of the Nigerian Consumer. *Indian J. Econ. Bus.*, *11*, 139–155.

Ileoje, N. P. (2001). *A New Geography of Nigeria* (New Revised Edition). Ibadan, Nigeria: Longman Publishers.

Ingram, S. J. I. (2011). *From Food Production to Food Security: Developing interdisciplinary, regional-level research*. Wageningen University. Retrieved from edepot.wur.nl/176450

International Dietary Data Expansion Project. (2015). *Dietary diversity*. Available on https://inddex.nutrition.tufts.edu/data4diets/data-source/dietary-diversity

International Food Policy Research Institute. (2005). *Land rights and common properties*. Washington, DC: International Food Policy Research Institute.

International Fund for Agricultural Development. (2010). *Annual Report 2010*. Retrieved from https://www.ifad.org/documents/38714170/39309757/annual+report+2010+e.pdf/da09a822-4ed9-4106-a646-6947fbfa5be7

International Labour Organization - ILO. (2016). *Women at work: Trends 2016*. International Labour Office. Retrieved from http://www.ilo.org/wcmsp5 /groups/public/—-dgreports/—-dcomm/—-publ/documents/publication/wcms_457317.pdf

International Task Force on Commodity Risk Management in Developing Countries (ITF). (1999). Dealing with commodity price volatility in developing countries: A proposal for a market based approach. Discussion paper for the roundtable on commodity risk management in developing countries, World Bank.

IPCC. (2015). *Agriculture, Forestry and Land Use. Fifth Assessment Report*. Retrieved from: https://www.ipcc.ch/site/assets/uploads/2018/02/ipcc_wg3_ar5_chapter11.pdf

IPES-Food. (2015). *The new science of sustainable food systems: Overcoming barriers to food systems reform*. Retrieved from http://www.ipes-food.org/images/Reports/IPES_report01_1505_web_br_pages.pdf

IPES-Food. (2017). *Too big to feed: Exploring the impacts of mega-mergers, concentration, concentration of power in the agri-food sector*. Retrieved from: http://www.ipes-food.org/_img/upload/files/Concentration_FullReport.pdf

Islam, K., Jashimuddin, M., Nath, B., & Nath, T. (2018). Land use classification and change detection by using multi-temporal remotely sensed imagery: The case of Chunati wildlife sanctuary, Bangladesh. *The Egyptian Journal of Remote Sensing and Space Sciences*, *21*(1), 37–47. doi:10.1016/j.ejrs.2016.12.005

Ismail, M. (2018). Food security is key to SDGs success. *The ASEAN Post*. Retrieved from https://theaseanpost.com/article/food-security-key-sdgs-success

Ja'afar-Furo, M. R., Bello, K., & Sulaiman, A. (2011). Assessment of the prospects of value addition among small-scale rural enterprises in Nigeria: Evidence from North-eastern Adamawa State. *Journal of Development and Agricultural Economics*, *3*(3), 144–149.

Jabo, M. S. M., Mansor, I. S., Nasir, S. M., Mahir, A. A., & Maikasuwa, A. M. (2014). Examining the Incidence, Depth and Severity of Food Insecurity among rural Households in Nigeria. *Journal of Economics and Sustainable Development*, *5*(26).

Jacobs, S. (2000). Zimbabwe: Why Land Reform is a Gender Issue. *Sociological Research Online*, *5*(2), 301–314. doi:10.5153ro.485

Jakle, J. A., & Wilson, D. (1992). *Derelict Landscapes: The Wasting of America's Built Environment*. Lanham, MD: Rowman & Littlefield Publishers.

Jarvis, A., Reuter, H. I., Nelson, A., & Guevara, E. (2008). *Hole-filled seamless SRTM data V4. International Centre for Tropical Agriculture*. CIAT.

Jasim, M. S., Norsida, M., Ahmad, H. L., Majeed, H. S., Salim, H., Nolila, M. N., & Bassim, H. K. (2016). A Review: Training Requirement of Agriculture Extension Officers in Iraq. *Asian Journal of Applied Sciences*, *9*(2), 34–40. doi:10.3923/ajaps.2016.34.40

Jatto, N. A., & Okebiorun, E. O. (2017). Value Addition in Cassava Processing: Evidence from Women in Ilesa West Local Government Area of Osun State. *Agriculture and Food Sciences Research*, *4*(1), 30–36. doi:10.20448/journal.512.2017.41.30.36

Jejeebhoy, S. J., & Sathar, Z. A. (2001). Women's Autonomy in India and Pakistan: The Influence of Religion and Region. *Population and Development Review*, *27*(4), 687–712. doi:10.1111/j.1728-4457.2001.00687.x

Jiang, Q., Ma, E., Zhan, J., & Shi, N. (2015). Seasonal and interannual variation in energy balance in the semiarid grassland area of China. *Advances in Meteorology*, *2015*, 2015. doi:10.1155/2015/120620

Jibowo, A. A., & Ajayi, A. O. (2011). Agricultural Extension in Nigeria: History of Agricultural Extension in Nigeria. Agricultural Extension Society of Nigeria (AESON).

Jibril, M. S., & Liman, H. M. (2014). Land use and land cover change detection in Ilorin, Nigeria using satellite remote sensing. *Journal of Natural Sciences Research*, *4*, 123–129.

Jonckheere, S., Musirimu, E., & Liversaage, H. (2013, April). *Legal empowerment to secure women's land rights in Burundi*. [Paper presentation]. The Annual World Bank Conference on Land and Poverty. The World Bank, Washington DC, United States.

Jørgen, C. (2008). The determinants of migrant remittances. *Oxford Review of Economic Policy*, *24*(3), 581–598. doi:10.1093/oxrep/grn022

Jost, C.C. Kristjanson, P., & Ferdous, N. (2015). *Lessons in Theory of Change: Gender and Inclusion*. CCSL Learning Brief No. 14.

Juma, C., & Ojwang, J. (1996). *In land we trust: Environment, private property and constitutional change*. Nairobi: Initiative Publishers.

Jurgilevich, A., Birge, T., Kentala-Lehtonen, J., Korhonen-Kurki, K., Pietikainen, J., Saikku, L., & Schösler, H. (2016). Transition towards circular economy in the food system. *Sustainability*, *8*(1), 1–14. doi:10.3390u8010069

Kabeer, N. (2012). *Women's Economic Empowerment and Inclusive Growth: Labour Markets and Enterprise Development*. SIG Working Paper 2012/1. London, UK: Department for International Development (DFID) and Ottawa, Canada: International Development Research Centre (IDRC) – Supporting Inclusive Growth (SIG) program.

Kabunga, N.S., Dubois, T., & Qaim, M. (2014). Impact of tissue culture banana technology on farm household income and food security in Kenya. *Journal of Food policy, 45*, 25-34.

Kachika, T. (2010). *Land Grabbing in Africa: A review of the impacts and the possible policy response.* Available at: http://www.oxfam.org.uk/resources/learning/landrights/downloads/land_grabbing_in_africa_impacts_&_policy_responses.pdf

Kachingwe, N. (2012). *From Under Their Feet: A Think Piece on the Gender Dimensions of Land Grabs in Africa.* Johannesburg, South Africa: Action Aid International.

Kaloustian, F. C., Jaffrey, L., & Newman, I. (2009). *Land Grabbing for Food and Fuel Outsourcing: a rising Threat to the Right to Food.* International Human Right Clinic and University of San Francisco School of Law.

Karamba, W. R., Quiñones, E. J., & Winters, P. (2011). Migration and food consumption patterns in Ghana. *Food Policy*, *36*(1), 41–53. doi:10.1016/j.foodpol.2010.11.003

Katz, E., & Chamorro, J. S. (2002). *Gender, Land Rights and the Household Economy in Rural Nicaragua and Honduras. Paper prepared for USAID/Broadening Access and Strengthening Input Market Systems (BASIS) Collaborative Research Support Program*. CRSP.

Kawarazuka, N., & Béné, C. (2011). The potential role of small fish species in improving micronutrient deficiencies in developing countries: Building evidence. *Public Health Nutrition*, *14*(11), 1927–1938. doi:10.1017/S1368980011000814 PMID:21729489

Kay, S. (2016). *Land grabbing and land concentration in Europe*. A research brief. Transnational Institute for HOTL. Available at: www.tni.org/files/publication-down loads/landgrabbingeurope_a5-2.pdf

Kearney, J. (2010). Food consumption trends and drivers. *Philosophical Transactions of the Royal Society of London. Series B, Biological Sciences*, *365*(1554), 2793–2807. doi:10.1098/rstb.2010.0149 PMID:20713385

Kehinde, A. L., & Aboaba, K. O. (2016). Analysis of value addition in the processing of cassava tubers to "garri" among cottage level processors in southwestern Nigeria. Invited poster presented at the 5th International Conference of the African Association of Agricultural Economists, Addis Ababa, Ethiopia.

Kennedy, G. L. (2009). *Evaluation of dietary diversity scores for assessment of micronutrient intake and food security in developing countries*. Wageningen University.

Keshtkar, H., & Voigt, W. (2016). A spatiotemporal analysis of landscape change using an integrated Markov chain and cellular autoamta models. *Modeling Earth Systems and Environment*, *§§§*, 2.

Ke, X., Ma, E., & Yuan, Y. (2014). Scenario simulation of the influence of land use change on the regional temperature in a rapidly urbanizing region: A case study in southern-Jiangsu, China. *Advances in Meteorology*, *2014*, 2014. doi:10.1155/2014/159724

Kidane, H., Alemu, Z. G., & Kundhlande, G. (2005). Causes of household food insecurity in Koredegaga peasant association, Oromiya zone, Ethiopia. *Agrekon*, *44*(4), 543–560.

Kim, G., Miller, P. A., & Nowak, D. J. (2018). Urban vacant land typology: A tool for managing urban vacant land. *Sustainable Cities and Society*, *36*, 144–156. doi:10.1016/j.scs.2017.09.014

Kim, I., Le, Q., Park, S., Tenhumen, J., & Koellner, T. (2014). Driving Forces in Archetypical Land-Use Changes in a Mountainous Watershed in East Asia. *Land (Basel)*, *3*(3), 957–980. doi:10.3390/land3030957

King, C. (2016). *Food insecurity and housing instability in vulnerable families. Review of Economics of the Household, published online ahead of print*. Lincoln, NE: Nutrition and Health Sciences - Faculty Publications.

Klarin, T. (2018). The Concept of Sustainable Development: From its Beginning to the Contemporary Issues. *Zagreb International Review of Economics & Business*, *21*(1), 67–94. doi:10.2478/zireb-2018-0005

Kobayashi, M., Msangi, S., Batka, M., Vannuccini, S., Dey, M. M., & Anderson, J. L. (2015). Fish to 2030: The Role and Opportunity for Aquaculture. *Aquaculture Economics & Management*, *19*(3), 282–300. doi:10.1080/13657305.2015.994240

Kohler, N. (1999). The relevance of Green Building Challenge: An observer's perspective. *Building Research and Information*, *27*(4/5), 309–320. doi:10.1080/096132199369426

Kohler, N., & Hassler, U. (2002). The building stock as a research project. *Building Research and Information*, *30*(4), 226–236. doi:10.1080/09613210110102238

Kosciejew, M. (2011). *The New Scramble for Africa: Land-grabbing Exploitation Presented as Economic Development*. Retrieved from http://www.caaglop.com/robbenisland-blog/economy/the-new-scramble-for-africa-land-grabbing-exploitation-presented-as-economic-development/

Kremer, P., Hamstead, Z. A., & McPhearson, T. (2013). A social-ecological assessment of vacant lots in New York City. *Landscape and Urban Planning*, *120*, 218–233. doi:10.1016/j.landurbplan.2013.05.003

Kreuter, U. P., Harris, H. G., Matlock, M. D., & Lacey, R. E. (2001). Change in ecosystem service values in the San Antonio area, Texas. *Ecological Economics*, *39*(3), 333–346. doi:10.1016/S0921-8009(01)00250-6

Kumar, S., Radhakrishnan, N., & Mathew, S. (2014). Land use change modeling using a Markov model and remote sensing. *Geomatics, Natural Hazards & Risk*, *5*(2), 145–156. doi:10.1080/19475705.2013.795502

Kuokkanen, A., Mikkilä, M., Kuisma, M., Kahiluoto, H., & Linnanen, L. (2017). The need for policy to address the food system lock-in: A case study of the Finnish context. *Journal of Cleaner Production*, *140*, 933–944. doi:10.1016/j.jclepro.2016.06.171

Kurekova, L. (2011). *Theories of migration: Conceptual review and empirical testing in the context of the EU East-West flows*. Academic Press.

Kuuire, V. Z., Mkandawire, P., Arku, G., & Luginaah, I. (2013). Abandoning farms in search of food. Food remittance and household food security in Ghana. *African Geographical Review*, *32*(2), 125–139. doi:10.1080/19376812.2013.791630

Kuwornu, J. K. M., Suleyman, D. M., & Amegashie, D. P. K. (2013). Analysis of Food Security Status of Farming Households in the Forest Belt of the Central Region of Ghana. *Russian Journal of Agricultural and Socio-Economic Sciences*, *1*(13), 17. doi:10.18551/rjoas.2013-01.04

Lambin, E., & Geist, H. (Eds.). (2006). *Land-use and land-cover change – local processes and global impacts*. Berlin: Springer-Verlag. doi:10.1007/3-540-32202-7

Lambin, E., Geist, H., & Lepers, E. (2003). Dynamics of land-use and land-cover change in tropical regions. *Annual Review of Environment and Resources*, *28*(1), 205–241. doi:10.1146/annurev.energy.28.050302.105459

Lambrecht, I., Vanlauwe, B., & Maertens, M. (2014). *What is the sense of gender targeting in agricultural extension programs? Evidence from eastern DR Congo*. Leuven, Belgium: University of Leuven, Department of Earth and Environmental Sciences, Division of Bioeconomics (Bioeconomics Working Paper 2014/4).

Landesa (2012). *Women's secure rights to land: Benefits, barriers and best practices. Center for Women's Land Rights Issue Brief.* Seattle, WA. Retrieved from https://cdn.landesa.org/wp-content/uploads/Landesa-Women-and-Land-Issue-Brief.pdf

Lang, T. (2009). *Food security and sustainability: The perfect fit*. Retrieved from http://www.sd-commission.org.uk/data/files/publications/SDCFoodSecurityPositionPaper.pdf

Lang, T., Barling, D., & Caraher, M. (2009). *Food policy: integrating health, environment and society*. Oxford: Oxford University Press. doi:10.1093/acprof:oso/9780198567882.001.0001

Langworthy, M., Frankenberger, T., Bonnard, P., Onwwubah, C., & Green, H. (2003). Food Access Indicator Review: A Food and Nutrition Technical Assistance Project. Academic Press.

Lawal, A. S., & Okeowo, T. A. (2014). Effects of rural urban migration on labour supply in cocoa production in ondo east Local Government Area of Ondo State. *Academic Journal of Interdisciplinary Studies*, *3*(7), 25–29.

Lawson, L. J. (2005). *City bountiful: a century of community gardening*. Berkeley: University of California Press.

Leggett, J. A., & Carter, N. T. (2012). *Rio+20: The United Nations Conference on Sustainable Development*. Congressional Research Service.

Legislation Information System, Official Gazette Date: 13.08.2016 Number of Official Gazette: 29800, No. 4/1 On Communication for Regulation of Fishing Fishing (Communication No: 2016/35). (n.d.). Retrieved from https://www.mevzuat.gov.tr/Metin.Aspx?MevzuatKod=9.5.22750&MevzuatIliski=0&sourceXmlSearch=ticari%20ama%C3%A7l%C4%B1

Lele, U., Masters, W. A., Kinabo, I., Meenakshi, J., Ramaswami, B., Tagwireyi, J., . . . Goswami, S. (2016). Measuring Food and Nutrition Security: An independent technical assessment and user's guide for existing indicators. http://www.fao.org/fileadmin/user_upload/fsin/docs/1_FSIN-TWG_UsersGuide_12June2016.compressed.pdf.

Lever, J., Sonnino, R., & Cheetham, F. (2019). Reconfiguring local food governance in an age of austerity: Towards a place-based approach? *Journal of Rural Studies*, *69*, 97–105. doi:10.1016/j.jrurstud.2019.04.009

Levidow, L. (2015). European transitions towards a corporate-environmental food regime: Agroecological incorporation or contestation? *Journal of Rural Studies*, *40*, 76–89. doi:10.1016/j.jrurstud.2015.06.001

Levien, M. (2011). Special Economic Zones and Accumulation by Dispossession in India. *Journal of Agrarian Change*, *11*(4), 454–483. doi:10.1111/j.1471-0366.2011.00329.x

Li, F., Ye, Y. P., Song, B. W., Wang, R. S., & Tao, Y. (2014). Assessing the changes in land use and ecosystem services in Changzhou municipality, People's Republic of China, 1991–2006. *Ecological Indicators*, *42*, 95–103. doi:10.1016/j.ecolind.2013.11.012

Li, R. Q., Dong, M., Cui, J. Y., Zhang, L. L., Cui, Q. G., & He, W. M. (2007). Quantification of the impact of land-use changes on ecosystem services: A case study in Pingbian County, China. *Environmental Monitoring and Assessment*, *128*(1-3), 503–510. doi:10.100710661-006-9344-0 PMID:17278007

Liu, P., Karoma, S., Arias, P., & Hallam, D. (2013). *Trends and Impacts on Foreign Investments in developing country agriculture. Evidence from case studies*. Rome, Italy: FAO.

Liu, Y., Njai, R. S., Greenlund, K. J., Chapman, D. P., & Croft, J. B. (2014). Relationships between Housingand food insecurity, frequent mental distress, and insufficient sleep amongadults in12 US states, 2009. *Preventing Chronic Disease*, *11*, 130334. doi:10.5888/pcd11.130334 PMID:24625361

Li, Z., Deng, X., Wu, F., & Hasan, S. S. (2015). Scenario analysis for water resources in response to land use change in the middle and upper reaches of the Heihe River Basin. *Sustainability*, *7*(3), 3086–3108. doi:10.3390u7033086

Long, H. L., Liu, Y. Q., Hou, X. G., Li, T. T., & Li, Y. R. (2014). Effects of land use transitions due to rapid urbanization on ecosystem services: Implications for urban planning in the new developing area of China. *Habitat International*, *44*, 536–544. doi:10.1016/j.habitatint.2014.10.011

Longtau, S. R. (2003). African Agriculture: Nigeria Case Study Report on Rice Production. Prepared by Eco-System Development Organsiation (EDO), Jos, Nigeria for Overseas Development Institute (ODI).

Lonza, L., Hass, H., Maas, H., Reid, A., & Rose, K. D. (2011). *EU Renewable Energy Targets in 2020: Analysis of Scenarios for Transport*. Joint Scientific and Technical Report. Retrieved from http://ies.jrc.ec.europa.eu/uploads/jec/JECBiofuels%20Report_2011_PRINT.pdf

Lorenzo, C., Vermeulen, S., Leonard, R., & Keeley, J. (2009). *Land grab or development opportunity? Agricultural Investment and International land deals in Africa*. London: IIED/FAO/IFAD.

Lovell, S. T. (2010). Multifunctional urban agriculture for sustainable land use planning in the United States. *Sustainability*, *2*(8), 2499–2522. doi:10.3390u2082499

Luginaah, I., Weis, T., Galaa, S., Nkrumah, K. M., Bezner-Kerr, R., & Bagah, D. (2009). Enviroment, migration and food security in the Upper West Region of Ghana. In *Environment and health in Sub-Saharan African: Managing an emerging crises* (pp. 25–38). Springer. doi:10.1007/978-1-4020-9382-1_2

Lundius, R. V. (2009). *Smallholder Agriculture and Food Security in the 21st Century*. IFAD.

Ma, C., Zhang, G. Y., Zhang, X. C., Zhao, Y. J., & Li, H. Y. (2012). Application of Markov model in wetland change dynamics in Tianjin Coastal Area, China. *Procedia Environmental Sciences*, *13*, 252–262. doi:10.1016/j.proenv.2012.01.024

Magurran, A. E., & McGill, B. J. (2011). *Biological diversity: frontiers in measurement and assessment*. Oxford: Oxford University Press.

Maharjan, A., Bauer, S., & Knerr, B. (2013). International migration, remittances and subsistence farming: Evidence from Nepal. *International Migration (Geneva, Switzerland)*, *5*(1), 249–263. doi:10.1111/j.1468-2435.2012.00767.x

Mahoi, I. (2015). *An Assessment on Gender Equity and Land Ownership, and it's Impact on Food Security in Sierra Leone*. Universita' Cattolica Del Sacro Cuore, Milano. Retrieved from https://tesionline.unicatt.it/bitstream/10280/12223/1/tesiphd_completa_mahoi.pdf

Maji, A. T., Ukwungwu, M. N., Nahemiah, D., & Bakare, S. O. (2017). Rice: History, Research & Development in Nigeria. In D. Nahemiah, I. Nkama, A.T. Maji, M.N. Ukwungu (Eds.), Rice in Nigeria: Traditional Recipes and Research Needs (pp. 1-10). RONAB GRAPHIX PRINTS.

Makutsa, P. (2010). *Land Grab in Kenya: Implications for Smallholder Farmers*. Nairobi, Kenya: Eastern Africa Farmers Federation.

Mallach, A. (2012). *Laying the Groundwork for Change: Demolition, Urban Strategy, and Policy Reform*. Washington, DC: Brookings Institution.

Mango, N., Zamasiya, B., Makate, C., Nyikahadzoi, K., & Siziba, S. (2014). Factors influencing household food security among smallholder farmers in the Mudzi district of Zimbabwe. *Development Southern Africa*, *31*(4), 625–640. doi:10.1080/0376835X.2014.911694

Maphosa, F. (2007). Remittances and development: The impact of migration to South Africa on rural livelihood in Southern Zimbabwe. *Development Southern Africa*, *24*(1), 123–136. doi:10.1080/03768350601165942

Marks, G. (1992). Structural Policy in the European Community. In A. Sbragia (Ed.), *Europolitics: Institutions and Policymaking in the "New" European Community*. Washington, DC: Brookings Institute.

Marsden, T. (2013). From post-productionism to reflexive governance: Contested transitions in securing more sustainable food futures. *Journal of Rural Studies*, *29*, 123–134. doi:10.1016/j.jrurstud.2011.10.001

Martin, M. (2005). *Africa's external constraints: What developed countries should do. Africa in the World Economy. The National, Regional and International Challenges*. Retrieved from w.w.w.fondad.org

Martini, R., & Lindberg, C. (2013). *Fishing for Tomorrow: Managing fisheries for sustainable development. Coherence for Development, Better Policies for Better Lives, (2)*. OECD.

Mason, K. O., & Smith, H. L. (2003). *Women's Empowerment and Social Context: Results from Five Asian Countries. Gender and Development Group*. Washington, DC: World Bank.

Mathey, J., & Rink, D. (2010). Urban wastelands—A chance for biodiversity in cities? In *Urban Biodiversity and Design* (pp. 406–424). Oxford, UK: Blackwell. doi:10.1002/9781444318654.ch21

Maxwell, S., & Frankenberger, T. (1992). *Household Food Security: Concepts, Indicators and Measurements: A Technical Review*. New York: UNICEF and IFAD.

Maziya, M., Mudhara, M., & Chitja, J. (2017). What factors determine household food security among smallholder farmers? Insights from Msinga, KwaZulu-Natal, South Africa. *Agrekon*, *56*(1), 40–52. doi:10.1080/03031853.2017.1283240

Mbah, E. N., Ezeano, C. I., & Agada, M. O. (2016). Effects of rural-urban youth migration on farm families in Benue state, Nigeria. *International Journal of Agricultural Research, Innovation and Technology*, *6*(1), 14–20. doi:10.3329/ijarit.v6i1.29207

McArt, S. H., Cook-Patton, S. C., & Thaler, J. S. (2012). Relationships between arthropod richness, evenness, and diversity are altered by complementarity among plant genotypes. *Oecologia*, *168*(4), 1013–1021. doi:10.100700442-011-2150-6 PMID:22002039

McElvenney, L. (2012). *Can Africa Feed the World? Land Acquisition as a Solution to Global Food Security* (Master Thesis). LUCSUS Lund University Centre for Sustainability Studies.

McGuire, V. C. (2007, Sept 2). Replacing neglect with peach trees. *The New York Times.*

McKinsey. (2014). *Overcoming obesity: An initial economic analysis*. Retrieved from: https://www.mckinsey.com/industries/healthcare-systems-and-services/our-insights/how-the-world-could-better-fight-obesity

McMillan, T. (2008, May 7). Urban farmers' crops go from vacant lot to market. *The New York Times.*

MEA. (2005). *Ecosystems and human well-being: Biodiversity synthesis*. Washington, DC: World Resources Institute, Island Press.

Meerza, S. I. A. (2010). Rural Urban Migration and its Consequences on Rural Children. In Mini The impact of Rural-Urban Migration and Rural Economy in Rural Village. Academic Press.

Meeske, M. (2018). *Rural-Urban migration, food consumption pattern and trend in the Local food system: A case study for Vietnam*. Wageningen University, Department of Social Sciences.

Melesse, M. B., Berg, M., Béné, C., Brauw, A., Brouwer, I. D., & Remans, R. (2017). *Improving Diets through Food Systems: Metrics and Methods for Analysis*. Unpublished Article.

Mendes, C. (2015). *Policy Coherence for Sustainable Development Toolkit Module 2: Food security*. Retrieved from https://www.oecd.org/gov/pcsd/Thematic%20module%20on%20Food%20Security_DRAFT.pdf

Me-Nsope, N. (2017). Gender Practice in Food Systems Innovations: Approaches, Lessons, and Challenges. Michigan State University.

Merian Research. (2009). *The Ventures of Land Grabbing. The Involvement of European Financial Companies in Large Scale Land Acquisition abroad*. European Union.

Metcalf, S. S., & Widener, M. J. (2011). Growing Buffalo's capacity for local food: A systems framework for sustainable agriculture. *Applied Geography (Sevenoaks, England)*, *31*(4), 1242–1251. doi:10.1016/j.apgeog.2011.01.008

Meybeck, M., Green, P., & Vorosmarty, C. J. (2001). A new typology for mountains and other relief classes: An application to global continental water resources and population distribution. *Mountain Research and Development*, *21*(1), 34–45. doi:10.1659/0276-4741(2001)021[0034:ANTFMA]2.0.CO;2

Mgbenka, R. N., & Mbah, E. N. (2016). A Review of Smallholder Farming in Nigeria: Need for Transformation. *International Journal of Agricultural Extension and Rural Development Studies*, *3*(2), 43–54.

Michael, J., Hiroyuki, T., Kwabena, G. B., & Oluyemisi, K. S. (2013). *Policy Options for Accelerated Growth and Competitiveness of the Domestic Rice Economy in Nigeria: Nigeria Strategy Support Program II*. Retrieved from www.ifpri.org

Miggiano, L. (2010). *Links between land tenure security and food security (Annex III). Contribution of the International Land Coalition to the updating of the High Level Task Force's (HLTF) Comprehensive Framework for Action (CFA).* Rome: International Land Coalition.

Miller, D. D., & Welch, R. M. (2013). Food system strategies for preventing micronutrient malnutrition. *Food and Agriculture Organization of the United Nations*. Available online at http://www.fao.org/fileadmin/templates/esa/Papers_and_documents/WP_13_06_Miller_Welch.pdf

Mini, S. E. (2001). *The impact of rural-urban migration on rural economy in rural village*. Retrieved from www. geofileonline.com

Miruka, M., Njuki, J., Starr, L., Kruger, E., & Hillenbrand, E. (2016). Addressing the multidimensional nature of gender dynamics in agriculture: Measuring gender relations and women's empowerment. In *Transforming Gender and Food Security in the Global South* (pp. 21-46). Routledge Studies in Food, Society and Environment. Retrieved from www.routledge.com/books/series/RSFSE/

Mirutse, D., Gebregiorgis, H., & Selam Ataklt, H. (2006). *Female-headed households and livelihood intervention in four selected Weredas in Tigray, Ethiopia. DCG report, vol.44.* Retrieved from https://www.utviklingsfondet.no/dcg/assets/documents/Report-44.pdf

Mishra, V., & Ray, R. (2006). *Dietary pattern, calorie intake and undernourishment: the Vietnamese experience.* Retrieved from https://www.researchgate.net/publication/237123306

Mitchell, R., & Hanstad, T. (2004). *Small home garden plots and sustainable livelihoods for the poor. Food and Agriculture Organization of the United Nations Livelihood Support Programme (LSP).* Rome: Food and Agriculture Organization of the United Nations.

Mittal, A. (2009). *The Great Land Grab: Rush for World's Farmland Threatens Food Security For the Poor*. Oakland: The Oakland Institute.

Moital, F., Almeida, M., & Pinto-Correia, T. (2012). "Proposta de diversificação de actividades nas explorações agrícolas. In PROVE (Ed.), *Promover, Vender, Participar e Diversificar. Relatório final,* [Promote, Sell, Participate and Diversify. Final report]. ICAAM / University of Évora and Monte-ACE. Retrieved from: http://www.prove.com.pt/www/wst/files/I167-05XDRAFTXXRELATOXXRIOX27JAN.PDF

Mok, H. F., Williamson, V. G., Grove, J. R., Burry, K., Barker, S. F., & Hamilton, A. J. (2014). Strawberry fields forever? Urban agriculture in developed countries: A review. *Agronomy for Sustainable Development, 34*(1), 21–43. doi:10.100713593-013-0156-7

Momodu, A.S., Akinbami, C.A.O., & Obisanya, J.F. (2011). Achieving food security and climate change mitigation through development in rural Nigeria: Gender perspective. *African Journal of Environmental Science and Technology, 5*(10), 834–854.

Momoh, P., & Kormawa, N. (2002). Need assessment study for market-driven agricultural technology transfer. *IITA*. Retrieved from www.https://books.google.com.ng/books?

Moser, C. (1998). *Unit 1: A conceptual framework for gender analysis and planning*. International Labour Organization. Retrieved from http://www.ilo.org/public/english/region/asro/mdtmanila/training/unit1/moserfw.htm

Moursi, M. M., Arimond, M., Dewey, K. G., Trèche, S., Ruel, M. T., & Delpeuch, F. (2008). Dietary diversity is a good predictor of the micronutrient density of the diet of 6-to 23-month-old children in Madagascar. *The Journal of Nutrition, 138*(12), 2448–2453. doi:10.3945/jn.108.093971 PMID:19022971

Mukasa, A., & Salami, A. (2016). A Gender Equality in Agriculture: What are really the Benefits for sub Saharan Africa? *Chief Economist Complex, AEB, 7*(3).

Mukudi, E. (2003). Nutrition status, education participation, and school achievement among Kenyan middle-school children. *Nutrition (Burbank, Los Angeles County, Calif.), 19*(7–8), 612–616.

Murisa, T. (2008). *Agrarian Reforms in Southern Africa: Contradictions of Neo-liberal Prescriptions*. Extrait du suite Foundation Gabriel Peri. Retrieved from http://www.gabrielperi.fr/

Mutisya, M., Ngware, M. W., Kabiru, C. W., & Kandala, N. (2016). The Effect of Education on Household Food Security in Two Informal Urban Settlements In Kenya: A Longitudinal Analysis. *Food Security, 8*(4), 743–756. doi:10.100712571-016-0589-3

Myers, D., & Wyatt, P. (2003). Rethinking urban capacity: identifying and appraising vacant buildings. In *Proceedings of the European Real Estate Society Conference*. ERES.

Myers, D., & Wyatt, P. (2004). Rethinking urban capacity: Identifying and appraising vacant buildings, *Building Research &amp. Information, 32*(4), 285–292. doi:10.1080/0961321042000221061

Nahuelhual, L., Carmona, A., Aguayo, M., & Echeverria, C. (2014). Land use and ecosystem services provision: A case study of recreation and ecotourism opportunities in southern Chile. *Landscape Ecology, 29*(2), 329–344. doi:10.100710980-013-9958-x

Najmuddin, O., Deng, X., & Siqi, J. (2017). Scenario analysis of land use change in Kabul River Basin–A river basin with rapid socio-economic changes in Afghanistan. *Physics and Chemistry of the Earth Parts A/B/C, 101*, 121–136. doi:10.1016/j.pce.2017.06.002

Nanyang, G. M. (2015). *Industrial Cassava Processing machine from China*. Retrieved from http://www.cassavamachine.cn/ProInfo_4.html

Napoli, M. (2011). Towards a Food Insecurity Multidimensional Index (FIMI). *Master*. In *Human Development And Food Security, (2010/2011)*. Roma: Roma Tre University.

National Bureau of Statistics. (2007). *Agricultural Survey Report 1994/95– 2005/06*. Abuja: National Bureau of Statistics.

National Commission on Urban Problems. (1968). *Land Use in 106 Large Cities*. Washington, DC: Three Land Research Studies.

National Institutes of Health. (2019). *Important Nutrients to Know: Proteins, Carbohydrates, and Fats*. Retrieved from https://www.nia.nih.gov/health/important-nutrients-know-proteins-carbohydrates-and-fats

National Land Use Database (2000) *Previously Developed Land (PDL): Vacant Sites and Derelict Sites, and Other Previously Developed Land and Buildings that May be Available for Redevelopment, Data Specification, NLUD – a partnership Rethinking urban capacity.* Author.

Naylor, R. L., Hardy, R. W., Bureau, D. P., Chie, A., Elliott, M., Farrell, A. P., ... Nichols, P. D. (2009, Sept. 8). Feeding Aquaculture in An Era of Finite Resources. *Proceedings of the National Academy of Sciences, 106*(36), 15103-15110. 10.1073/pnas.0905235106

NBS. (2016). *National population estimates*. National Bureau of Statistics, FCT.

NCRI. (2007). *Definition of Ofada rice qualities through varietal identification and testing*. NCRI.

NCSS Statistical Software NCSS. (2019). *Two-stage least squares*. Retrieved from https://www.ncss-wpengine.netdna-ssl.com/wp content/themes/ncss/pdf/Procedures/NCSS/Two Stage_Least_Squares.pdf

Németh, J., & Langhorst, J. (2014). Rethinking Urban Transformation: Temporary Uses for Vacant Land. *Cities (London, England)*, *40*, 143–150. doi:10.1016/j.cities.2013.04.007

Newbold, T., Hudson, L., Hill, S., Contu, S., Lysenko, I., Senior, R., ... Purvis, A. (2015). Global effects of land use on local terrestrial biodiversity. *Nature*, *520*(7545), 45–50. doi:10.1038/nature14324 PMID:25832402

Newman, G. D., Bowman, A. O. M., Jung Lee, R., & Kim, B. (2016). A current inventory of vacant urban land in America. *Journal of Urban Design*, *21*(3), 302–319. doi:10.1080/13574809.2016.1167589

NGCC. (2014). 30 meter global land cover dataset (GlobeLand30). http://www. globallandcover.com/GLC30Download/index.aspx, Accessed date: 18 February 2016.

Nguyen, D. L., Grote, U., & Nguyen, T. T. (2017). Migration and rural household expenditures: A case study from Vietnam. *Economic Analysis and Policy*, *56*, 163–175. doi:10.1016/j.eap.2017.09.001

Nguyen, M. C., & Winters, P. (2011). The Impact of migration on food consumption patterns: The case of Vietnam. *Food Policy*, *36*(1), 71–87. doi:10.1016/j.foodpol.2010.11.001

NIMET. (2015). Abuja, Nigeria: Nigerian Meteorological Agency.

Njoku, J. D., Ebe, T. E., & Pat-Mbano, E. C. (2010). Detection and mapping of land use and land cover classes of a developing city in south-eastern region of Nigeria using multi-band digital remotely-sensed data. *African Research Review*, *4*(3), 315–330. doi:10.4314/afrrev.v4i3.60272

Norris, M., & Winston, N. (2009). Rising second home numbers in rural Ireland: Distribution, drivers and implications. *European Planning Studies*, *17*(9), 1303–1322. doi:10.1080/09654310903053448

Northam, R. M. (1971). Vacant Urban Land in the American City. *Land Economics*, *47*(4), 345–355. doi:10.2307/3145071

NPC. (2006). *Nigeria National Census Report*. Abuja: National Population Commission.

NRC. (2014). *Advancing Land Change Modeling: Opportunities and Research Requirements*. Washington, DC: The National Academic Press.

Nugent, R. (2000). The impact of urban agriculture on the household and local economies. In Growing cities. Growing Food, (pp 67–97). Feldafing: Dtsch Stift int Entw.

Nwakor, F. N., Ifenkwe, G. E., & Asumughha, G. N. (2011). Root and Tuber Crops Research for Food Security and Empowerment. In C.O. Amadi, K.C. Ekwe, G. O. Chukwu, A. O. Olojede., & C.N. Egesi (Eds.), Potentials in Adoption of Improved Cassava Varieties for increased Food Security in Nigeria (pp. 563-571). National Root Crops Research Institute Umudike.

Nwalieji, H. U., & Uzuegbunam, C. O. (2012, December). Effect of climate change on rice production in Anambra state, Nigeria. *Journal of Agricultural Extension*, *16*(12), 81–91.

Nzelu, C. (2017, Dec 13). CBN: Local Rice Production Records 70% Increase. *This Day Newspaper.* Retrieved from https://www.thisdaylive.com

O' Bowman, A. M., & Pagano, M.A. (2004). Terra Incognita: Vacant Land and UrbanStrategies. Washington, DC: Georgetown University Press.

O'Sullivan, D., Evans, T., Manson, S., Metcalf, S., Ligmann-Zielinska, A., & Bone, C. (2016). Strategic directions for agent-based modeling: Avoiding the YAAWN syndrome. *Journal of Land Use Science*, *11*(2), 177–187. doi:10.1080/1747423X.2015.1030463 PMID:27158257

Oakland Institute. (2011). *Understanding land investment deals in Tanzania. Country Report* (p. 59). Oakland: The Oakland Institute.

Obayelu, O. A., & Idowu, O. O. (2019). Dietary diversity status of rural households in Nigeria: A gendered perspective. *Economia Agro-Alimentare*, *21*(3). doi:10.3280/ECAG2019-003003

Obibuaku, L. O. (1983). Agricultural Extension as a Strategy for Agricultural Transformation. Nsukka: University of Nigeria Press.

OCHA. (2012). Gender Definitions and Mandates 2. *OCHA Gender Toolkit*. Available at: https://www.unocha.org/sites/dms/Documents/GenderToolkit1_2_GenderDefinitionsandMandates.pdf

Odebode, S. O. (2008). Appropriate Technology for Cassava Processing in Nigeria: User's Point of View. *Journal of International Women's Studies*, *9*(3), 269–286.

Odeny, M. (2013). *Improving Access to Land and Strengthening Women's Land Rights in Africa.* Paper presented at the World Bank conference on Land and Poverty, Washington, DC.

ODI (Overseas Development Institute), (1997, Feb.). *Global hunger and food security after the World Food Summit.* Briefing Paper 1997, (1). London: ODI.

Odozi, J., Awoyemi, T. T., & Omonona, B. T. (2010). Household poverty and inequality: The implication of migrants. *Journal of Economic Policy Reform*, *13*(2), 191–199. doi:10.1080/17487871003700788

Odunze, D. I. (2019). A review of the Nigerian Agricultural Promotion Policy (APP, 2016-2020): Implications for entrepreneurship in the agribusiness sector. *International Journal of Agricultural Policy and Research*, *7*(3), 70–79.

Odurukwe, S. N., Matthews-Njoku, E. C., & Ejiogu-Okereke, N. (2006). Impacts of the women-in-agriculture (WIA) extension programme on women's lives; implications for subsistence agricultural production of women in Imo State, Nigeria. *Livestock Research for Rural Development*, *18*(2). Retrieved from http://www.lrrd.org/lrrd18/2/odur18018.htm

OECD. (2019). OECD Environmental Performance Reviews: Turkey 2019. OECD Environmental Performance Reviews. Paris: OECD Publishing.

Oginni, A., & Adebamowo, M. (2013). An evaluation of the socio-cultural effects of climate change on vulnerable Africa: Making a case for urgent action towards adaptation in Nigeria. *Br. J. Arts Soc. Sci.*, *13*, 141–149.

Ojo, G. U., & Offiong, R. A. (2018). Land grabbing: Implications of 1978 Land Use Act on rural livelihood sustainability in Cross River State, Nigeria. *International Journal of Development and Economic Sustainability*, *6*(5), 60–69.

Ojogho, O. (2010). Determinants of food insecurity among arable farmers in Edo State, Nigeria. *Agricultural Journal*, *5*(3), 151–156. doi:10.3923/aj.2010.151.156

Okoro, D., & Ujah, O. C. (2009). *Agricultural Policy and Budget Analysis in Nigeria (1999 –2007): Perspectives and Implications for SLSSFAN Project States*. OXFAM.

Okoruwa, V. O., Ojo, O. A., Akintola, C. M., Ologhobo, A. D., & Ewete, F. K. (2008). Post- harvest Grain Management Storage Techniques and Pesticides Use by Farmers in South- west Nigeria. *Journal of Economics and Rural Development*, *18*(1), 53–72.

Oladimeji, Y. U. (2017). Food Production trend in Nigeria and Malthus theory Of Population: Empirical evidence from RiceProduction. *Nigerian Journal of Agriculture, Food and Environment, 13*(1), 126–132.

Oladimeji, Y. U., & Ajao, A. M. (2014). Structure and determinants of profit efficiency among small scale rice farmers in Kwara State, Nigeria. *International Journal of Applied Research and Technology, 3*(11), 27–33.

Olagunju, F. I. (2008). Economics of Palm Oil Processing in South Western Nigeria. *International Journal of Agricultural Research, 1*(2), 69–77.

Olagunju, F. I., Oke, J. I., Babatunde, R. O., & Ajiboye, A. (2012). Determinants of food security in Ogbomoso Metropolis of Oyo State. *Nigeria. Production Agriculture and Technology, 8*(1), 111–124.

Olaoye, T. A., Kayode, A. O., & Oladiran, J. O. (2016). Determinants of food security in Ogbomoso Agricultural zone. *International Journal of Research and Development Organization, 2*(1).

Olawale, S. (2018). *List of Rice Producing States in Nigeria*. Retrieved from https://naijaquest.com/rice-producing-states-in-nigeria/

Olayide, D. W. (2009). Rural-urban migration and agriculture development in Nigeria. *Art Soc. Sci. International Research Journal, 1*(30), 44–57.

Olowa, O. W. (2009). The Impact of migrant remittances on household welfare in rural Nigeria. *Social Sciences, 4*(1), 59–64.

Olowa, O. W., Awoyemi, T. T., Shittu, M. A., & Olowa, O. A. (2013). Effects of Remittances on Poverty among Rural Households in Nigeria European. *Journal of Sustainable Development, 2*(4), 263–284.

Olowa, O. W., Awoyemi, T. T., Shittu, M. A., & Olowa, O. A. (2013). Effects of remittances on poverty among rural households in Nigeria. *European Journal of Sustainable Development, 2*(4), 263–284.

Onuekwusi, G. C., Odoemelam, L. E., & Kanu, R. I. (2017). Capacity Building on Cocoyam Value Additions Training for Rural Women in Abia State, Nigeria. *Journal of Agriculture Research and Technology, 6*(4), 2–6.

Onwualu, A. P. (2012). *Agricultural sector and national development: focus on value chain approach*. Presented at the 5th Edition of the Annual Lecture of Onitsha Chamber of Commerce at Sharon house, GRA, Onitsha.

Onyemauwa, C. S. (2012). Analysis of Women Participation in Cassava Production and Processing in Imo State, Southeast Nigeria. *Journal of Economics and Sustainable, 3*(5), 81–90.

Orewa, S. I., Adakaren, B., Ilechie, C. O., & Obulechei, S. (2009). An Analysis of the profitability of using the NIFOR small scale palm oil processing equipment (SSPE). *American-Eurasian Journal of Agronomy, 2*(3), 192–300.

Oseni, G., Corral, P., Goldstein, M., & Winters, P. (2015). Explaining gender differentials in agricultural production in Nigeria. *Agricultural Economics, 46*(3), 285–310. doi:10.1111/agec.12166

Osondu, C. K., & Ibezim, C. O. (2001). The impact of rural-urban migration on plantation Agriculture in the Niger-Delta Region, Nigeria. *Journal of Rural Social Sciences, 1*(20).

Oteng, J. W., & Sant'Anna, R. (2002). *Rice production in Africa: current situation and issues*. Retrieved from http://www.fao.org/docrep/003/x2243t/x2243t00.htm#TopOfPage

Oxfam International. (2012). *Our Land, Our Lives: Time Out on the Global Land Rush*. Oxfam Briefing Note 5. Retrieved from https://www.oxfam.org/sites/www.oxfam.org/files/bn-land-livesfreeze-041012-en_1.pdf

Oyediran, W. O. (2016) Assessment of Paddy Rice (*Ofada*) Processing in ensuring Rural Women livelihood sustainability in Ogun State, Nigeria. *International Journal for Research in Agricultural and Food Science,* 82-92.

Oyenuga, V. A. (1967). Agriculture in Nigeria. Rome, Italy: FAO.

Oyinloye, R. O., & Oloukoi, J. (2012). Spatio-temporal assessment and mapping of the landuse landcover dynamics in the central forest belt of southwestern Nigeria Research. *Journal of Environment and Earth Science*, *4*, 720–730.

Pant, L. P. (2016). Paradox of mainstreaming agroecology for regional and rural food security in developing countries. *Technological Forecasting and Social Change*, *111*, 305–316. doi:10.1016/j.techfore.2016.03.001

Parsa, V. A., & Salehi, E. (2016). Spatio-temporal analysis and simulation pattern of land use/cover changes, case study: Naghadeh, Iran. *Journal of Urban Management*, *5*(2), 43–51. doi:10.1016/j.jum.2016.11.001

Parsons, K., & Hawkes, C. (2018). *Connecting food systems for co-benefits: How can food systems combine diet-related health with environmental and economic policy goals? Policy Brief 31*. Copenhagen: World Health Organisation. Retrieved from http://www.euro.who.int/__data/assets/pdf_file/0007/387070/policy-brief-31-austria-eng.pdf?ua=1

Parsons, K., Hawkes, C., & Wells, R. (2019). What is the food system? A Food policy perspective. In *Rethinking Food Policy: A Fresh Approach to Policy and Practice*. London: Centre for Food Policy. Available online at https://www.city.ac.uk/__data/assets/pdf_file/0008/471599/7643_Brief-2_What-is-the-food-system-A-food-policy-perspective_WEB_SP.pdf

Pascual, U., Muradian, R., Brander, L., Gómez-Baggethun, E., Martín-López, B., Verma, M., ... Polasky, S. (2010). The Economics of Valuing Ecosystem Services and Biodiversity. In P. Kumar (Ed.), *The Economics of Ecosystems and Biodiversity: The Ecological and Economic Foundations*. London: Earthscan.

Peck, J., Theodore, N., & Brenner, N. (2009). Neoliberal urbanism: Models, moments, mutations. *SAIS Review (Paul H. Nitze School of Advanced International Studies)*, *29*(1), 49–66.

Pedro, S. (2019). Food Governance: Multistakeholder Decision- Making Mechanisms in Food Systems. In W. Leal (Ed.), Encyclopaedia of the UN Sustainable Development Goals, (Vol 2: Zero Hunger). Springer Publishing

Pedro, S. (2018). Family Farming, the Environment and the Global Food Chain. In P. Cooney & W. Sacher (Eds.), *Environmental Impacts of Transnational Corporations in the Global South. Research in Political Economy* (Vol. 33). Emerald Publishing Limited. doi:10.1108/S0161-723020180000033008

Peng, J., Wang, Y. L., Wu, J. S., Yue, J., Zhang, Y. A., & Li, W. F. (2006). Ecological effects associated with land-use change in China's southwest agricultural landscape. *International Journal of Sustainable Development and World Ecology*, *13*(4), 315–325. doi:10.1080/13504500609469683

Perez-Escamilla, R. (2017). Food Security and the 2015–2030 Sustainable Development Goals: From Human to Planetary Health. *Current Developments in Nutrition*, *1*(7), e000513. doi:10.3945/cdn.117.000513 PMID:29955711

Pettit, H. (2014). Cities without Suburbs: A Census 2010 Perspective. *European Planning Studies*, *22*(9), 1989–1991. doi:10.1080/09654313.2014.924219

PIND. (2011). *Cassava Value Chain Analysis in the Niger Delta*. A Report by Foundation for Partnership Initiative in the Niger Delta.

Pinstrup-Andersen, P., & Watson, D. D. (2011). *Food policy for developing countries: the role of government in global, national, and local food systems*. Ithaca: Cornell University Press.

Piyathamrongchai, K., & Batty, M. (2007). Integrating Cellular Automata and Regional Dynamics Using Gis. In E. Koomen, J. Stillwell, A. Bakema, & H. Scholten (Eds.), *Modelling Land-Use Change: Progress and Applications*. Dordrecht: Springer.

Plummer, M. L. (2009). Assessing benefit transfer for the valuation of ecosystem services. *Frontiers in Ecology and the Environment*, *7*(1), 38–45. doi:10.1890/080091

Polack, E., Cotula, L., & Côte, M. (2013). *Accountability in Africa's Land Rush: What Role for Legal Empowerment?* London: IIED/IDRC.

Polasky, S., Nelson, E., Pennington, D., & Johnson, K. A. (2011). The impact of land-use change on ecosystem services, biodiversity and returns to landowners: A case study in the State of Minnesota. *Environmental and Resource Economics*, *48*(2), 219–242. doi:10.100710640-010-9407-0

Portela, R., & Rademacher, I. (2001). A dynamic model of patterns of deforestation and their effect on the ability of the Brazilian Amazonia to provide ecosystem services. *Ecological Modelling*, *143*(1-2), 115–146. doi:10.1016/S0304-3800(01)00359-3

Porter, J. R., Xie, L., Challinor, A. J., Cochrane, K., Howden, S. M., Iqbal, M. M., ... Travasso, M. I. (2014). Food security and food production systems. In *Climate Change 2014: Impacts, Adaptation, and Vulnerability. Part A: Global and Sectoral Aspects. Contribution of Working Group II to the Fifth Assessment Report of the Intergovernmental Panel on Climate Change* (pp. 485–533). Cambridge, UK: Cambridge University Press.

Practvuthy, M. (2011). *Land Acquisition by Non-Local Actors and Consequences for Local Development: Impact of Economic Land Concessions on the Livelihoods of Indigenous Communities in Northeast Provinces of Cambodia*. Royal University of Phnon Penh, Mekong Institute of Cambodia.

Pratley, J. E. (2017). *Agriculture: from macho to gender balance*. Paper for 18th Australian Society of Agronomy Conference, Ballarat, Australia.

Premium Times. (2015, Oct 19). *Nigerian women contribute 70 per cent of agricultural workforce – AfDB*. Retrieved from https://www.premiumtimesng.com/news/top-news/191777-nigerian-women-contribute-70-per-cent-of-agricultural-workforce-afdb.html

Pro-poor Opportunities in Commodity and Service Markets Project (Propcom). (2012). Retrieved from http://www.propcom.org/

Qui, P., Yang, Y., Zhang, J., & Ma, X. (2011). Rural-to-urban migration and its implication for new cooperative medical scheme coverage and utilization in China. *BMC Public Health*, *11*(1), 520. doi:10.1186/1471-2458-11-520 PMID:21718506

Quisumbing, A. R., Brown, L. R., Feldstein, H. S., Haddad, L. J., & Pena, C. (1995). *Women: The Key to Food Security*. Washington, DC: International Food Policy Research Institute.

Rafiee, V., & Stenberg, E. (2018). *Land grabbing and its implications on rural livelihoods in Ghana and Ethiopia: A comparative study. Södertörns högskola*. Institutionen för Naturvetenskap.

Ramankutty, N., Hertel, T., Lee, H. & Rose, S. (2007). *Global spatial data of 18 Agro-ecological Zones (AEZs)*. Academic Press.

Randazzo, T., & Piracha, M. (2014). *Remittances and Household Expenditure Behaviour in Senegal*. IZA Discussion Paper No. 8106.

Rathna. (2016). *Rice milling technology*. New Rathna Rice (pvt) ltd. Retrieved from http://newrathnarice.com

Reardon, T., Codron, J., Busch, L., Bingen, R., & Harris, C. (1999). Global Change. IN Agri-food Grades and Standards: Agribusiness Strategic Responses. In Developing Countries. International Food and Agribusiness Management Review, 2, 1-15.

Reddy, S. S., Ram, P. R., Sastry, T. V. N., & Devi, I. B. (2004). *Agricultural Economics*. New Delhi: Oxford and IBH Publishing Co. Pvt Ltd.

Renting, H., & Wiskerke, J. (2010). *New Emerging Roles for Public Institutions and Civil Society in the Promotion of Sustainable Local Agro-Food Systems*. Paper presented at 9th European IFSA Symposium, Vienna, Austria.

Republic of Turkey, Ministry of Development. (2016). Report on Turkey's Initial Steps towards the Implementation of the 2030 Agenda for Sustainable Development. Ankara: Author.

Richards, M. (2013). Social and environmental impacts of agricultural large scale land acquisitions in Africa-with a focus on West and Central Africa. Rights and Resources Initiative.

Richardson, L., Loomis, J., Kroeger, T., & Casey, F. (2015). The role of benefit transfer in ecosystem service valuation. *Ecological Economics*, *115*, 51–58. doi:10.1016/j.ecolecon.2014.02.018

Rick, H., Bernard, M., & Rembold, F. (2014). *APHLIS – Postharvest cereal losses in Sub-Saharan Africa, their estimation, assessment and reduction*. A Technical Report for European Commission. Retrieved from https://www.google.com.ng/url?sa=t&rct=j&q=&esrc=s&source=web&cd=11&cad=rja&uact=8&ved=2ahUKEwjK5IG38_bmAhUhsaQKHf41AhEQFjAKegQIAxAC&url=https%3A%2F%2Fpublications.jrc.ec.europa.eu%2Frepository%2Fbitstream%2FJRC92152%2Flbna26897enn.pdf&usg=AOvVaw2TZCnu3x18iFqEqZ97J1hQ

Rose, D., & Charlton, K. E. (2002). Prevalence of household food poverty in South Africa: Results from a large, nationally representative survey. *Public Health Nutrition*, *5*(3), 383–389. doi:10.1079/PHN2001320 PMID:12003648

Rosol, M. (2005). Community gardens—A potential for stagnating and shrinking cities? Examples from Berlin. *Die Erde*, *136*(2), 165–178.

Rubin, D., Manfre, C., & Barrett, K. N. (2009). *Promoting Gender Equitable Opportunities in Agricultural Value Chains: A Handbook*. Prepared under the Greater Access to Trade Expansion project, Women in Development IQC (Contract No. GEW-I-00-02-00018-00, Task Order No. 02). Washington, DC: USAID. Retrieved from http://www.culturalpractice.com/site/wpcontent/downloads/4-200916.pdf

Ruel, M. T. (2003). Operationalizing dietary diversity: A review of measurement issues and research priorities. *The Journal of Nutrition*, *133*(11), 3911S–3926S. doi:10.1093/jn/133.11.3911S PMID:14672290

Rupprecht, C. D. D., & Byrne, J. A. (2014). Informal urban greenspace: A typology and trilingual systematic review of its role for urban residents and trends in the literature. *Urban Forestry & Urban Greening*, *13*(4), 597–611. doi:10.1016/j.ufug.2014.09.002

Russell, N., Karlsson, K., Ashby, J., & Mascarenhas, M. (2015). *Change in the Making: Progress Reports on CGIAR Gender Research: Issue No. 1. Toward gender-equitable control over productive assets and resources*. Cali, Colombia: CGIAR Gender and Agriculture Research Network, CGIAR Consortium Office and International Center for Tropical Agriculture.

Rwanga, S. S., & Ndambuki, J. M. (2017). Accuracy Assessment of Land Use/Land Cover Classification Using Remote Sensing and GIS. *International Journal of Geosciences*, *8*(04), 611–622. doi:10.4236/ijg.2017.84033

Ryder, J., Iddya, K., & Ababouch, L. (2014). Assessment and Management of Seafood Safety and Quality Current Practices and Emerging Issues. *FAO Fisheries and Aquaculture Technical Paper*, 574.

Samie, A., Deng, X., Jia, S., & Chen, D. (2017). Scenario-based simulation on dynamics of land-use-land-cover change in Punjab Province, Pakistan. *Sustainability*, *9*(8), 1285. doi:10.3390u9081285

Sampson, O. M. (2014). *Women's Contribution to Local Economic Development: A study of Women in Cassava Production and Processing in Central Tongu District* (Thesis). Kwame Nkuumah University of Science and Technology.

Saraux, A., Tobón, G. J., Benhamou, M., Devauchelle-Pensec, V., Dougados, M., Mariette, X., ... Schaeverbeke, T. (2013). Potential classification criteria for rheumatoid arthritis after two years: Results from a French multicenter cohort. *Arthritis Care and Research*, *65*(8), 1227–1234. doi:10.1002/acr.21982 PMID:23401491

Savage, M., Dougherty, B., Hamza, M., Butterfield, R., & Bharwani, S. (2009). Socioeconomic Impacts of Climate Change in Afghanistan. Department of International Development & Stockholm Environment Institute.

Schaffnit, C. (2012). *Foreign investment in farmland: Frankfurt*. Deutsche Bank, AG.

Schilling, J. & Logan, J. (2008). Greening the Rust Belt: A green infrastructure model for right-sizing America's shrinking cities. *Journal of the American Planning Association, 74*, 451-466.

Schmelzkopf, K. (1995). Urban community gardens as contested space. *Geographical Review*, *85*(3), 364–381. doi:10.2307/215279

Schroter, D., Cramer, W., Leemans, R., Prentice, I. C., Araujo, M. B., Arnell, N. W., ... Zierl, B. (2005). Ecosystem service supply and vulnerability to global change in Europe. *Science*, *310*(5752), 1333–1337. doi:10.1126cience.1115233 PMID:16254151

Schutter, O. (2014). The Reform of the Committee on World Food Security: The Quest for Coherence in Global Governance. In *Rethinking Food Systems. Structural Challenges, New Strategies and the Law* (pp. 219–238). Dordrecht: Springer. doi:10.1007/978-94-007-7778-1_10

Schutzbank, M. H. (2012). *Growing vegetables in Metro Vancouver: An urban farming census (Unpublished master's thesis)*. Vancouver: University of British Columbia. Retrieved from https://circle.ubc.ca/handle/2429/43559

Searchinger, T., Hanson, C., Ranganathan, J., Lipinski, B., Waite, R., Winterbottom, R., & Heimlich, R. (2013). *Creating a sustainable food future: Interim findings. A menu of solutions to sustainably feed more than 9 billion people by 2050.* Retrieved from https://www.wri.org/sites/default/files/wri13_report_4c_wrr_online.pdf

Shaw, W. (2007). *Migration in Africa: A review of the economic literature on international migration in 10 countries.* Washington, DC: Development Prospects Group, The World Bank.

SIDA. (2015). *Women and food security. Gender tool Box (Brief).* https://www.sida.se/contentassets/3a820dbd152f4fca98bacde8a8101e15/women-and-food-security.pdf

Simon, G. A. (2018). Definition of Food Security, The Four Dimensions of Food Security, Food Insecurity, Food Sovereignty, Food Systems (Master in human Development and Food Security). Roma Tre University.

Sims, R., Flammini, A., Puri, M., & Bracco, S. (2015, Nov.). *Opportunities for Agri-Food Chains To Become Energy-Smart.* FAO and USAID. Retrieved from http://www.fao.org/3/a-i5125e.pdf

Sindayigaya, W. (2012). *Foreign investment in agriculture-land grabbing*. PELUM Network.

Singh, R. B., Woodhead, T., & Papademtriou, M. K. (2003). *Strategies to Sustain and Enhance Asia Pacific Rice Production.* Retrieved from www.fao.org/docrep/006/y475Ie0t.htmHTopOfpage

Singh, S., Srivastava, P., Pandey, A., & Gautam, S. (2013). Integrated assessment of groundwater influenced by a confluence river system: Concurrence with remote sensing and geochemical modelling. *Water Resources Management*, *27*(12), 4291–4313. doi:10.100711269-013-0408-y

Sinha, P., & Kumar, L. (2013). Markov land cover change modeling using pairs of time-series satellite images. *Photogrammetric Engineering and Remote Sensing*, *79*(11), e1051. doi:10.14358/PERS.79.11.1037

Siqwana-Ndulo, N. (2013). The Informal Sector in South Africa: Women Street Traders in Durban. Consultancy Africa Intelligence. Retrieved from http://www.polity.org.za/article/the-informal-sector-in-south-africa-women-street-traders-in-durban-part-2-2013-05-14

Skees, J. R., Hartell, J., & Hoa, J. (2006). Agricultural Commodity Markets and Trade: New Approaches to Analyzing Market Structure and Instability. In A. Sarris & D. Hallam (Eds.), *Weather and Index-Based Insurance for Developing Countries: Experience and Possibilities*. Edward Elgar and Food and Agriculture Organization of the United Nations.

Skeldon, R. (1997). Rural-to-urban migration and its implications for poverty alleviation. *Asia-Pacific Population Journal*, *12*(1), 3–16. doi:10.18356/cd2c964e-en PMID:12292421

Sleeter, B. M., Sohl, T. L., Loveland, T. R., Auch, R. F., Acevedo, W., Drummond, M. A., ... Stehman, S. V. (2013). Land-cover change in the conterminous United States from 1973 to 2000. *Global Environmental Change*, *23*(4), 733–748. doi:10.1016/j.gloenvcha.2013.03.006

Small, J. (1997). Women's Land Rights: a case study from the Northern Transvaal. In S. Meer (Ed.), *Women, Land and Authority: Perspectives from South African*. London: Oxfam.

Smith, J. P., Li, X., & Turner, B. L. II. (2017). Lots for greening: Identification of metropolitan vacant land and its potential use for cooling and agriculture in Phoenix, AZ, USA. *Applied Geography (Sevenoaks, England)*, *85*, 139–151. doi:10.1016/j.apgeog.2017.06.005

Smith, L. C., & Subandoro, A. (2007). *Measuring Food Security Using Household Expenditure Surveys. Food Security in Practice Technical Guide Series*. Washington, D.C.: International Food Policy Research Institute; doi:10.2499/0896297675

Smith, M. D., Roheim, C. A., Crowder, L. B., Halpern, B. S., Turnipseed, M., Anderson, J. L., ... Selkoe, K. A. (2010). Sustainability and Global Seafood. *Science*, *327*(5967), 784–786. doi:10.1126cience.1185345 PMID:20150469

Smith, M., Pointing, J., & Maxwell, S. (1993). *Household Food Security, Concepts and Definitions: An Annotated Bibliography*. Brighton: Institute of Development Studies, University of Sussex.

Solon, M., Mink, P., Dernini, S., Bortoletti, M., & Lomax, J. (2019). The One Planet Sustainable Food Systems Programme as a Multi-stakeholder Platform for a Systemic Approach. In B. Burlingame & S. Dernini (Eds.), *Sustainable diets: Linking Nutrition and Food Systems*. Wallingford: C.A.B. International. doi:10.1079/9781786392848.0178

Somado, E. A., Guei, R. G., & Nguyen, N. (2010). *Nguyen Overview: Rice in Africa*. Retrieved from www.africarice,org

Song, W., & Deng, X. Z. (2017). Land-use/land-cover change and ecosystem service provision in China. *The Science of the Total Environment*, *576*, 705–719. doi:10.1016/j.scitotenv.2016.07.078 PMID:27810757

Song, W., Deng, X., Yuan, Y., Wang, Z., & Li, Z. (2015). Impacts of land-use change on valued ecosystem service in rapidly urbanized North China Plain. *Ecological Modelling*, *318*, 245–253. doi:10.1016/j.ecolmodel.2015.01.029

Spore, C. T. A. (2018). *A promising new continent-wide trade agreement for Africa: Trade and Marketing*. Retrieved from https://spore.cta.int/en/marketing/all/article/a-promising-new-continent-wide-trade-agreement-for-africa-sid0343958eb-d79b-4a03-9d46-44dfe69f9188b

Stahre, M., VanEenwyk, J., Siegel, P., & Njai, R. (2015). Housing insecurity and the association withhealth outcomes and unhealthy behaviors, Washington State, 2011. *Preventing Chronic Disease, 12*, 140511. doi:10.5888/pcd12.140511 PMID:26160295

Sterpka, M. (2009, July 6). *Cleveland's for-profit urban gardens are growing*. Retrieved from https://www.cleveland.com/metro/2009/07/clevelands_forprofit_urban_gar.html

Stewart, W. P., Gobster, P. H., Rigolon, A., Strauser, J., Williams, D. A., & van Riper, C. J. (2019). Resident-led beautification of vacant lots that connects place to community. *Landscape and Urban Planning, 185*, 200–209. doi:10.1016/j.landurbplan.2019.02.011

Sule, A. M., Ogunwale, S. A., & Atala, T. K. (2002). Factors affecting adoption of fishing innovation among fishing entrepreneurs in Jebba Lake Community. *Journal of Agricultural Extension, 6, 48-60.*

Sunderland, T., Powell, B., Ickowitz, A., Foli, S., Pinedo-Vasquez, M., Nasi, R., & Padoch, C. (2013). *Food Security and Nutrition: The Role Of Forests*. Discussion Paper. CIFOR. Retrieved from https://www.cifor.org/publications/pdf_files/WPapers/DPSunderland1301.pdf

Sustainable Development Goals Knowledge Platform. (2015). *Transforming our world: the 2030 Agenda for Sustainable Development*. Retrieved from https://sustainabledevelopment.un.org/post2015/transformingourworld

Syms, P. (1998) *Re-using Brownfield Sites: The Problem of Releasing Brownfield Land for Development.* Planning and Property Paper No. PP6, Centre for Regional Economic and Social Research, Sheffield Hallam University.

Syms, P. (1994). The funding of developments on derelict and contaminated sites. In R. Ball & A. Pratt (Eds.), *Industrial Property Policy and Economic Development* (pp. 63–82). London: Routledge.

Syms, P. (1999). Redeveloping brownfield land: The decision making process. *Journal of Property Investment & Finance, 17*(5), 481–500. doi:10.1108/14635789910294903

Syms, P. (2001). *Releasing Brownfields*. London: RICS Foundation, Royal Institution of Chartered Surveyors.

Tacon, A. G. J., & Metian, M. (2013). Fish Matters: Importance of Aquatic Foods in Human Nutrition and Global Food Supply. *Journal Reviews in Fisheries Science, 21*(1), 22–38. doi:10.1080/10641262.2012.753405

Tadlock, C. (2002). *Value-Added Agricultural Enterprisesin Rural Development Strategies*. The Library of Congress.

Taiwo, K. A., & Fasoyiro, S. B. (2015). Women and cassava processing in Nigeria. *International Journal of Development Research, 5*(2), 3513–3517.

Taiwo, K. A., Osunbitan, A. J., Sunmmonu, T. O., Ajayi, M. O., & Ajibola, O. O. (2001). Technology choice and technical capacity in gari production. *Food Reviews International, 17*(1), 89–107. doi:10.1081/FRI-100000516

Taylor, J. R., & Lovell, S. T. (2012). Mapping public and private spaces of urban agriculture in Chicago through the analysis of high-resolution aerial images in Google Earth. *Landscape and Urban Planning, 108*(1), 57–70. doi:10.1016/j.landurbplan.2012.08.001

Tendall, D. M., Joerin, J., Kopainsky, B., Edwards, P., Shreck, A., Le, Q. B., ... Six, J. (2015). Food system resilience: Defining the concept. *Global Food Security, 6*, 17–23. doi:10.1016/j.gfs.2015.08.001

The Economist. (2017). *The Global Food Security Index 2017: Measuring food security and the impact of resource risks is the sixth edition of an Economist Intelligence Unit study.* Retrieved from http://foodsecurityindex.eiu.com/Home/DownloadResource?fileName=EIU%20Global%20Food%20Security%20Index%20-%202017%20Findings%20%26%20Methodology.pdf

The World Bank. (2019). Retrieved from https://www.worldbank.org/en/country/turkey/overview

Theting, H., & Brekke, B. (2010. *Land investments or land grab: A critical view from Tanzania and Mozambique*. Available at: http://www.spireorg.no/files/spire/documents/Land_investment_or_land_grab08112010.pdf

Thorpe, J. (2013). *Sugar rush, land rights and the supply chains of the biggest food and beverage companies*. Oxford, UK: OXFAM GB for Oxfam International.

Tijani, B. A., & Ndukwe, O. K. (2012). Assessment of household food security situation in Marte Local Government Area of Borno State, Nigeria. *Maiduguri Journal of Arts and Social Science*, *10*(1), 47–60.

Tijani, S. A., & Umoh, I. E. (2016). Attitude of Women Cassava Farmers towards Participation in Growth Enhancement Support Scheme in Oyo state, Nigeria. *Journal of Agricultural Extension*, *20*(2), 89. doi:10.4314/jae.v20i2.7

Tlusty, M. F. (2012). Environmental improvement of seafood through certification and ecolabelling: Theory and analysis. *Fish and Fisheries*, *13*(1), 1–13. doi:10.1111/j.1467-2979.2011.00404.x

Torheim, L. E., Ouattara, F., Diarra, M. M., Thiam, F. D., Barikmo, I., Hatløy, A., & Oshaug, A. (2004). Nutrient adequacy and dietary diversity in rural Mali: Association and determinants. *European Journal of Clinical Nutrition*, *58*(4), 594–601. doi:10.1038j.ejcn.1601853 PMID:15042127

Törhönen, M., & Palmer, D. (2004, April). *Land administration in post conflict Cambodia*. [Paper presentation]. The Symposium on Land Administration in Post Conflict Areas. Geneva, Switzerland.

Tortorello, M. (2012, Oct 31). Urban gardens grow everything except gardeners. *The New York Times*.

Tsegai, D. (2007). Migration as a household decision: What are the roles of income differences? Insights from the Volta Basin of Ghana. *European Journal of Development Research*, *19*(2), 305–326. doi:10.1080/09578810701289212

TÜDAV (Turkish Marine Research Foundation). (2017). *2017 Turkish Seas Report*. Retrieved from Retrieved from http://tudav.org/wp-content/uploads/2018/04/TUDAV_2017_Denizler_Raporu_s.pdf

Turkish Ministry of Agriculture and Forestry. (2019). *Fishery Statistics*. Retrieved from https://www.tarimorman.gov.tr/sgb/Belgeler/SagMenuVeriler/BSGM.pdf

Turkish Ministry of Environment and Urban Planning. (2018). *Sık Sorulan Sorular, Deniz Ve Kiyi Yönetimi* [Frequently Asked Questions, Sea and Marine Management]. Retrieved from https://csb.gov.tr/sss/deniz-ve-kiyi-yonetimi

TURKSTAT (Turkish Statistical Institute). (2019a). *Fishery Statistics*. Retrieved from http://www.tuik.gov.tr/PreTablo.do?alt_id=1005

TURKSTAT. (2019b, June). *Aquaculture (Amsterdam, Netherlands)*, *2018*, 12. Retrieved from http://tuik.gov.tr/PreHaberBultenleri.do?id=30697

Twene, S. K. (2016). *Land grabbing and rural livelihood sustainability: Experiences from the Buidam construction* (A published M.Sc Thesis). Available at: http://ir.knust.edu.gh/bitstream/123456789/10495/1/SAMUEL%20TWENE%20%28FINAL%20THESIS%29.pdf

Udoh, E. J. (2005). Demand and control of credit from informal sources by rice producing female farmers of Akwa Ibom State Nigeria. *Journal of Agriculture and Social Sciences*, *1*(2), 152–155.

Ugwu, D. S. (2009). Problems and Prospects of Commercial Small and Medium Scale Cocoa and Oil Palm Production in Cross River State, Nigeria. *Journal of Applied Sciences Research*, *5*(7), 827–832.

Ugwu, J. N., Mgbakor, M. N., & Chitor, C. (2014). Assessment of Value addition in Rice Production and Processing in Adani inUzouwani Local Government Area of Enugu state, Nigeria. *International Journal of Agricultural Science and Veterinary Medicine*, *2*(4), 54–68.

Umezulike, I. A. (2004). *The Land Use Act – More than two Decades After and Problems of Adaptive Strategies of Implementation. College of Law, Ugbinedion University*. Enugu: Snaap Press ltd.

UN Department of Economic and Social Affairs. (2019). News, Growing at a slower pace, world population is expected to reach 9.7 billion in 2050 and could peak at nearly 11 billion around 2100. Author.

UN Environment Programme (2016). *Collaborative framework for food systems transformation: A multi-stakeholder pathway for sustainable food systems*. Author.

UN General Assembly. (2015). *Resolution adopted by the General Assembly on 25 September 2015*. A/RES/70/1. Retrieved from https://www.un.org/en/development/desa/population/migration/generalassembly/docs/globalcompact/A_RES_70_1_E.pdf

UN. (2018). *United Nations Decade of Family Farming (2019–2028). Resolution adopted by the General Assembly on 20 December 2017*. Retrieved from: https://undocs.org/A/RES/72/239

UNCTAD. (2017). The Role Of Scıence, Technology And Innovatıon In Ensurıng Food Securıty By 2030. United Nations.

UNEP. (2016). *Food Systems and Natural Resources*. A Report of the Working Group on Food Systems of the International Resource Panel.

United Kingdom Department for Environment, Food Rural Affairs - UK DEFRA. (2016). *Environment Secretary Salutes Britain's Women Farmers*. London: Department for Environment, Food and Rural Affairs. Retrieved from www.gov.uk/government/news/environment-secretary-salutes-britains-womenfarmers

United Nations (UN). (1972). *Report of the United Nations Conference on the Human Environment*. Stockholm, 5-16 June 1972, A/CONF.48/14/Rev.1. UN.

United Nations Development Programme. (2015). Gender, Climate Change and Food Security. *Training Module*, *3*, 1–36.

United Nations Educational, Scientific and Cultural Organisation – UNESCO. (2003). *Baseline Definitions of Key Concept and Terms*. UNESCO's Gender Mainstreaming Implementation Framework. Retrieved from http://www.unesco.org/new/fileadmin/MULTIMEDIA/HQ/BSP/GENDER/PDF/1.%20Baseline%20Definitions%20of%20key%20gender-related%20concepts.pdf

United Nations Population Fund. (2014). *Adolescents, youth and the transformation of the future*. Retrieved from https://www.unfpa.org/sites/default/files/pub-pdf/EN-SWOP14-Report_FINAL-web.pdf

United Nations System High Level Task Force on Global Food Security. (2011). *Food and nutrition security: Comprehensive framework for action. Summary of the Updated Comprehensive Framework for Action*. Rome: UCFA.

United Nations. (1987). *Report of the World Commission on Environment and Development: Our Common Future*. Retrieved from https://sustainabledevelopment.un.org/content/documents/5987our-common-future.pdf

United Nations. (2002). *Report of the World Summit on Sustainable Development*. Johannesburg, South Africa: United Nations.

United Nations. (2012). *Back to Our Common Future Sustainable Development in the 21st century (SD21) project Summary for policymakers*. United Nations. Retrieved from https://sustainabledevelopment.un.org/content/documents/UN-DESA_Back_Common_Future_En.pdf

United Nations. (2015). *Transforming our world: The 2030 Agenda for Sustainable Development*. United Nations. Retrieved from https://www.un.org/pga/wp-content/uploads/sites/3/2015/08/120815_outcome-document-of-Summitfor-adoption-of-the-post-2015-development-agenda.pdf

United Nations. (2015). *Zero Hunger Challenge. Pathways to zero hunger: Transforming our Food Systems to Transform our World*. Rome: Committee on World Food Security.

United Nations. (2019). *World population prospects 2019*. New York: United Nations.

United States Agency for International Development - USAID. (2007). *Gender Terminology*. Retrieved from https://pdf.usaid.gov/pdf_docs/Pnadl089.pdf

United States Agency for International Development - USAID. (2010). *Guide to Gender Integration and Analysis: Additional Help for Automated Directive System (ADS)*. Washington, DC: United States Agency for International Development.

United States Agency for International Development - USAID. (2012). *Gender Equality and Female Empowerment Policy*. Washington, DC: USAID.

United States Agency for International Development - USAID. (2013). *Integrating Gender Equality and Female Empowerment in USAID's Program Cycle. ADS Chapter 205*. Washington, DC: USAID. Retrieved from https://www.usaid.gov/ads/policy/200/205

United States Agency for International Development - USAID. (2017). *ADS Chapter 205 Integrating Gender Equality and Female Empowerment in USAID's Program Cycle*. Retrieved from https://www.usaid.gov/sites/default/files/documents/1870/205.pdf

Uriarte, M., Yackulic, C., Lim, Y., & Arce-Nazario, J. (2011). Influence of land use on water quality in a tropical landscape: A multi-scale analysis. *Landscape Ecology*, *26*(8), 1151–1164. doi:10.100710980-011-9642-y PMID:26146455

USDA. (2018). *Nigerian Milled Rice Production by Year (1960 - 2018)*. Retrieved from https://www.indexmundi.com/agriculture/?country=ng&commodity=milledrice&graph=production-growth-rate

Uzodike, E. N. U. (1993). Women's Rights in Law and Practice: Property Rights. In A. O. Obilade (Ed.), *Women in Law*. Lagos: Southern University of Lagos.

van Berkum, S., Dengerink, J., & Ruben, R. (2018). *The food systems approach: sustainable solutions for a sufficient supply of healthy food*. Wageningen: Wageningen Economic Research, Memorandum 2018-064, 32.

Van Donge, J. K., & Pherani, L. (1999). Law and order as a development issue: Land conflicts and the creation of social order in Southern Malawi. *The Journal of Development Studies*, *36*(2), 373–399. doi:10.1080/00220389908422620

van Schrojenstein Lantman, J., Verburg, P. H., Bregt, A., & Geertman, S. (2011). Core principles and concepts in land-use modelling: A literature review. In E. Koomen & J. Borsboom-Van Beurden (Eds.), Land-Use Modelling in Planning Practice. Springer Science+Business Media B.V. doi:10.1007/978-94-007-1822-7_3

Vandermeulen, V., Gellynck, X., Huylenbroeck, G. V., Orshoven, J. V., & Bomans, K. (2009). Farmland for tomorrow in densely populated areas. *Land Use Policy*, *26*(4), 859–868. doi:10.1016/j.landusepol.2008.10.014

Varangis, P., Larson, D., & Anderson, J. (2002). *Agricultural Markets and Risks: Management of the Latter Not the Former.* World Bank, Policy Research Working Paper No. 2793.

Vargas-Lundius, R., Guillaume, L., Villarreal, M., & Osorio, M. (2008). *International migration, remittances and rural development*. Retrieved from http://www.fao.org/3/a-ak405e.pdf

Vaughan, K., Kaczynski, A., Wilhelm Stanis, S., Besenyi, G., Bergstrom, R., & Heinrich, K. (2013). Exploring the distribution of park availability, features, and quality across Kansas City, Missouri by income and race/ethnicity: An environmental justice investigation. *Annals of Behavioral Medicine*, *45*(S1), 28–38. doi:10.100712160-012-9425-y PMID:23334757

Veldés, A., Foster, W., Gustavo, A., Carlo, A., Covarrubias, K., Benjamin, D., ... Zezza, A. (2010). *A profile of rural poor. Background Paper for IFAD Rural Poverty Report 2010*. Rome, Italy: IFAD.

Veldkamp, A., & Lambin, E. F. (2001). Editorial: Predicting land-use change. *Agriculture, Ecosystems & Environment*, *85*(1-3), 1–6. doi:10.1016/S0167-8809(01)00199-2

Verbruggen, P., & Havinga, T. (Eds.). (2017). *Hybridization of food governance: Trends, types and results*. Cheltenham: Edward Elgar. doi:10.4337/9781785361708

Verburg, P. H., Eickhout, B., & Meijl, H. V. (2008). A multi-scale, multi-model approach for analyzing the future dynamics of European land use. *The Annals of Regional Science*, *42*(1), 57–77. doi:10.100700168-007-0136-4

Verburg, P. H., Kok, K., Pontius, R. G., & Veldkamp, A. (2006). Modeling land-use and land-cover change. In E. F. Lambin & H. J. Geist (Eds.), *Land-use and land-cover change: local processes and global impacts*. Berlin: Springer. doi:10.1007/3-540-32202-7_5

Viera, A. J., & Garrett, J. M. (2005). Understanding interobserver agreement: The kappa statistic. *Family Medicine*, *37*, 360–363. PMID:15883903

Vision. (2010). *Agriculture 1997 Final Report of the Sub Committee of Agricultural Economic Sector*. Federal Government of Nigeria.

Vliet, J. V. (2009). Assessing the Accuracy of Changesin Spatial Explicit Land Use Change Models. In *Proceedings of the 12th AGILE International Conference on Geographic Information Science, 2009 Germany*. Leibniz Universität Hannover.

Voicu, I., & Been, V. (2008). The effect of community gardens on neighboring property values. *Real Estate Economics*, *2*(2), 241–283. doi:10.1111/j.1540-6229.2008.00213.x

von Braun, J., & Meinzen-Dick, R. (2009). *Land grabbing by foreign investors in developing countries: Risks and opportunities. IFPRI Policy Brief 13*. Washington, DC: International Food Policy Research Institute.

Wahl, N., Jamnadass, R., Baur, H., Munster, C., & Liyama, M. (2009). *Economic Viability of Jatropha Plantations in Northern Tanzania-Assessing Farmers Prospects via Cost Benefit Analysis*. ICRAF Working Paper, World Agro forestry Centre.

Walker, R., & Walker, R. A. (1978). Two sources of uneven development under advanced capitalism: Spatial differentiation and capital mobility. *The Review of Radical Political Economics*, *10*(3), 28–37. doi:10.1177/048661347801000304

Wang, Y., Li, X., Zhang, Q., Lia, J., & Zhou, X. (2018). Projections of future land use changes: Multiple scenarios-based impacts analysis on ecosystem services for Wuhan city, China. *Ecological Indicators*, *94*, 430–445. doi:10.1016/j.ecolind.2018.06.047

Wang, Z., Zhang, B., Zhang, S., Li, X., Liu, D., Song, K., ... Duan, H. (2006). Changes of land use and of ecosystem service values in Sanjiang Plain, Northeast China. *Environmental Monitoring and Assessment*, *112*(1-3), 69–91. doi:10.100710661-006-0312-5 PMID:16404535

Waswa, L. M., Jordan, I., Herrmann, J., Krawinkel, M. B., & Keding, G. B. (2015). Community-based educational intervention improved the diversity of complementary diets in western Kenya: Results from a randomized controlled trial. *Public Health Nutrition*, *18*(18), 3406–3419. doi:10.1017/S1368980015000920 PMID:25857703

Waterhouse, R., Lauriciano, G., & Norfolk, S. (2010). *Social analysis of selected projects – Issues Note and Case Studies Large-Scale Land Acquisition for Agricultural Production Mozambique*. World Bank.

Wentling, T. L. (1992). *Planning for Effective Training: A Guide to Curriculum Development.* Rome: Food and Agriculture Organization of the United Nations.

West African Rice Development Authority (WARDA). (2003). *The Nigerian rice economy in a competitive world: constraints, opportunities and strategic choices: strategy for rice sector revitalization in Nigeria.* A paper on rice sector study implemented WARDA. Presented at a Technical Workshop 20-21 August, Ibadan, Nigeria.

Whitehead, A., & Tsikata, D. (2003, January). Policy Discourses on Women's Land Rights in Sub-Saharan Africa: The Implications of the Re-turn to the Customary. *Journal of Agrarian Change*, *3*(1&2), 67–112. doi:10.1111/1471-0366.00051

WHO. (2018). *2018 Global Nutrition Report*. Retrieved from: https://www.who.int/nutrition/globalnutritionreport/en/

Wiechman, T. (2007). Between Spectacular Projects and Pragmatic Deconstruction. The Future of Shrinking Cities: Problems, Patterns, & Strategies of Urban Transformation in a Global Context, 5–16.

Wikipedia. (2019). *Ijebu North Local Government Area in Ogun State*. Retrieved from https://en.wikipedia.org/wiki/Ijebu_North

Willett, W., Rockström, J., Loken, B., Springmann, M., Lang, T., Vermeulen, S., . . . Majele Sibanda, L. (2019). *Food in the Anthropocene: the EAT–Lancet Commission on healthy diets from sustainable food systems.* Retrieved from: https://www.thelancet.com/journals/lancet/article/PIIS0140-6736(18)31788-4/fulltext

Williams, J. (2010, December 1). Some landowners using community gardens to get a tax break. *The Vancouver Sun*. Retrieved from http://thetrac.ca/resources/docuploads/2010/12/02/ Some%20landowners%20using%20community%20gardens%20to%20get%20a%20tax%20break.pdf

Women Engage for a Common Future – WECF. (2018). *The Gender Impact Assessment and Monitoring Tool*. Retrieved from www.women2030.org

World Bank Report. (2005). *Empowering People by Transforming Institutions, Social Development in World Bank Operations*. Retrieved from http://documents. worldbank.org/curated/en/563751468780547056/Empowering-people-by-transforminginstitutions-social-development-in-World-Bank-operations.

World Bank Report. (2005). *Empowering People by Transforming Institutions, Social Development in World Bank Operations*. Retrieved from http://documents.worldbank.org/curated/en/563751468780547056/Empowering-people-by-transforming-institutions-social-development-in-World-Bank-operations

World Bank, Food and Agricultural Organization., & International Fund for Agricultural Development. (2009). Module 4: Gender Issues in Land Policy and Administration. Gender in Agriculture Sourcebook. Washington, DC: World Bank. Food and Agricultural Organization., & International Fund for Agricultural Development

World Bank. (1986). *Poverty and Hunger: Issues and Options for Food Security in Developing Countries*. Retrieved from http://documents.worldbank.org/curated/en/166331467990005748/pdf/multi-page.pdf

World Bank. (1998). *The assessment of vegetation and land use changes in Nigeria between 1976/78 and 1993/95*. Washington, DC: World Bank.

World Bank. (2010). *Rising Global Interest in Farmland: Can it Yield Sustainable and Equitable Benefits?* Washington, DC: World Bank.

World Bank. (2011). *Missing Food: The Case of Post-harvest Grain Losses in Sub-Saharan Africa*. World Bank Report No: 60371-AFR. Retrieved from www.worldbank.org/rural

World Bank. (2012). *Nigeria Agricultural Production: World Bank Indicators*. Washington, DC: World Bank.

World Bank. (2016). *Migration and Remittances. Factbook 2016, Global Knowledge Partnership on Migration and Development*. KNOMAD.

World Food Istanbul. (2018). *Infografik: Türkiye'de Balık ve Deniz Mahsülleri* [Infographics: Fish and Seafood in Turkey]. Retrieved from https://www.worldfood-istanbul.com/Fuar-Hakk%C4%B1nda/Sektorel-Haberler/Infografik-Turkiye-de-Bal%C4%B1k-ve-Deniz-Mahsulleri

World Wide Fund for Nature (WWF). (2016). *The Global Food System: An Analysis*. WWF Netherlands & Metabolic Lab.

Worldstage. (2018). *Nigeria: Federal Government to improve supply value chains of rice, cassava by partnering with the Nigerian Incentive-Based Risk Sharing System For Agricultural Lending (NIRSAL) in collaboration with the united nations development project(undp)*. Retrieved from https://www.worldstagegroup.com/category/business/agriculture

WRM. (2010, Dec). *World Rainforest Movement. Oil palm in Nigeria: shifting from smallholders and women to mass production*. Retrieved from https://oilpalminafrica.wordpress.com/2010/08/06/oil-palm-in-nigeria/

XLSTAT. (2019). *Two-Stage Least Squares Regression*. Retrieved from https://www.xlstat.com/en/solutions/features/two-stage-least-squares-regression

Yamano, T., & Deininger, K. (2005). *Land Conflict in Kenya: Causes; Impacts and Resolutions*. FASID Discussion Paper 2005-12-002, Foundation for Advanced Studies on International Development National Graduate Institute for Policy Studies, The World Bank.

Ye, Y. (2015). Global Fisheries, Current Situation and Challenges. In H. D. Smith, J. L. Suárez de Vivero, & T. S. Agardy (Eds.), *Routledge Handbook of Ocean Resources and Management, 1th. Ediiton*. Routledge. doi:10.4324/9780203115398-15

Yi, H., Güneralp, B., Filippi, A. M., Kreuter, U. P., & Güneralp, İ. (2017). Impacts of land change on ecosystem services in the San Antonio River Basin, Texas, from 1984 to 2010. *Ecological Economics*, *135*, 125–135. doi:10.1016/j.ecolecon.2016.11.019

Yıldırım, S., Yıldırım, D. Ç., & Gedikli, A. (2016). Sustainable Consumption Trends in the World in the Context of Green Economy and Sustainability. In M. M. Erdoğdu, T. Arun, & I. H. Ahmad (Eds.), *Handbook of Research on Green Economic Development Initiatives and Strategies* (pp. 66–84). IGI Global. doi:10.4018/978-1-5225-0440-5.ch004

Youdeowei, A., & Kwarteng, J. (2006). *Tool Kit for the Production of Agricultural Extension Materials: Guidebook*. CTA Publishing.

Yuan, Y., Zhao, T., Wang, W., Chen, S., & Wu, F. (2013). Projection of the spatially explicit land use/cover changes in China, 2010–2100. *Advances in Meteorology*, *2013*, 9. doi:10.1155/2013/908307

Yusuf, N., Olatunji, A. G., & Issah, M. (2014). Factors influencing foreign land acquisitions and sustainable development in Nigeria. In E. Osabuohien (Ed.), *In-country determinants and implications of foreign land acquisition* (pp. 214–236). IGI Global.

Zang, S. Y., Wu, C. S., Liu, H., & Na, X. D. (2011). Impact of urbanization on natural ecosystem service values: A comparative study. *Environmental Monitoring and Assessment*, *179*(1-4), 575–588. doi:10.100710661-010-1764-1 PMID:22013594

Zezza, A., Carletto, C., Davis, B., & Winters, P. (2011). Assessing the impact of migration on food and nutrition security. *Food Policy*, *36*(1), 1–6. doi:10.1016/j.foodpol.2010.11.005

Zhang, W., Chen, J., Liao, A., Han, G., Chen, X., Chen, L., ... Zhang, J. (2016). Geospatial knowledge-based verification and improvement of GlobeLand30. *Science China. Earth Sciences*, *59*(9), 1709–1719. doi:10.100711430-016-5318-4

Zhang, Y., Zhao, L., Liu, J., Liu, Y., & Li, C. (2015). The impact of land cover change on ecosystem service values in urban agglomerations along the coast of the Bohai Rim, China. *Sustainability*, *7*(8), 10365–10387. doi:10.3390u70810365

Zhao, B., Kreuter, U. L. B., Ma, Z., Chen, J., & Nakagoshi, N. (2004). An ecosystem service value assessment of land-use change on Chongming Island, China. *Land Use Policy*, *21*(2), 139–148. doi:10.1016/j.landusepol.2003.10.003

Zhen, L., Deng, X., Wei, Y., Jiang, Q., Lin, Y., Helming, K., ... Hu, J. (2014). Future land use and food security scenarios for the Guyuan district of remote western China. *Iforest-biogeosciences and Forestry*, *7*(6), 372–384. doi:10.3832/ifor1170-007

ZMO (The Chambers of Agricultural Engineers). (2017). *Genel Merkez, e-kütüphane, Su Ürünleri Raporu* [Headquarters, e-library, Fisheries Report]. Retrieved from http://www.zmo.org.tr/genel/bizden_detay.php?kod=27302&tipi=17&sube=010.02.2017

About the Contributors

Abiodun Elijah Obayelu is a seasoned scholar with over 10 years of experience in teaching, research, community service and consultancy. He possesses Bachelor degree in Agriculture (Agricultural Economics and Farm Management) and Master in Business Administration (MBA) from the University of Ilorin, Kwara State, Nigeria; M.Sc. and Ph.D degrees in Agricultural Economics from the University of Ibadan, Oyo State, Nigeria. Dr Obayelu currently lectures at the Federal University of Agriculture, Abeokuta, Nigeria. His research foci are: Agricultural Development, Food and Consumers Economics. Studies that deals with, and relates communities and households' socio-economic characteristics with development outcomes such as resilience of livelihood, economic welfare, impact assessment, and food security across cultures and countries have been the foci of Dr Obayelu's research endeavors over the years. These research foci are reinforced by the plethora of emerging evidences that point out agricultural transformation as the panacea for eliminating hunger and poverty in Africa. Dr Obayelu has experience in African Agricultural Research and Development contributing extensively to emerging issues in agriculture and agricultural innovation as part of his research and consultancies for international and national organizations. He has a lot of experience in baseline survey and impact evaluation. He also has a wide experience in data collection using both quantitative and qualitative techniques, participatory rural appraisal (PRA) using information gathering tools such as questionnaire, interview schedule, focus group discussion, pair ranking and use of key informants. Dr Obayelu's research foci provide a robust rationale for collaboration and multidisciplinary approaches in enquiry and have led to his participation as a team member (Agricultural and Food Economist) in a number of funded research and consultancy works. He has played one or more roles in the conceptualization and design of research, facilitation of field work and data collection, data analyses and report writing, facilitation of stakeholders meeting, liaison with government officials of the Federal Ministry of Agriculture, agencies and Non-Governmental Organization (NGOs) for possible mainstreaming of research projects and findings into agricultural development policy of Nigeria. Some of the research collaborations include: Michigan State University-Agricultural Policy Research in Africa (MSU-APRA) research on pathways to agricultural commercialization in Africa; socio-demographic changes and potential impact on nutrition in Africa (project on Sustainable Nutrition for Africa in the Years to Come (SUNRAY) Funded by the Commission of the European Communities Seventh Framework Programme); incentivising adoption of climate smart practices in cereals production in Nigeria: Socio-cultural and economic diagnosis, funded by Economic Community of West African States (ECOWAS); Regional Agency for Agriculture and Food (RAAF) and the Support Programme for Food and Nutrition Security in West Africa (PASANAO); livelihood impacts of improved on-farm water control in sub-Saharan Africa: an empirical investigation of three modes of smallholder agricultural water management in Nigeria, funded by the International

Water Management Institute, Colombo, Sri Lanka; impact evaluation of Nigerian Diary Development Program in Oyo and Kano States by Afridev Path International and Sahel Consulting Agriculture and Nutrition Limited, sponsored by Bill & Melinda Gates Foundation; impact assessment of GEP3-Cash Transfer Program by CAPRA, Canada, sponsored by UNICEF, Nigeria; diffusion strategy of Green Technology and Green Industry in Africa: a study of renewable energy technology market, and energy efficiency adoption in Cassava and Maize Processing Industries in Kenya and Nigeria by the United Nations University-Maastricht Economic and Social Research and Training Centre on Innovation and Technology (UNU-MERIT); diffusion and agricultural transformation: the case of NERICA in Nigeria"- a project funded by African Growth Initiative, Brookings Institution in collaboration with the Nigerian Institute of Social and Economic Research, Nigeria; competitive grants and commissioned research schemes funded by the Federal University of Agriculture, Abeokuta, Nigeria through the Institute of Food Security, Environmental Resources and Agricultural Research (IFSERAR) and Directorate of Grant Management. Dr Obayelu A. E. was a Ph.D. Doctoral Research Thesis Writing awardee of the African Economic Research Consortium (AERC) and International Development Research Centre (IDRC) in 2009 and had benefitted from several travel grants from different organizations such as: African Growth and Development Policy (AGRODEP) Consortium, UNICEF, United Nations University Institute for Natural Resources in Africa (UNU-INRA), Nigeria, Institute of Labour Economics (IZA), Research and Innovation Network for Europe and Africa (RINEA), to mention just a few. Dr. Obayelu is a member of African Growth and Development Policy (AGRODEP) Impact Evaluation Network (AIEN), African Association of Agricultural Economists (AAAE), Nigerian Association of Agricultural Economists (NAAE), Farm Management Association of Nigeria (FAMAN), Emerald Literati Network, African Economics Research Consortium Network, and the Nigerian Association of Evaluators (NAE). He is the editor of these books (i) Household Food Consumption Patterns and Food Security, published by Lap Lambert Academic Publishing, Germany in 2010 (ii) Food Systems Sustainability and Environmental Policies in Modern Economies, published by IGI International, USA in 2018. His research productivity is self-evident with over 90 peer-reviewed scientific publications in reputable local and international peered-reviewed journals, books and conference proceedings.

Oluwakemi Adeola Obayelu is a Lecturer in the Department of Agricultural Economics, University of Ibadan, Nigeria. She is an applied economist with research foci on gender, food and nutrition security and development issues. She is a member of many professional organisations and has published extensively in reputable peer-reviewed international journals.

* * *

Funmi Lydia Adeduntan is a post-graduate student in the Department of Agricultural Economics, University of Ibadan, Ibadan, Nigeria. Her research area is Welfare Economics.

Eleojo Grace Adejo hails from Ogugu, a community in Olamaboro Local Government Area of Kogi State. She Obtained B. Agric. Tech (Hons) in Agricultural Economics and Extension from Abubakar Tafawa Balewa University, Bauchi. She is a Ph.D candidate and research student in Department of Agricultural Economics and Extension, Kogi State University. She joined the service of Kogi State University since 2018 as an Assistant Lecturer with the Department of Agricultural Economics and Extension.

Patrick Emmanuel Adejo is a native of Anyigba a community in Dekina Local Government Area of Kogi State. He Obtained B.Agric. Tech (Hons) in Agricultural Economics and Extension and M.Sc. Agricultural Extension; both from Abubakar Tafawa Balewa University, Bauchi. He also holds Ph.D in Agricultural Extension and Communication from the Federal University of Agriculture, Makurdi. He joined the Services of Kogi State University, Anyigba in 2009. He has been involved in in-depth research in his field of specialization and has published extensively. He is currently the Head of Department (HOD) of the Department of Agricultural Economics and Extension, Faculty of Agriculture, Kogi State University.

Abimbola Oluyemisi Adepoju is a senior Lecturer in the Department of Agricultural Economics University of Ibadan, Oyo State Nigeria. She is an astute academic with research focus in Development and Welfare Economics. She is versatile and has a number of publications in both local and international learned journals on poverty, vulnerability, food security, social protection and other agriculture-related issues. She is a dynamic, resourceful, highly-motivated, proactive and performance-driven professional.

Rahman Ayodeji Adewole hails from Oyo Town in Oyo State, Nigeria where he attended Olivet Baptist High School. He obtained a Bachelor Degree in Agriculture from University of Ilorin in 2012 and proceeded to the University of Ibadan in 2016 where he obtained a Master Degree in Agricultural Economics in 2018. He is an avid advocate of Food Production and Security and has been fully involved in Agribusiness since 2010.

Rebecca Funmi Akinmulewo obtained her Master of Science degree in Agricultural Economics from the University of Ibadan. Her research foci are food policy and environmental economics.

Raheem O. Aminu is a graduate of Agricultural Economics and Farm Management from the Federal University of Agriculture Abeokuta. He has a veritable experience as a researcher, and this earned him the Chief O. Osunkeye Prize award for the Graduating Student with the Best Final Year Project in the Department of Agricultural Economics and Farm Management for the 2012/2013 set. His quest for self-actualization in academics made him applied and won the Chinese Government Scholarship to study a master's degree in Economics and Management at the China Agricultural University, which was completed in 2018. He also won the 3rd best presentation prize award at the 2018 International Student Conference on Environment and Sustainability (ISCES) Organized by United Nations Environment Programme and Tongji University Shanghai, P. R. China. Mr Aminu currently works as an Assistant Lecturer in the Department of Agricultural Economics and Farm Management, Federal University of Agriculture Abeokuta. His research interests lie in environmental, food and health economics.

Aisha Olushola Arowolo has 10 years of experience in teaching and research activities. She had a Bachelor of Agriculture in 2009 from the University of Agriculture, Abeokuta (UNAAB), Nigeria now known as the Federal University of Agriculture, Abeokuta (FUNAAB). She graduated with a First-Class honor and as the best graduating student of the Department of Agricultural Economics and Farm Management to which she received the UNAAB Senate Annual Prize, Chief O. Osunkeye Prize and Nigeria Deposit Insurance Corporation Prize. In 2010, she joined the services of the Federal University of Agriculture Abeokuta as a Junior Research Fellow and had her Master of Agriculture in Agricultural Economics in 2012 in the same university. She was an awardee of 2015 Chinese Government scholarship, which she used to pursue her PhD in Agricultural Economics and Management in the University of Chinese Academy of Sciences, Beijing, China, between 2015 and 2019. She is currently a Lecturer I in the Department of Agricultural Economics and Farm Management, Federal University of Agriculture, Abeokuta, Nigeria. Dr Arowolo has twenty-one (21) publications, 17 of which are journal articles, 3 chapters in book, and 1 conference proceedings. She is a member of the Farm Management Association of Nigeria (FAMAN) and a reviewer for many national and international journals. Her research interests are environmental and resource economics as well as health and welfare economics.

Simeon Olusola Ayansina is a Senior Lecturer in the Department of Agricultural Administration, Federal University of Agriculture, Abeokuta, Nigeria. Dr Ayansina obtained PhD. and MSc. Degrees from University of Ilorin and BSc. (Hons) Agricultural Education from University of Calabar all in Nigeria. He is a seasoned Academic and Administrator who has taught at different levels of education and currently teaching and conducting researches in Administrative areas such as Training and Development; Communication, Leadership, strategic planning, Management and Team Building in Organization. He is a member of different professional body, he supervises Post Graduate Research Works and has involved in many Multidisciplinary researches both within and outside the University.

Sarah Edewor is an early career researcher with a PhD from the Department of Agricultural Economics and Farm Management, Federal University of Agriculture Abeokuta, Nigeria. She is multidisciplinary research-oriented with research interest in development studies, international trade and environmental economics. She is highly proficiency in quantitative and qualitative analysis.

Shakirat Ibrahim is an agricultural economics expert with about 10 years of progressive teaching and research experience applying sound research methods, econometric methodologies and economic theories to research issues relating to climate change and agricultural households' welfare and food systems in Nigeria. Dr Ibrahim obtained her B. Agric and M. Agric degrees from the prestigious University of Agriculture, Abeokuta Nigeria in year 2008 and 2011 respectively. She has a PhD degree in Agricultural and Food Economics from the prestigious University of Reading, Reading United Kingdom in the year 2017. Dr Ibrahim's research-based degree trainings in world-class Universities in the field of environmental and welfare economics have equipped her with a great wealth of theoretical knowledge and research skills such as strong expertise in survey design and implementation, proficiency in data cleaning, management, practical application of econometric methods in data analysis using econometric software (Microsoft Excel, SPSS, STATA and LIMDEP), scientific writing and presentation. Currently, she lectures in the Department of Agricultural Economics and Farm Management, Federal University of Agriculture, Abeokuta Nigeria.

Merve Kaplan was born in 1986. She is an aquaculter engineer. She is preparing her master thesis in Kocaeli University since 2018. She is a master student at Kocaeli University in Turkey and also works as an aquaculture engineer in the Provincial Directorate of Food, Agriculture and Livestock in Kocaeli, Turkey.

Huynh Viet Khai obtained a Ph.D. degree and researched as a postdoctoral fellow at Kyushu University, Japan. Presently, he is an associate professor and the head of the Environmental and Resource Economics Department at the College of Economics, Can Tho University. His research involves productive efficiency of agricultural production, water pollution, the demand of agricultural insurance, the solution of poverty reduction, organic production, biodiversity conservation, as well as the economic valuation of non-market goods such as externality of agriculture and forest, waste management, environmental policy, the applications of contingent valuation and choice modeling methods.

Idumah Felix Oaikhena is a senior scientist in Forestry Research Institute of Nigeria, Ibadan Nigeria. He is a graduate of Economics and has higher degrees in Agricultural Economics.

Agatha O. Ogbe is an early career researcher. She holds a B. Agric and M.Sc degrees in Agricultural Economics. She is a doctoral student in the Department of Agricultural Economics and Farm Management, Federal University of Agriculture Abeokuta. Her research interest has revolved round trade and related issues, agricultural production, Agricultural marketing and gender issues. She has worked as a research assistant during her post-graduate studies where she assisted in preparing manuscripts for scholar submission, peer-reviewing of scientific journals and writing reports of research conducted.

Elizabeth Funmilola Okedeji is a graduate from the Department of Agricultural Extension Management, Faculty of Agricultural Sciences at the National Open University of Nigeria Awa Campus Ogun State.

Titilope Omolara Olarewaju is a socio-economic researcher with interest in environmental and resource economics. She received her PhD in Agricultural economics with specialty in Environmental and Resource Economics. She is concerned with sustainable use of natural resources for poverty reduction and food security. She is presently employed in Forestry Research Institute of Nigeria and works as an extension officer and researcher. She is involved in outreach programmes to farmers, students and also mentors industrial attachment students and corp members on national youth service corps. She lectures and supervises undergraduates from tertiary institutions across the country.

Oke Olugbenga is a zealous researcher with lecturing experiences at the Federal College of Forestry, Ibadan, a graduate from University of Agriculture, Abeokuta, Ogun State, Nigeria. Currently working with forestry research institute of Nigeria as a researcher.

Agatha Oseghale is a Lecturer I in the Department of Agricultural Economics and Farm Management, Federal University of Technology Minna, Niger State, Nigeria. She obtained her Master Degree in Agricultural Economics and Farm Management in 2011 and her PhD Degree in 2018 from the Federal University of Agriculture, Abeokuta, Ogun State, Nigeria. Her research interest includes Environmental and Resource Economics, Development Economics, Agricultural Marketing and Agricultural Production Economics. She teaches and supervises undergraduate and postgraduate students and has over 15 publications in local and international journals. She is a member of the Nigerian Association of Agricultural Economists(NAAE), Agricultural Society of Nigeria and Farm Management Association of Nigeria (FAMAN).

Funminiyi Peter Oyawole holds a Master degree in Agricultural Economics from the Federal University of Agriculture, Abeokuta, Nigeria. His research interests are in sustainable development, poverty reduction, food security and rural development. He has participated in a number of funded research studies. He has presented research findings at many conferences including International Association of Agricultural Economists Inter-Conference symposium and African Conference of Agricultural Economists.

Sérgio Pedro is a Research Fellow of the Ecology and Society Lab (ECOSOC, CES) of the Centre for Social Studies (CES) of the University of Coimbra. Scientific Manager for Univ. Coimbra´s Strategic Area - Environment, Natural Resources and Agrifood. PhD candidate in Food Governance at CES. MSc in Public Law (specialization in Human Rights and Public Policy) and BA Degree (2012) in Law. He has over 8 years of international and national experience in project management, network coordination, specialized in legal consultancy and research within the third sector and academia. His thematic interest as a professional and citizen focuses on the transdisciplinary intersection between environment, political economy, human rights, public policy and sociology of law.

Olusola Jamiu Saliu hails from Aiyetoro Gbede in Ijumu Local Government Area of Kogi State. He acquired B.Sc (HONS) in Agricultural Extension Services and M.Sc in Agricultural Extension Services from the University of Ibadan in 1986 and 1995 respectively. He also obtained Ph.D in Agricultural Extension from the Federal University of Agriculture Makurdi in 2010. Prof. O.J. Saliu taught for some years in a Secondary School before he joined the Department of Agricultural Economics and Extension, Kogi State University Anyigba in the year 2000. In the University, he has taught and supervised students at both undergraduate and postgraduate levels and has published widely. He is a member of Kogi State University Senate Committee on affiliation, Senate Committee on Curriculum Development, University Council Committee on Internally Generated Revenue and a member of Committee of Deans and Directors. He rose through the ranks of Academic Ladder to become a professor in the year 2017. He was the Head of Department of Agricultural Economics and Extension between 2012 and 2014, Director, Centre for Pre-Degree and Diploma Studies between 2014 and 2019 and currently the Director of Academic Planning and Development, Kogi State University. He is happily married with children. He is a devout Muslim.

Fatai Abiola Sowunmi (Ph.D) is a lecturer in the Department of Agricultural Economics, University of Ibadan, Ibadan, Nigeria. His research areas are: Welfare Economics, Environmental Economics and Quantitative Techniques. He was a Visiting Scholar at the United Nations University-Institute for Natural Resources in Africa (2014). He consults for many International Organisations: IFAD, IITA and African Centre for Shared Development Capacity Building. Dr. Sowunmi has many publications in reputable Local and International Journals.

José G. Vargas-Hernández, M.B.A., Ph.D., Member of the National System of Researchers of Mexico and a research professor at University Center for Economic and Managerial Sciences, University of Guadalajara. Professor Vargas-Hernández has a Ph. D. in Public Administration and a Ph.D. in Organizational Economics. He has undertaken studies in Organisational Behaviour and has a Master of Business Administration, published four books and more than 200 papers in international journals and reviews (some translated to English, French, German, Portuguese, Farsi, Chinese, etc.) and more than 300 essays in national journals and reviews. He has obtained several international Awards and recognition.

Seda Yıldırım graduated from Kocaeli University, Department of Business Administration in 2006. She completed her postgraduate and doctoral education at the Department of Business Administration, Kocaeli University, Production Management and Marketing. She worked as a research assistant in Kocaeli University between 2010-2013 and as an assistant professor in Recep Tayyip Erdoğan University between 2013-2016. Since 2017, she has been working as an Associate Professor at Namık Kemal University. Her current research interests include sustainable development, green marketing, sustainable consumption, Islamic economics, marketing culture, and values. She works as an editorial board member and arbitrator in various journals. She also studies as a member of the organizing committee of various international congresses (İCOMEP, ENSCON, İSEFE).

Index

2030 Sustainable Development Goals 16, 19, 22-23, 36

A

Agricultural land 5, 7, 86, 109, 111, 114, 117-120
Agricultural productivity 89, 95-96, 100, 176, 182-183, 185-187, 189, 191, 210, 218
Agriculture 2-3, 5, 7-8, 10, 17-18, 21-22, 25, 36, 38-40, 43, 46-50, 60-61, 63, 65-69, 84-85, 109-110, 112, 117-119, 128-129, 131, 133, 135, 138, 152, 163-164, 168, 175, 182, 184-188, 197, 199, 217
Asymmetries 181, 189-191

C

Cassava 114, 127-136, 138, 142-144, 166, 171, 221
Committee on World Food Security (CFS) 36, 65
Constraints 42-43, 128, 130, 144, 151, 159, 183, 189, 191
CVM 74

D

Developing countries 17, 23, 31, 73, 109-112, 114, 117-120, 138, 163-164, 169, 176, 181, 184, 197

E

Ecosystem Service Value 94, 97-98, 100
Empowerment 46, 163, 165, 177, 182-185
Extension 80, 129, 151, 153, 159, 169, 175-176, 187, 189, 197, 201, 204, 208, 218

F

Farming households 163, 165-166, 168-169, 173-174, 210, 216-218, 225, 228, 233
Food access 5, 47, 198
Food and Agriculture Organization 18, 22, 25, 36, 65, 164, 168
food and nutritional security 66-67, 172
Food availability 5, 7, 26, 68
Food consumption pattern 216, 218-219, 223, 229
Food governance 60-70
Food group 216, 223, 225, 228, 233
Food inaccessibility 7
Food Insecurity Access Scale 197, 199
Food production 1-4, 6, 18, 46-50, 63, 65, 97, 110, 112, 118-120, 133, 165, 169, 181-182
Food Security 1-10, 16, 18-19, 21-23, 25-26, 30-31, 36, 46-47, 51, 61, 63, 65-66, 68, 84, 109-112, 114, 118-119, 128-129, 131, 135, 163-168, 172, 174-177, 182-185, 197-199, 204-205, 208, 210, 218-219, 229
Food sustainability 17
Food system failure 6-7
Food unacceptability 8
Food unavailability 7

G

Gender Equality 182-185
Gender Mainstreaming 181
Governance structures 38-39, 49-50

H

Healthy diet 185

I

Inclusiveness 181

L

Land rights 163-169, 171-172, 174, 176, 182
Land tenure 118, 164-165, 171, 173-174, 182
Land-use Change 83, 85-86, 88-90, 92, 94

M

Markov Model 89
Mechanized 117, 135, 151-152, 156-157
Millennium Development Goals 17-18, 24, 36
Moderately Food Insecure 167-168, 172, 175-176, 199, 201, 204-206

N

Nested Logit 197, 200, 207

O

Oil palm 131, 144, 151-152, 154, 157, 159, 221
Ownership rights 163, 166, 168-169, 172-174, 176-177

P

Pearson correlation 154, 156-157, 159
Productive Inputs 182, 184, 187, 204, 208

R

Regression 76, 85, 90-92, 163, 166, 168, 173-174, 199, 219-221
Remittances 197-202, 204-208, 210, 218, 225, 233
Rice production 76, 80, 128, 130-136, 138
Risk management 73
Rural Households 129, 197-202, 204, 206-207, 210, 217
Rural poor 109, 112, 218
Rural-Urban Migration 216-218

S

Seafood Security 16-17, 22-24, 26, 30-31, 36
Seafood Sustainability 16-17, 23, 30, 36
Severely Food Insecure 197, 199, 201-202, 204-207, 214
Shannon index 219
Socio economic characteristics 154, 156-157
Sustainable Development 3, 16-19, 21-24, 30-31, 36, 41, 44-45, 60, 69, 109, 164, 182, 184
sustainable development goals 16-19, 21-24, 30-31, 36, 60, 69, 109, 184
sustainable food systems 1-3, 5, 9, 60-62, 68-70, 109-110, 181, 183, 185-186, 191

T

Turkish seafood market 16, 24, 26, 29-30
Two-stage least squares 219, 223, 229

U

Urban agriculture 38-40, 43, 46-50
Urban farming 39, 43, 46-50
Urban vacant spaces 38-39, 46-49
Use rights 163, 167, 169, 172-173

V

Value addition 127-131, 133-136, 138, 142-145
Value chain 4, 127-129, 135-136, 140, 142

W

Willingness to pay 74, 78, 80, 138

www.ingramcontent.com/pod-product-compliance
Lightning Source LLC
LaVergne TN
LVHW080536300926
849162LV00007B/1622